Communications
in Computer and Information Science          295

Dickson Lukose   Abdul Rahim Ahmad
Azizah Suliman (Eds.)

# Knowledge Technology

Third Knowledge Technology Week, KTW 2011
Kajang, Malaysia, July 18-22, 2011
Revised Selected Papers

 Springer

Volume Editors

Dickson Lukose
MIMOS Berhad, Knowledge Technology
57000 Kuala Lumpur, Malaysia
E-mail: dickson.lukose@mimos.my

Abdul Rahim Ahmad
Universiti Tenaga Nasional
College of Information Technology
43000 Kajang, Malaysia
E-mail: abdrahim@uniten.edu.my

Azizah Suliman
Universiti Tenaga Nasional
College of Information Technology
43000 Kajang, Malaysia
E-mail: azizah@uniten.edu.my

ISSN 1865-0929                           e-ISSN 1865-0937
ISBN 978-3-642-32825-1                   e-ISBN 978-3-642-32826-8
DOI 10.1007/978-3-642-32826-8
Springer Heidelberg Dordrecht London New York

Library of Congress Control Number: 2012944572

CR Subject Classification (1998): I.2.6, I.2.10-11, I.2.4, H.3, I.4, H.4.2, H.2.8, J.3

*Typesetting:* Camera-ready by author, data conversion by Scientific Publishing Services, Chennai, India

Printed on acid-free paper

Springer is part of Springer Science+Business Media (www.springer.com)

# Preface

The Knowledge Technology Week (KTW) is the premier platform for researchers and practitioners of artificial intelligence and semantic technology in Malaysia. The KTW series aims at stimulating research by promoting the exchange and cross-fertilization of ideas among different branches of artificial intelligence and semantic technology. It also provides a common forum for researchers and practitioners in various fields of artificial intelligence and semantic technology to exchange new ideas and share their experience. KTW 2011 was held in Kajang, Malaysia, hosted by the College of Information Technology (COIT), Universiti Tenaga Nasional (UNITEN) on the Putrajaya Campus.

KTW 2011 consisted of the following co-located events:

- Conferences:
  - Third Malaysian Joint Conference on Artificial Intelligence (MJCAI 2011)
  - Third Semantic Technology and Knowledge Engineering Conference (STAKE 2011)
- Workshops:
  - International Workshop on Semantic Agents (IWSA 2011)
  - Workshop on Semantic Augmented Reality (WSAR 2011)
  - Workshop on Blended Learning in Higher Education and the Workplace: Tools, Practices and Experiences (WBL 2011)
- Tutorials:
  - Hybrid Metaheuristic, from Optimization to Modeling by Bong Chin Wei
  - Prolog Rule Language for Semantic Technology by Sheng-Chuan Wu
  - Setting up a Linked Open Data Infrastructure by Johannes Keizer
  - Knowledge Design Patterns by John Sowa
  - Searching, Orienteering and Finding by Andreas Dengel
  - Building Multi-Agent Systems Using Win-Prolog by Mohd Sharifuddin Ahmad
- Other Events:
  - Poster Sessions
  - PhD Students Symposium
  - Artificial Intelligence Software Demonstration
  - Artificial Intelligence Special Sessions on Agriculture, Econmics, Health and Medical Education.

This volume contains selected papers from the proceedings of the Third Malaysian Joint Conference on Artificial Intelligence (MJCAI 2011), the Third Semantic Technology and Knowledge Engineering (STAKE 2011), and the International Workshop on Semantic Agents (IWSA 2012). KTW 2011 received a total of 105 submission from seven countries. From these, 29 regular papers (28%) and nine short papers (9%) were accepted for publication in this volume. We also included five invited papers from the keynote speakers. All submitted papers were referred by three or more reviewers selected by the Program Committee members. The reviewers' evaluations and comments were carefully examined by the core members of the Program Committee to ensure fairness and consistency in the paper selection. After the conference, another round of paper revision and review was conducted to ensure all comments during the presentation of the paper was taken into account and incorporated in the final version of the papers.

The technical program comprised two days of workshops and tutorials, followed by paper, poster and demo sessions, invited talks, and three special sessions. The keynote speakers of KTW 2011 were:

- Andreas Dengel, Scientific Director, German Research Center for Artificial Intelligence (DFKI) Germany
- John Sowa, President, VivoMind Intelligence Inc., USA
- Dietrich Albert, Professor of Psychology, University of Graz, Austria
- Sheng-Chuan Wu, Vice President, Franz Inc. USA
- Klaus Tochtermann, Director of Leibniz Information Center for Economics, Kiel, Germany
- Johannes Keizer, Office of Knowledge Exchange, Research and Extension, Food and Agriculture Organization of the United Nation (FAO)
- Eric Tsui, Associate Director, Knowledge Management Research Center, The Hong Kong Polytechnic University, Hong Kong
- Erich Weichselgartner, Deputy Scientific Director, Leibniz Center for Psychological Information and Documentation, Trier, Germany

The success of a conference depends on the support and cooperation of many people and organizations; KTW 2011 was no exception. KTW 2011 was supported by the Ministry of Science, Technology and Innovation (MOSTI), MIMOS Berhad, Universiti Tenaga Nasional, Leibniz Information Centre for Economics, Franz Inc., Sabah Economic Development and Investment Authority, Infovalley Sdn. Bhd., and NOTA ASIA Sdn. Bhd.

We would like to take this opportunity to thank the authors, the Program Committee members, reviewers, and fellow members of the various Conference and Workshop Committees for their time and efforts spent in making KTW 2011 a successful and enjoyable conference.

Finally, we thank Springer, its computer science editor, Alfred Hofmann, Anna Kramer and Leonie Kunz, for their assistance in publishing the KTW 2011 proceedings as a volume in the *Communications in Computer and Information Science* series.

July 2011                                                        Dickson Lukose
                                                          Abdul Rahim Ahmad
                                                              Azizah Suliman

# Organization

## Organizing Committee

### General Chairs

Dickson Lukose                     MIMOS Berhad, Malaysia

### Program Chairs

Abdul Rahim Ahmad                  Universiti Tenaga Nasional, Malaysia
Azizah Suliman                     Universiti Tenaga Nasional, Malaysia
Karthigayan Muthukaruppan          MIMOS Berhad, Malaysia
Patricia Anthony                   Universiti Malaysia Sabah, Malaysia

### Local Arrangements Chairs

Lai Weng Kin                       MIMOS Berhad, Malaysia
Mohd Zaliman Mohd. Yusof           Universiti Tenaga Nasional, Malaysia
Norziana Jamil                     Universiti Tenaga Nasional, Malaysia

### PhD Symposium Chairs

Mohammad Reza Beik Zadeh           MIMOS Berhad
Asmidar Abu Bakar                  Universiti Tenaga Nasional, Malaysia

### Tutorial Chairs

Kow Weng Onn                       MIMOS Berhad, Malaysia
Chen Soong Der                     Universiti Tenaga Nasional, Malaysia

### Workshop Chairs

Arun Anand Sadanandan              MIMOS Berhad, Malaysia
Mohammed Azlan Mohamed
  Iqbal                            Universiti Tenaga Nasional, Malaysia

### Demo Chairs

Tan Sieow Yeek                     MIMOS Berhad, Malaysia
Alicia Tang Yee Chong             Universiti Tenaga Nasional, Malaysia

## Publicity Chairs

| | |
|---|---|
| Aziza Mamadolimova | MIMOS Berhad, Malaysia |
| Norziana Jamil | Universiti Tenaga Nasional, Malaysia |

## Special Session Chairs

| | |
|---|---|
| Ng Pek Kuan | MIMOS Berhad, Malaysia |
| Tan Sieow Yeek | MIMOS Berhad, Malaysia |

## Sponsorship Chairs

| | |
|---|---|
| Ng Pek Kuan | MIMOS Berhad, Malaysia |
| Tan Sieow Yeek | MIMOS Berhad, Malaysia |
| Badariah Solemon | Universiti Tenaga Nasional, Malaysia |

## Committee Secretaries

| | |
|---|---|
| Rokiah Bidin | MIMOS Berhad, Malaysia |
| Tan Chin Hooi | Universiti Tenaga Nasional, Malaysia |
| Majdi Bseiso | Universiti Tenaga Nasional, Malaysia |

# Local Organizing Committee

| | |
|---|---|
| Benjamin Min Xian Chu | MIMOS Berhad, Malaysia |
| Fadzly Zahari | MIMOS Berhad, Malaysia |
| Low Loi Ming | Universiti Tenaga Nasional, Malaysia |
| Uwe Heinz Rudi Dippel | Universiti Tenaga Nasional, Malaysia |

# Program Committee

| | |
|---|---|
| Abdul Rahim Ahmad | Universiti Tenaga Nasional |
| Ahsan Morshed | United Nations Food and Agriculture Organization, Italy |
| Ajith Abraham | Machine Intelligence Research Labs |
| Aldo Gangemi | CNR-ISTC |
| Alessandro Agostini | Prince Mohammad Bin Fahad University |
| Ali Selamat | Universiti Teknologi Malaysia |
| Andrea Turbati | University of Rome |
| Ansgar Bernardi | German Research Center for Artificial Intelligence (DFKI) |
| Armando Stellato | University of Rome, Tor Vergata |
| Arouna Woukeo | University of Southampton |
| Asmidar Abu Bakar | Universiti Tenaga Nasional |
| Azizah Suliman | Universiti Tenaga Nasional |
| Bernhard Schandl | University of Vienna |
| Bong Chin Wei | MIMOS Bhd. |

| | |
|---|---|
| Burley Zhong Wang | Sun-Yat-Sen University |
| Cheah Yu-N | Universiti Sains Malaysia |
| Cora Beatriz Excelente-Toledo | LANIA |
| Daniel Bahls | Leibniz Information Centre for Economics |
| Dickson Lukose | MIMOS Bhd. |
| Didier Stricker | German Research Center for Artificial Intelligence (DFKI) |
| Dimitrios Skoutas | Technical University of Crete |
| Dominik Slezak | Infobright Inc. |
| Edna Ruckhaus | Andrés Bello Catholic University |
| Emilio Corchado | University of Burgos |
| Eric Tsui | The Hong Kong Polytechnic University |
| Fitri Suraya Mohamad | Universiti Malaysia Sarawak |
| Gianluca Moro | University of Bologna |
| Hon Hock Woon | MIMOS Bhd. |
| Ilya Zaihrayeu | University of Trento, Italy |
| Ioan Toma | STI Innsbruck |
| Jason Teo | Universiti Malaysia Sabah |
| Jessie Hey | University of Southampton |
| John Goodwin | Ordnance Survey |
| Joyce El Haddad | University Paris-Dauphine |
| Jun Shen | University of Wollongong |
| Kang Byeong Ho | University of Tasmania |
| Karthigayan Muthukaruppan | MIMOS Bhd. |
| Kiu Ching Chieh | MIMOS Bhd. |
| Lim Chee Peng | Universiti Sains Malaysia |
| Lim Tek Yong | Multimedia University |
| Magdy Aboul-Ela | French University |
| Mandava Rajeswari | University of Science Malaysia |
| Maria Esther Vidal | Universidad Simón Bolívar |
| Matteo Palmonari | Universita Degli Studi di Milano |
| Maurice Pagnucco | The University of New South Wales |
| Md. Nasir Sulaiman | Universiti Putra Malaysia |
| Mian Muhammad Awais | Lahore University of Management Sciences |
| Michael Granitzer | Know Center |
| Minghua He | Aston University |
| Minjie Zhang | University of Wollongong |
| Mohammad Reza Beik Zadeh | MIMOS Bhd. |
| Mohd Sharifuddin Ahmad | Universiti Tenaga Nasional |
| Mukhtar Rana | University of Hail |
| Neubert Joachim | Leibniz Information Centre for Economics |
| Norazila Abd. Aziz | Universiti Malaysia Sarawak |
| Norbert Luttenberger | Christian-Albrechts-Universität zu Kiel |
| Panagiotis Karras | National University of Singapore |
| Patrice Boursier | Université de La Rochelle |

| | |
|---|---|
| Patrick Hanghui Then | Swinburne University of Technology Sarawak Campus |
| Quan Bai | Auckland University of Technology |
| Rayner Alfred | Universiti Malaysia Sabah |
| Rene Witte | Concordia University |
| Reto Krummenacher | Siemens AG |
| Samhaa El-Beltagy | Nile University |
| Saravanan Muthaiyah | Multimedia University |
| Sarvapali D. Ramchurn | University of Southampton |
| Sebastian Tramp | University of Leipzig |
| Sebti Foufou | Qatar University |
| Shahrul Azman Mohd Noah | Universiti Kebangsaan Malaysia |
| Shyamala Doraisamy | Universiti Putra Malaysia |
| Sim Kim Lau | University of Wollongong |
| Soon Lay Ki | Multimedia University |
| Srinath Srinavasa | International Institute of Information Technology |
| Stefan Mandl | University of Augsburg |
| Stefano Fantoni | University of Augsburg |
| Sundar Raju Raviraja | University of Malaya |
| Sung-Kook Han | Wonkwang University |
| Syed Malek F.D. Syed Mustapha | Asia e University |
| Tristan Ling | University of Tasmania |
| Umar Moustafa Al-Turky | The King Fahd University of Petroleum and Minerals |
| Ventzeslav Valev | Bulgarian Academy of Sciences |
| Viet Dung Dang | Google Inc. |
| Vijay Dalani | GE Global Research |
| Wayne Wobcke | The University of New South Wales |
| Werner Kießling | University of Augsburg |
| Wilmer Pereira | Universidad Católica Andrés Bello |
| Yudith Cardinale | Universidad Simón Bolívar |
| Zili Zhang | Deakin University |

## Additional Reviewers

| | |
|---|---|
| Ahamad Tajudin Khader | Universiti Sains Malaysia |
| Albena Strupchanska | Concordia University |
| Benjamin Adrian | German Research Center for Artificial Intelligence |
| Bradford Heap | The University of New South Wales |
| David Rajaratnam | The University of New South Wales |
| Gunnar Grimnes | German Research Center for Artificial Intelligence |
| Hafsa Yazdani | Prince Mohammad Bin Fahad University |

| Kok Soon Chai | Universiti Sains Malaysia |
| Kyongho Min | University of Tasmania |
| Lahouari Ghouti | King Fahd University of Petroleum and Minerals |
| Majdi Beseiso | Universiti Tenaga Nasional |
| Munir Zaman | Universiti Sains Malaysia |
| Nur'Aini Abdul Rashid | Universiti Sains Malaysia |
| Timothy Cerexhe | The University of New South Wales |

## Editorial Board

| Abdreas Dengel | German Research Center for Artificial Intelligence, Germany |
| Eric Tsui | The Hong Kong Polytechnic University, Hong Kong |
| John F. Sowa | VivoMind Intelligence, Inc., USA |
| Klaus Tochtermann | Leibniz Information Centre For Economics, Germany |
| Sheng-Chuan Wu | Franz Inc., USA |
| Dietrich Albert | University of Graz, Austria |
| Erich Weichselgartner | The Leibniz-Institute for Psychology Information, Germany |
| Patricia Anthony | Universiti Malaysia Sabah, Malaysia |
| Abdul Rahim Ahmad | Universiti Tenaga Nasional, Malaysia |
| Azizah Suliman | Universiti Tenaga Nasional, Malaysia |
| Dickson Lukose | MIMOS Berhad, Malaysia |

## Sponsoring Organizations

 Ministry of Science, Technology and Innovation, Malaysia

 MIMOS Berhad, Malaysia

 Universiti Tenaga Nasional, Malaysia

 Leibniz Information Centre for Economics, Germany

 Franz Inc., USA

 Sabah Economic Development and Investment Authority, Malaysia

 Infovalley Sdn. Bhd., Malaysia

 NOTA ASIA Sdn. Bhd., Malaysia

# Table of Contents

## Invited Papers

Touch & Write: Penabled Collaborative Intelligence .................. 1
  *Andreas Dengel, Marcus Liwicki, and Markus Weber*

A Framework for Knowledge Sharing and Interoperability
in Agricultural Research for Development ......................... 11
  *Caterina Caracciolo and Johannes Keizer*

E-Learning Based on Metadata, Ontologies and Competence-Based
Knowledge Space Theory ......................................... 24
  *Dietrich Albert, Cord Hockemeyer, Michael D. Kickmeier-Rust,
  Alexander Nussbaumer, and Christina M. Steiner*

Linked Library Data: Offering a Backbone for the Semantic Web ....... 37
  *Joachim Neubert and Klaus Tochtermann*

The CRM of Tomorrow with Semantic Technology .................. 46
  *Sheng-Chuan Wu*

## Regular Papers

Identifying Metabolic Pathway within Microarray Gene Expression
Data Using Combination of Probabilistic Models .................... 52
  *Abdul Hakim Mohamed Salleh and Mohd Saberi Mohamad*

A Transparent Fuzzy Rule-Based Clinical Decision Support System
for Heart Disease Diagnosis ..................................... 62
  *Adel Lahsasna, Raja Noor Ainon, Roziati Zainuddin, and
  Awang M. Bulgiba*

Knowledge Modeling for Personalized Travel Recommender .......... 72
  *Atifah Khalid, Suriani Rapa'ee, Norlidza Mohd Yassin, and
  Dickson Lukose*

Formulating Agent's Emotions in a Normative Environment .......... 82
  *Azhana Ahmad, Mohd Sharifuddin Ahmad,
  Mohd Zaliman Mohd Yusoff, and Moamin Ahmed*

Service Oriented Architecture for Semantic Data Access Layer ......... 93
  *Chee Kiam Lee, Norfadzlia Mohd Yusof, Nor Ezam Selan, and
  Dickson Lukose*

A Novel Image Segmentation Technique for Lung Computed
Tomography Images . . . . . . . . . . . . . . . . . . . . . . . . . . . . . . . . . . . . . . . . . . .    103
    *Chin Wei Bong, Hong Yoong Lam, and Hamzah Kamarulzaman*

Implementation of a Cue-Based Aggregation with a Swarm Robotic
System . . . . . . . . . . . . . . . . . . . . . . . . . . . . . . . . . . . . . . . . . . . . . . . . . . . . . . .    113
    *Farshad Arvin, Shyamala C. Doraisamy, Khairulmizam Samsudin,
Faisul Arif Ahmad, and Abdul Rahman Ramli*

Parallel Web Crawler Architecture for Clickstream Analysis . . . . . . . . . . .    123
    *Fatemeh Ahmadi-Abkenari and Ali Selamat*

Extending Information Retrieval by Adjusting Text Feature Vectors . . . .    133
    *Hadi Aghassi and Zahra Sheykhlar*

Tracking Multiple Variable-Sizes Moving Objects in LFR Videos Using
a Novel Genetic Algorithm Approach . . . . . . . . . . . . . . . . . . . . . . . . . . . . .    143
    *Hamidreza Shayegh Boroujeni, Nasrollah Moghadam Charkari,
Mohammad Behrouzifar, and Poonia Taheri Makhsoos*

From UML to OWL 2 . . . . . . . . . . . . . . . . . . . . . . . . . . . . . . . . . . . . . . . . . . .    154
    *Jesper Zedlitz, Jan Jörke, and Norbert Luttenberger*

Completeness Knowledge Representation in Fuzzy Description Logics . . .    164
    *Kamaluddeen Usman Danyaro, Jafreezal Jaafar, and
Mohd Shahir Liew*

Random Forest for Gene Selection and Microarray Data
Classification . . . . . . . . . . . . . . . . . . . . . . . . . . . . . . . . . . . . . . . . . . . . . . . . . .    174
    *Kohbalan Moorthy and Mohd Saberi Mohamad*

A Novel Fuzzy HMM Approach for Human Action Recognition
in Video . . . . . . . . . . . . . . . . . . . . . . . . . . . . . . . . . . . . . . . . . . . . . . . . . . . . . .    184
    *Kourosh Mozafari, Nasrollah Moghadam Charkari,
Hamidreza Shayegh Boroujeni, and Mohammad Behrouzifar*

An Image Annotation Technique Based on a Hybrid Relevance
Feedback Scheme . . . . . . . . . . . . . . . . . . . . . . . . . . . . . . . . . . . . . . . . . . . . . . .    194
    *Mehran Javani and Amir Masoud Eftekhari Moghadam*

Using Semantic Constraints for Data Verification in an Open World . . . .    206
    *Michael Lodemann, Rita Marnau, and Norbert Luttenberger*

Ontology Model for National Science and Technology . . . . . . . . . . . . . . . . .    216
    *Michelle Lim Sien Niu, Nor Ezam Selan, Dickson Lukose, and
Anita Bahari*

Norms Detection and Assimilation in Multi-agent Systems:
A Conceptual Approach ......................................... 226
    *Moamin A. Mahmoud, Mohd Sharifuddin Ahmad, Azhana Ahmad,*
    *Mohd Zaliman Mohd Yusoff, and Aida Mustapha*

Model Based Human Pose Estimation in MultiCamera Using Weighted
Particle Filters ...................................... 234
    *Mohammad Behrouzifar, Hamidreza Shayegh Boroujeni,*
    *Nasrollah Moghadam Charkari, and Kourosh Mozafari*

Route Guidance System Based on Self-Adaptive Algorithm ............ 244
    *Mortaza Zolfpour-Arokhlo, Ali Selamat,*
    *Siti Zaiton Mohd Hashim, and Md Hafiz Selamat*

Location Recognition with Fuzzy Grammar ........................ 254
    *Nurfadhlina Mohd Sharef*

High Level Semantic Concept Retrieval Using a Hybrid Similarity
Method.......................................... 262
    *Sara Memar Kouchehbagh, Lilly Suriani Affendey,*
    *Norwati Mustapha, Shyamala C. Doraisamy, and*
    *Mohammadreza Ektefa*

A Conceptualisation of an Agent-Oriented Triage Decision Support
System ......................................... 272
    *Shamimi Halim, Muthukaruppan Annamalai,*
    *Mohd Sharifuddin Ahmad, and Rashidi Ahmad*

Evaluating Multiple Choice Question Generator...................... 283
    *Sieow-Yeek Tan, Ching-Chieh Kiu, and Dickson Lukose*

Ontology Model for Herbal Medicine Knowledge Repository ........... 293
    *Supiah Mustaffa, Ros'aleza Zarina Ishak, and Dickson Lukose*

A Data Structure between Trie and List for Auto Completion ......... 303
    *Vooi Keong Boo and Patricia Anthony*

Agent for Mining of Significant Concepts in DBpedia ................ 313
    *Vooi Keong Boo and Patricia Anthony*

Prediction of Protein Residue Contact Using Support Vector
Machine ......................................... 323
    *Weng Howe Chan and Mohd Saberi Mohamad*

Optimized Local Protein Structure with Support Vector Machine
to Predict Protein Secondary Structure .......................... 333
    *Yin Fai Chin, Rohayanti Hassan, and Mohd Saberi Mohamad*

# Short Papers

Ensuring Security and Availability of Cloud Data Storage Using Multi
Agent System Architecture ........................................ 343
*Amir Mohamed Talib, Rodziah Atan, Rusli Abdullah, and
Masrah Azrifah Azmi Murad*

Multiple Object Classification Using Hybrid Saliency Based
Descriptors ...................................................... 348
*Ali Jalilvand and Nasrollah Moghadam Charkari*

Parameter Estimation for Simulation of Glycolysis Pathway by Using
an Improved Differential Evolution ............................... 352
*Chuii Khim Chong, Mohd Saberi Mohamad, Safaai Deris,
Yee Wen Choon, and Lian En Chai*

A Multi-objective Genetic Algorithm for Optimization Time-Cost
Trade-off Scheduling .............................................. 356
*Hadi Aghassi, Sedigheh Nader Abadi, and Emad Roghanian*

Finding Web Document Associations Using Frequent Pairs of Adjacent
Words ........................................................... 360
*Jason Yong-Jin Tee, Lay-Ki Soon, and Bali Ranaivo-Malançon*

Representing E-Mail Semantically for Automated Ontology Building ... 364
*Majdi Beseiso, Abdul Rahim Ahmad, and Roslan Ismail*

A Nodal Approach to Agent-Mediation in Personal Intelligence ........ 368
*Shahrinaz Ismail and Mohd Sharifuddin Ahmad*

Benchmarking the Performance of Support Vector Machines
in Classifying Web Pages ......................................... 375
*Wein-Pei Wong, Ke-Xin Chan, and Lay-Ki Soon*

Conic Curve Fitting Using Particle Swarm Optimization: Parameter
Tuning .......................................................... 379
*Zainor Ridzuan Yahya, Abd Rahni Mt Piah, and Ahmad Abd Majid*

**Author Index** ..................................................... 383

# Touch & Write:
# Penabled Collaborative Intelligence

Andreas Dengel[1,2], Marcus Liwicki[1], and Markus Weber[1,2]

[1] German Research Center for AI (DFKI GmbH), Trippstadter Straße 122, 67663
Kaiserslautern, Germany,
`firstname.lastname@dfki.de`,
`http://www.dfki.de/~lastname`
[2] Knowledge-Based Systems Group, Department of Computer Science,
University of Kaiserslautern, P.O. Box 3049, 67653 Kaiserslautern

**Abstract.** Multi-touch (MT) technology becomes more and more famous and several frameworks have been proposed in the last decade. All of them focus on the support of touch input, gestures, and objects. Recently, however, a new generation of MT-tables emerged, which allows for pen-input in addition to the touch paradigm. These devices, such as the *Touch & Write* Table of DFKI, consider multiple pen interaction. In this paper we propose a software development kit (SDK) which integrates the basic processing of the pen input. Moreover, the *Touch & Write* SDK includes handwriting recognition and geometric shape detection. Using writing mode detection the SDK automatically applies the correct recognition component based on features extracted from the pen data.

**Keywords:** Software architecture, SDK, Touch & Write, sketch-based interface.

## 1 Introduction

Usability is the main aim of Human-Computer Interaction (HCI) research. It is essential to design interfaces that allow users to intuitively work and interact with applications even for first time users. Appropriate metaphors and devices have to be used to allow easy and fast interaction.

Almost three decades after the computer mouse started its triumph with the Apple Macintosh, it is about time for a next generation of HCI. The mouse does not completely reflect the interactions paradigms of the real world. This new generation seems to be found in MT tabletop environments, which provide hands on experience and offer a wider domain of usage scenarios. However, current MT solutions, such as the Microsoft Surface[1] or DiamondTouch [5], lack in a way of intuitively switching between two important modes: moving the objects and editing their content, i.e., drawing or writing.

---

[1] `http://www.microsoft.com/surface`: Last accessed 05/22/2010.

D. Lukose, A.R. Ahmad, and A. Suliman (Eds.): KTW 2011, CCIS 295, pp. 1–10, 2012.

To cope with this problem, we have recently proposed a tabletop solution which also integrates pen input. The *Touch & Write* Table [13] integrates the Anoto technology[2] on top of the usual frustrated total internal reflection (FTIR) proposed by Han [8]. Another table integrating touch and pen input was proposed in the Flux project [12], where a similar hardware is used.

In this paper we propose an SDK which enables developers to use touch and write input metaphors without the need of in-depth knowledge about pattern recognition and handwriting recognition. All these techniques are integrated into the framework and the developer just has to handle events which are triggered by a separate engine.

The rest of this paper is organized as follows. First, Section 2 presents the *Touch & Write* SDK with more detail. Next, Section 3 introduces an application which makes use of the SDK and allows architects to annotate and retrieve architectural floor plans. Subsequently, Section 4 introduces an application for shopping in smart environments and Section 5 presents an application for video annotation and retrieval. Finally, Section 6 concludes this paper and draws some outlook for future work.

## 2    *Touch & Write* SDK

### 2.1    Previous Frameworks

Using sketch-based interface is a promising approach in several domains as the usage of pen gives a user more freedom and a precise tool. Examples for sketch-based systems are COMIC system [14], Simulink's[3] Sim-U-Sketch [10], and recently an application to draw UML diagrams [7]. Unfortunately, most of the existing frameworks for multi-touch tables focus only on touch and object support.

MT4j [11] is designed for rapid development of graphically rich applications on a variety of contemporary hardware, from common PCs and notebooks to large-scale ambient displays, as well as different operating systems. The framework has a special focus on making multi-touch software development easier and more efficient.

Echtler introduced libTisch [6], a layered architecture which allows easy integration of existing software, as several alternative implementations for each layer can co-exist.

pyMT [1] is based on the Python programming language, thus it is portable and offers a broad framework. It comes with native support for many multi-touch input devices, a growing library of multi-touch aware widgets, hardware accelerated OpenGL drawing, and an architecture that is designed to let you

---

[2] Anoto pen http://www.anoto.com/: Last accessed 04/02/2010.

[3] Simulink is an environment for multi-domain simulation and Model-Based Design for dynamic and embedded systems.
http://www.mathworks.com/products/simulink/: Last accessed 06/20/2010.

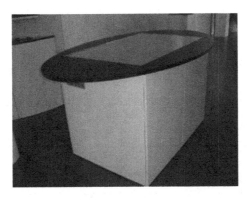

**Fig. 1.** Picture of the Touch&Write table

focus on building custom and highly interactive applications as quickly and easily as possible.

Microsoft Windows 7 (Windows Presentation Foundation) offers a comprehensive API for multi-touch support. As the framework is based on Microsoft Windows technology cross-platform portability is of course an issue. Similarly, Adobe Flash/Air also has begun to integrate multi-touch support, but here, there are severe performance issues, as full screen updates are notoriously performance intensive in Flash. Further releases will bring some improvements.

However, none of the above mentioned frameworks offers pen support and further analysis of the pen data. We propose the integration of pen devices and pattern recognition techniques.

Our research is mainly motivated by the fact that in several domains, such as design or engineering, sketches are used to exchange ideas in a collaborative environment. Therefore supporting pen devices is a promising approach. A pen offers more freedom for designers, they can just draw as on usual paper. Furthermore it does not limit their creativity, since the designer does not need to think about how he or she puts the information on the paper, there is no need to go through various context menus and dialogs. Unfortunately, if the information is just drawn on normal paper it is hard to transfer it back to the digital world.

The main idea of the *Touch & Write* table and SDK is to offer an innovative new platform for creating applications that users find natural to use. The table seamlessly integrates the paper world into the digital world. Editing, arranging and writing tasks can be easily performed by using hands and the pen. The SDK goes beyond traditional MT-SDKs by including automated pen-input analyses and interpretation.

## 2.2   *Touch & Write* Components

The *Touch & Write* table, illustrated in Figure 1, combines the paradigm of multi-touch input devices and a multi-pen input device. The table is a novel rear-projection tabletop which detects touching by using FTIR and a high resolution

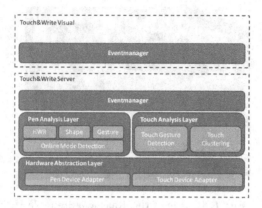

**Fig. 2.** *Touch & Write* SDK overview

pen technology offered by Anoto Group AB. This section shortly summarizes the main technical setup of the table surface.

Although touch and write interaction is very useful for several collaborative designing tasks, to the authors' knowledge there is no SDK available. Therefore we propose an SDK which integrates the novel input device, the pen. The *Touch & Write* SDK is an approach to simplify the development for a tabletop device with pen-support. Figure 2 illustrates the components of the *Touch & Write* SDK.

The *Touch & Write* SDK is divided into two parts, the *Touch & Write* Server and the *Touch & Write* Visual system. The basic idea is that the server part is always running on the system as service and handles the input devices. In contrast to the server component, the visual component is intended to be replaceable. Furthermore it should not be bound to any specific programming language. Therefore we use inter-process communication between the subsystems via a network protocol on the local machine. Each application using an implementation of the *Touch & Write* Visual component, has to register at the Event Manager of the server component.

The first layer of the server architecture is a hardware abstraction layer that offers adapters for the touch and pen input device. In the current development phase just the hardware of the *Touch & Write* table is supported. For the touch input the open source software Community Core Vision (CCV)[4] is employed which uses the TUIO protocol [9] and the on-line pen data is send via bluetooth. The adapter analyzes the streamed pen data and the coordinates are handed to the next layer.

The second layer is split up in two parts, the Pen Analysis Layer and Touch Analysis Layer. In this paper we focus on the pen data analysis, since the touch analysis can be done like in the above mentioned references. While the user

---

[4] http://ccv.nuigroup.com/: Last accessed 05/22/2010.

interacts with the system the pen data is cached and whenever the user stops to interact for a short interval $T_{delay}$ the on-line mode detection is triggered. Depending on the detection result, further analysis is applied on the cached on-line pen data. Either the handwriting recognition is used to detect written text, the shape detection subsystem analysis the drawing and recognizes geometrical shapes, or the gesture recognizer is used to recognize a gesture.

Finally, the Event Manager will be informed about the result of the analysis. There is a general distinction between so called Low-Level Events and High-Level Events. Low-Level Events directly transmit the raw data, i.e., the screen coordinates extracted from the on-line pen data and the touch positions received CCV. High-Level Events contain the results of the analysis component, e.g., handwriting recognition results, detected shapes, or recognized gestures.

The event manager of the server component is responsible for distributing the events to the registered applications. Low-Level Events have to be delivered in real-time, giving the user immediate feedback, such as rendering the pen trajectory or for interaction with touched objects. While High-Level-Events are fired whenever the user stops the interaction with the system and the analysis is triggered.

As mentioned above, the visual component of the SDK is programming language independent. Each component needs to implement the events, and to receive them from server event manager by registering. Currently, there exist two implementations for the visual component. The first is based on a Java-based game engine called jMonkeyEngine[5]. There we have extended the jMonkeyEngine's input framework to seamlessly integrated pen and touch input. The second implementation is based on Adobe FLEX. An example how this works is shown at `http://www.touchandwrite.de/`.

# 3 a.Scatch - Sketch-Based Retrieval for Architectural Floor Plans

Recently, we have developed a system for architectural floor plan retrieval which builds on the *Touch & Write* SDK. It is a sketch-based approach to query a floor plan repository of an architect. The architect searches for semantically similar floor plans just by drawing a new plan. An algorithm extracts the semantic structure sketched on the *Touch & Write* table and compares the structure of the sketch with the ones from the floor plan repository (see Fig. 3). More details on this application can be found in [16].

Experiments investigating the performance of the Floor Plan Detection revealed that our approach outperforms most existing approaches [3,2]. The detection rate on 80 images was more than 90% in many cases all plans were correctly recognized. Also the retrieval performance is close to 100% and it works very fast [16].

---

[5] `http://www.jmonkeyengine.com/`: Last accessed 05/22/2010.

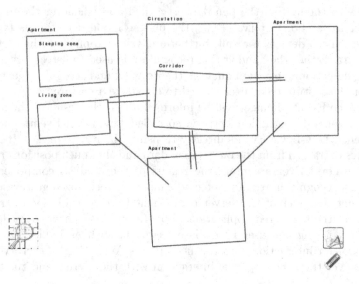

**Fig. 3.** a.SCatch retrieval interface

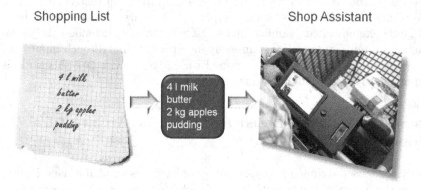

**Fig. 4.** SmartSL system overview

## 4   SmartSL - The Smart Shopping List

Besides the system for floor plan retrieval, we recently proposed a novel system which automatically extracts the intended items to buy from a handwritten shopping list. This intelligent shopping list relies on an ontology of the products which is provided by the shopping mall.

The work flow of the SmartSL system is illustrated in Fig. 4. The handwritten strokes are recognized by *Touch & Write* SDK components and then parsed in order to detect the amount and the desired item. This is then matched to the underlying ontology and the intended order is recognized. Finally, the user can use a specially equipped shopping cart to find the project in the marked and not to forget any item. Such carts exist in the Innovative Retail Laboratory (IRL) [15] located in St. Wendel. The IRL, a research laboratory of DFKI in cooperation with GLOBUS SB-Warenhaus Holding, illustrates how a shopping center in the future might look like.

Our current prototype works on an ontology of 300 products. In our real-world experiments we asked 20 persons to write shopping lists without any constrains. We have found out that in most cases the participants were happy with the retrieved results.

# 5  CoVidA - Pen-Based Collaborative Video Annotation

As an application for collaborative work, we have developed a pen-based annotation tool for videos. Annotating videos is an exhausting task, but it has a great benefit for several communities, as labeled ground truth data is the foundation for supervised machine learning approaches. Thus, there is need for an easy-to-use tool which assists users with labeling even complex structures. For outlining and labeling the shape of an object, we make use of the *Touch & Write* SDK.

## 5.1  Interaction

First, we introduce the touch-based gestures for the interaction with the video.

The video media can be controlled via touch on the *Video Controls* and a drag gesture on the *Slider* (see below the video in Figure 5 where the 42 % is indicating the current slider position). The *Video Controls* provide the main functionality to navigate in the active video (fast backward or forward, play, pause, or stop). To navigate fast through the video the user can also use the slider by tapping to the desired position or fast searching via dragging.

Videos, annotation list and annotation fields can be manipulated via touch gestures. The standard functionalities are resizing, moving, or rotation (these gestures are similar to the well-known gestures on small Touch displays). These touch gestures provide a simple access for end-users. These manipulation operations are important as we are dealing with an environment for multiple users. As several users can be located around the tabletop, they need the functionality to adjust the videos to their needs. The touch-based gestures are detected by the *Touch & Write* server and triggered in the visual component via High-Level-Events. The *CoVidA* application receives these events and processes the event data to manipulate the videos in excepted ways.

## 5.2  Annotations

Annotations were performed by using the Anoto pen. Similar to a real pen, the anoto pen is used for outlining the desired shapes and for writing the annotation terms. Here we differentiate between a *frame annotation* and an *object annotation*.

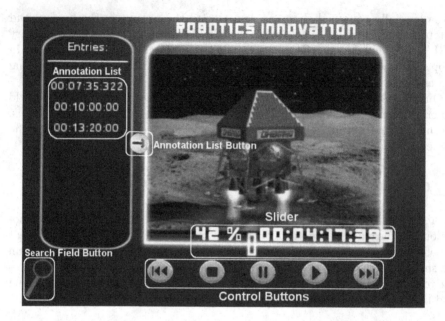

**Fig. 5. Touch-Based Interactions**: Users can interact with the buttons and annotation entries with simple single touches

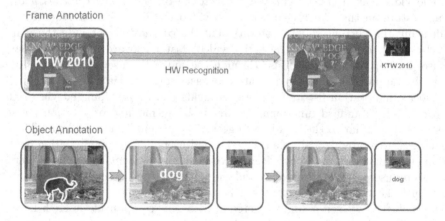

**Fig. 6.** Video Annotation Process

A *frame annotation* is associated with the whole frame and an *object annotation* is associated to a specific object which is outlined by a drawn shape. The user can annotate whole frames fast and simple by writing on the video using the pen device. The online mode detection classifies the pen input as text and the handwriting recognition is triggered. *CoVidA* now receives the recognition results as a ranked word list of the recognition results. The received top result is stored as description for the whole frame after touching on the save button (see Figure 6).

If the user starts with outlining an object in the image, the online mode detection detects a shape drawing. Then this shape is send to the *CoVidA* application. Thus it is assumed that the user aims for an object annotation (see Figure 6, Bottom and By having this kind of automatism, users are not forced to switch the current mode manually.

### 5.3   Evaluation

The *CoVidA* system has been evaluated in a user study. The evaluation had three main goals:

1. Investigate the ease of interaction with video data in *CoVidA*.
2. Possibility to work with *CoVidA* without detailed explanation.
3. Annotation of complex structures using a pen device.

The *CoVidA* software was evaluated against the *LableMe* application which is a well-recognized application by the MIT.

In our experiments we show that especially for complex structures the usage of a pen device improves the effectiveness of the outlining process. Furthermore, the toolkit was very intuitive and most users, even experts favored *CoVidA* over *LableMe*[6].

## 6   Conclusions

*Touch & Write* is a robust means for multi-touch and pen input providing the basis for a whole bunch of experimental research. It allows an intuitive switch between the modes object manipulation, and content editing. Online streaming provides direct feedback from the various applications. In this paper we have presented three examples where touching and writing are successfully combined in different application areas.

## References

1. PyMT project homepage (2010), http://pymt.txzone.net/
2. Ahmed, S., Weber, M., Liwicki, M., Dengel, A.: Improved automatic analysis of architectural floor plans. In: Proc. 10th ICDAR (September 2009) (to appear)
3. Ahmed, S., Weber, M., Liwicki, M., Dengel, A.: Text/graphics segmentation in architectural floor plans. In: Proc. 10th ICDAR (September 2009) (to appear)
4. Brooke, J.: SUS: A quick and dirty usability scale. In: Jordan, P.W., Weerdmeester, B., Thomas, A., Mclelland, I.L. (eds.) Usability Evaluation in Industry. Taylor and Francis, London (1996)

---

[6] This has been found by applying the System Usability Scale (SUS) [4].

5. Dietz, P., Leigh, D.: Diamondtouch: a multi-user touch technology. In: UIST 2001: Proceedings of the 14th Annual ACM Symposium on User Interface Software and Technology, pp. 219–226. ACM, New York (2001)
6. Echtler, F., Klinker, G.: A multitouch software architecture. In: Proceedings of NordiCHI 2008 (October 2008)
7. Frisch, M., Heydekorn, J., Dachselt, R.: Investigating multi-touch and pen gestures for diagram editing on interactive surfaces. In: ITS 2009: Proceedings of the ACM International Conference on Interactive Tabletops and Surfaces, pp. 149–156. ACM, New York (2009)
8. Han, J.Y.: Low-cost multi-touch sensing through frustrated total internal reflection. In: UIST 2005: Proceedings of the 18th Annual ACM Symposium on User Interface Software and Technology, pp. 115–118. ACM, New York (2005)
9. Kaltenbrunner, M., Bovermann, T., Bencina, R., Costanza, E.: Tuio - a protocol for table based tangible user interfaces. In: Proceedings of the 6th International Workshop on Gesture in Human-Computer Interaction and Simulation (GW 2005), Vannes, France (2005)
10. Kara, L.B., Stahovich, T.F.: Sim-u-sketch: a sketch-based interface for simulink. In: AVI, pp. 354–357 (2004)
11. Laufs, U., Ruff, C.: MT4J project homepage (2010), http://www.mt4j.org/
12. Leitner, J., Powell, J., Brandl, P., Seifried, T., Haller, M., Dorray, B., To, P.: Flux: a tilting multi-touch and pen based surface. In: CHI EA 2009: Proceedings of the 27th International Conference Extended Abstracts on Human Factors in Computing Systems, pp. 3211–3216. ACM, New York (2009), http://mi-lab.org/wp-content/blogs.dir/1/files/publications/int125-leitner.pdf
13. Liwicki, M., El-Neklawy, S., Dengel, A.: Touch & Write - A Multi-Touch Table with Pen-Input. In: Proceedings International Workshop on Document Analysis Systems (2010) (to appear)
14. Os, E.D., Boves, L.: Towards ambient intelligence: Multimodal computers that understand our intentions. In: Proc. eChallenges, pp. 22–24 (2003)
15. Spassova, L., Schöning, J., Kahl, G., Krüger, A.: Innovative Retail Laboratory. In: Roots for the Future of Ambient Intelligence. European Conference on Ambient Intelligence (AmI 2009), Salzburg, Austria (November 2009)
16. Weber, M., Liwicki, M., Dengel, A.: a.SCAtch - A Sketch-Based Retrieval for Architectural Floor Plans. In: 12th International Conference on Frontiers of Handwriting Recognition (2010)

# A Framework for Knowledge Sharing
# and Interoperability in Agricultural
# Research for Development*

Caterina Caracciolo and Johannes Keizer

Food and Agriculture Organization of the United Nations (FAO)
v.le Terme di Caracalla 1
00154 Rome, Italy
{Caterina.Caracciolo,Johannes.Keizer}@fao.org

**Abstract.** In an ideal world  all data would be produced using open formats
and would be linked directly to other related data on the web. This would give
the possibility for service providers to set up information systems by mixing
and matching data from different distributed repositories. A scenario like this is
no science fiction. Nevertheless most data (of all kinds) resides in database and
repository silos, and efforts to create one stop access to distributed data lack
functionalities, robustness or sustainability. The CIARD  initiative is working
to make agricultural research information publicly available and accessible to
all, by acting on both those issues. Among its actions are advocating and promoting open access, improving applicability and enabling effective use of data
and information in agricultural research and innovation. In this paper we
present the CIARD initiative and concentrate on FAO's contribution to it. We
present the Linked Data approach, the vocabulary editor VocBench, the domain
specific tagger AgroTagger and the RING registry of services and tools.

**Keywords:** Interoperability, Information sharing, CIARD, FAO, Linked Data.

## 1 Information in Agriculture: Where Are We?

Scholarly communications through articles and conferences with collateral exchange
of limited datasets have been the means of data sharing in the past. However, scientific publication and data production is growing at a much faster rate than ever before.
Figure 1 charts the number of articles indexed by MEDLINE from 1950 to 2010. The
steep increase that has begun in 1995, the internet era, is clearly visible.

But the production of peer-reviewed scholarly journals is only the tip of the iceberg.
Behind growing numbers of scholarly publications there is a growing amount of scientific data, such as experimental data, published as data sets with their associated metadata and quality indicators. Furthermore, scholarly papers are no longer the only way in
which scientific information is exchanged. Document repositories of white papers or
technical reports are widely used, and so are other web-based repositories, that may be

---

* http://www.ciard.net

D. Lukose, A.R. Ahmad, and A. Suliman (Eds.): KTW 2011, CCIS 295, pp. 11–23, 2012.

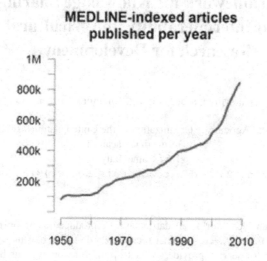

**Fig. 1.** MEDLINE-indexed articles published per year. Source: http://altmetrics.org/manifesto/

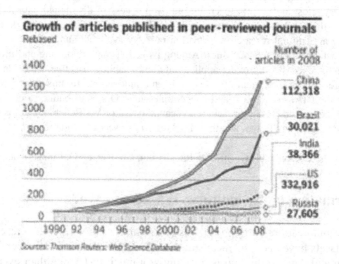

**Fig. 2.** Growth of articles pubished since 1990. See the outstanding growth of China, Brazil, India. Source Thomson Reuters. Web of Science Database.

considered as Knowledge 'derivatives' such as collections of descriptions of agricultural technologies, learning object repositories, expertise databases, etc. Researchers are also using more social platforms such as blogs[1] to discuss results before they are published in scholarly journals or after they have been published. Personal communications are now often made more general and turned into community communications, thanks to

---

[1] See as an example ScienceBlogs http://scienceblogs.com/

blogs and professional platforms that may have access to directories of peoples and institutions. This growth of scientific output calls for a growth of the instruments at disposal of scientists to orientate themselves into this wealth of information.

Another factor to take into account is the expansion of the scientific world to include a few larger countries with so-called "emerging economies", especially China, Brazil and India, into the scientific mainstream. This phenomenon is especially clear since the year 2000, as it is shown in Figure 2. For example, taking 1990 as a base, Brazil has increased its scientific production by 800% and China by 1200%. The growth of these key new players in the scholarly communication arena has made interoperability a more important global issue, with special regard to the handling of languages other than English.

Data on publication rates in agriculture are not easily available, but it is quite clear that the trend will be similar, although perhaps less dramatic, to the other life sciences which are monitored by MEDLINE[2]. Agricultural scientists should be responsible for actively ensuring that the new knowledge generated from their research is easily accessible and so easily taken up by their colleagues and the agricultural community. This issue is also vital in the face of rapidly shifting challenges such as climate change, trans-boundary pests and diseases, and agricultural trade tariffs.

Some of the principal partners in the CIARD initiative have organized an e-consultation with the goal of identifying how to enhance the sharing and interoperability of information for global agricultural research for development. This event was followed by an expert consultation in Beijing in June 2011 with the same title. These two events are exercises to describe the current status and analyze the needs for tools, standards and infrastructures, leading on to defining future actions.

## 2     Data Sharing and Interoperability: Where Are We?

The IEEE definition for interoperability refers to the ability of two systems or components to exchange information and use the information that has been exchanged. In practice, the actions that require some degrees of interoperability include the transferring of data from one repository to another, the harmonization of different data and metadata sets, the creation of a virtual research environments, the creation of documents from distributed data sets, the reasoning over distributed datasets, and the creation of new information services using distributed data sets. This list, although not inclusive, shows that the interoperability of systems implies "interoperability" of data. Nowadays, this implication is widely acknowledged and we can claim that we are in a data-centered world.

Data sets (consider for example those mentioned in the previous section) are often connected in a way or the other, as in the case of two scientific papers about the same experiment, or the scientific data and the published paper based on that data. However, much of that connection is often implicit. One of the reasons of such a limited availability of connections between data sets is that each data set usually has an

---

[2] http://www.nlm.nih.gov/

independent life cycle, with big variety in the way data is structured and accessed. This is a well-known problem among researchers and practitioners in the area of relational databases.

There are several interesting examples of successful data exchange between distributed datasets, and some of them in the area of agricultural research and innovation. A common characteristic of most examples is that they are based on specific ad-hoc solutions more than on a general principle or architecture. Usually, they rely on pre-existing agreement between the people or organizations involved in the exchange, which leads to the sharing of data through web services, or accessible APIs.

Some well known initiatives try to achieve data sharing for scholarly publications, mostly organized by the publishers[3]. Other initiatives concentrate on document repositories and open access. Certainly, the best known of the initiatives dealing with document repositories is the Open Archives Initiative[4] (OAI), which promotes interoperability standards to facilitate the efficient dissemination and sharing of content.

Initiatives related to data sharing are also happening in specific domains. The Open Geospatial Consortium[5] (OGC) represents a strong community in the area of geospatial information, which promotes geospatial and location standards. It has spurred important open source projects such as GeoServer[6]. Considerable cooperation among international organizations has focused around the problem of sharing statistical data, such as census data and time-series, leading to initiatives such as the Statistical Data and Metadata Exchange[7] (SDMX) and the Data Documentation Initiative[8] (DDI), an XML based standard for statistical data exchange, and a specification for capturing metadata about social science data, respectively. These standards are embraced by various international organizations, including the World Bank[9], the International Monetary Fund[10], United Nations Statistical Division[11], and the Organization for Economic Co-operation and Development[12]. Similar experience is the Gene Ontology Consortium[13], which promotes the standardization of the representation of gene and gene product attributes across species and databases. Specific initiatives are Singer System[14] and GeoNetwork[15] in the area of genetic resources and georeferenced maps respectively.

Apart from the virtuous examples represented by the initiatives just mentioned, the general problem regarding the actual way information systems developers access,

---

[3] For example, Science Direct by Elsevier http://www.sciencedirect.com/
[4] http://www.openarchives.org/
[5] http://www.opengeospatial.org/
[6] http://geoserver.org/display/GEOS/Welcome
[7] http://sdmx.org/
[8] http://www.ddialliance.org
[9] http://www.worldbank.org/
[10] http://www.imf.org/external/index.htm
[11] http://unstats.un.org/unsd/default.htm
[12] http://www.oecd.org/
[13] GeneOntology Consortium http://www.geneontology.org/
[14] Singer System http://singer.cgiar.org/
[15] GeoNetwork http://www.fao.org/geonetwork/srv/en/main.home

collect and mash up data from distributed sources is still to be solved. Mainly, data sources are as disconnected and unorganized as before. The problem then is how to connect diverse yet intrinsically connected data sets, and organize efficient collaboration processes.

## 3    Interoperable Is Not Centralized. Or, the Linked Data Approach

One approach to data sharing consists in making all interested parties use the same platform. In other words, through centralization. Quite naturally, interesting degrees of interoperability may be achieved in this way. *Facebook* and *Google* are the largest examples of centralized systems that allow easy sharing of data among users, and some interoperability of services. Uniform working environments (software & database schemas) help create interoperability despite physically distributed information repositories. But there are social, political and practical reasons why centralization of repositories or unification of software and working tools will not happen. This does not mean that a certain concentration of data in specific servers on the web or a common set of software tools is useless. It just means that this cannot be the unique road to allow sharing and interoperability.

The alternative to centralization of data or unification of working environments is the development of **standard ways to encode, transmit and process data**, to make distributed data sets interoperable and shareable. This is exactly the idea of the "semantic web", the idea of a web of global interoperability, launched more than 10 years ago. Since then, a number of standards promoted by the W3C have contributed to the achievement of that goal. The underlying idea of these standards is that interoperability is achieved when machines understand the meaning of data and are able to programmatically process it.

The latest paradigm for interoperability on the web is the data publication style called **Linked Data**[16] [1] [2]. Instead of pursuing ad hoc solutions for the exchange of specific data sets, the concept of linked data establishes the possibility to express structured data in a way that it can be linked to other data sets that are following the same principle. This style of publishing data is gaining much attention and new services such as the New York Times and the BBC are already publishing data that way. Some governments too, are pushing heavily to publish administrative information as Linked Data.

Linked Data focuses on data and its links to other pieces of data, independently of their physical location around the world. Linked Data is then about using the web to link together what was not previously linked, and exploit the potentiality of web publication. It is the equivalent for data of what an HTML document (the hypertext of the early days of the internet) is for text. But, unlike HTML documents, linked data may be used programmatically by machines, and so exploited within web applications.

More than a specific technology, Linked Data is a web publication style based on the use of a few web standards and protocols. Primarily, Linked Data is based on the

---

[16] Linked Data - Connect Distributed Data across the Web http://linkeddata.org/

Resource Description Framework[17] (RDF), which facilitates the exchange of structured information regardless of the specific structure in which they are expressed at source. Any database can be expressed using RDF, but also structured textual information from content management systems can be expressed in RDF. Once data is expressed as RDF and exposed through a web server with a SPARQL[18] end-point, the web client may be presented with either a human-readable (HTML) [3], or a machine readable version of the data, depending on the request. The RDF version of the data makes it understandable and processable by machines, which are able to mash-up data from different sites. The advantage of data mesh-up based on Linked Data is that data is exposed with uniform interfaces. On the contrary, in common data mesh-up techniques, ad-hoc procedures are required for each data source.

There are now mainstream open source data management tools like Drupal[19] or Fedora commons[20] which includes already RDF as the way to present data. The other important standards for Linked Data is the URIs to name and locate entities in the web, and the HTTP protocol to exchange web content through web browsers.

Once these three mechanisms (URIs to locate and name data objects, RDF to encode them, and HTTP to move them around in the web) are implemented, data can be linked to any other relevant data set published using the same technologies and protocols. And the whole thing, the resulting linked data, may be exploited in applications.

Let us suppose that a data owner publishes a data set about sardine captures as linked data. Then it could link its data to the a third party data set, say a reliable source of information about biological profiles of fish species. The resulting linked data may be of interest to humans, but also programmatically used in applications exploring the connection between fish capture and the climate.

When Linked Data is published with an explicit open license, or when it is at least openly accessible from a network point of view (not behind an authorization check or paywall), then it is Linked Open Data. The Linked Open Data project[21] collects information about data sets published that way.

FAO, as well as many other large data producers (BBC, the Library of Congress, the US government, to name just a few), has started publication of its data as Linked Open Data. The very first data set published by FAO is its multilingual thesaurus AGROVOC[22] [3], available in over 20 languages and covering all subject areas related to FAO interest. To date, AGROVOC is linked to ten relevant thesauri and vocabularies in specialized domains, and a few more are in the pipeline.[23] Fig. 3 shows an intuitive view of the fact that AGROVOC is linked to other vocabularies, by

---

[17] Resource Description Framework http://www.w3.org/RDF/

[18] http://www.w3.org/TR/rdf-sparql-query/

[19] http://drupal.org/

[20] http://fedora-commons.org/

[21] http://richard.cyganiak.de/2007/10/lod/

[22] http://aims.fao.org/website/AGROVOC-Thesaurus/sub

[23] For the entire list of vocabularies connected to AGROVOC; and a human-readable version of the connected data sets see: http://aims.fao.org/standards/agrovoc/linked-open-data

representing three concepts belonging to three vocabularies (NALT[24], Eurovoc[25], and GEMET[26]). That figure also captures the fact that once various vocabularies are linked, one also gains the access to the document repositories associated to them. With the publication of AGROVOC as linked data, not only has FAO exposed its first data set in the linked data world, but the largest data set about agriculture is now out there for public use.

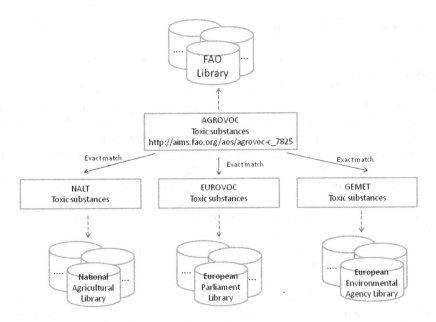

**Fig. 3.** AGROVOC concepts are linked to other vocabularies (solid lines). Vocabularies are then connected to one or more document repositories (dashed lines) because they index the documents contained in them.

We envisage that this is the way to go to achieve interoperability of data in the agricultural domain. In our view, the four key elements to make this happen are:

1. common standard vocabularies to allow for the use of RDF in specific domains and facilitate automatic data linking (thesauri, authority files, value vocabularies)

2. appropriate tools to store, manage and exploit data in RDF, including content management systems, RDF wrappers for legacy systems, and RDF based information extractors

3. a common way for users to know what vocabularies, data sets and tools are available

---

[24] http://agclass.nal.usda.gov/

[25] http://eurovoc.europa.eu/

[26] http://www.eionet.europa.eu/gemet

4. people able to understand these mechanisms, to produce linked data and exploit it in relevant applications. This implies that training and capacity development activities are needed.

This is, in a nutshell, the perspective endorsed by CIARD, which we explain in detail in the next section.

# 4     The CIARD Perspective

The Coherence in Information for Agricultural Research for Development (CIARD) [27] was established in 2008 with the aim of making research information and knowledge accessible to all [4]. Through consultations, virtual and face-to-face, with a wide range of organizations it was agreed that data and information sharing needs to be enhanced and improved starting from within organizations at national, regional and global level. In 2011, two such international consultations took place, to analyze and understand the state of the art in outputs of agricultural research. The result was a plan of actions, to implement the idea that data should be available and shareable avoiding central repositories or standardized software, but rather enabling horizontal communication. Interoperability is then a notion interpreted at the data level, as interoperability of distributed data sets. In this sense, the CIARD perspectives couples very well with the Linked Data perspective introduced previously.

Data should be made available using formats suitable for the web and for common consumption, such as RDF, in combination with standard vocabularies. In case of data born in other formats, it will have to be made available in a web-oriented format by means of suitable software, then stored. Tools, software and services will have to be available to institutions and users in a common and accessible place. Also, data storage will have to be possible. Finally, it is important that awareness and capacities are well developed in the community of users.

In this section we point out at these components individually, and try to provide the reader with a concise, high level summary of the CIARD view. For a more detailed account of that, the interested reader should consult the CIARD website[28]. Note that there is no a priority ranking in our listing, but rather an attempt to provide a narrative line to our view.

1. **Tools to convert existing data into formats suitable to sharing.** Most data repositories will keep their current format and structure for a long time. In principle, this is not a problem, also because many of these systems have been created to optimize data processing. However, this implies that data should be converted in order to be sharable over the web and accessed by other parties in a programmatic way. Various technologies to convert data born in other formats into RDF are available. Also, Linked Data is a publication style that can already be easily

---

[27] http://ciard.net

[28] Core documents can be accessed at: http://www.ciard.net/ciard-documents, while the entire repository is available at: http://www.ciard.net/repository

adopted. However, such processes of conversions are usually performed according to ad-hoc procedures. Our goal is to streamline these processes, and make available to the community of information producers in agriculture suitable methodologies to address these problems. In this way we can produce "triple stores" of RDF data that interlink distributed information sources and make them accessible from aggregations sides or specialized services.

2. **Vocabularies.** The notion of standard vocabulary is needed in all information tagging activities, including the traditional activity of document indexing. In fact, that is the reason why vocabularies like AGROVOC were originally developed. Beyond AGROVOC, a number of vocabularies, thesauri and specialized glossaries are available in the agricultural domain. They need not only to be mapped and linked, but also to be made available for the purpose of data markup. Tools such as Open Calais[29] and AgroTagger (see next section) are examples of software for data markup: based on specific vocabularies, they use entries in the vocabularies to mark up documents. We would also mention another important application of standard vocabularies in information sharing, since data cannot be interoperable without an explicit qualification of its intended meaning. Specifically, being RDF essentially a data model based on the triple structure of "subject-predicate-complement", we need to be clear about what is the intended meaning of the predicates used. Lists of predicates are also called vocabularies, and are essential to information sharing.

3. **Data Storage on the cloud.** Data for the web, will have to be stored in a way that is accessible to the world. This requires some infrastructure that may be costly for many institutions involved into data production. We see here a need that could be addressed by CIARD, by providing an infrastructure available to the community for the storage of their data. This will be a sort of "cloud storage" of data.

4. **Tools for data processing and storing.** A selection of the tools currently available for the publishing of interoperable data should be made, according to the requirements and possibilities of the actors represented by CIARD. What we need is then a suite of tools and services to process and store data, that can be installed and used at institutional levels.

5. **Interfaces to distributed triple stores.** Once data is made available for consumption, it is important that interfaces are also available to make it exploited by applications. The CIARD partners should produce a library of such interfaces with inbuilt data selectors.

6. **Registry for data and software.** We need a one stop access to existing services from the community. The registry of tools and services developed within CIARD, called RING[30], should become a hub of data streams, from which the community can channel their data.

7. **Awareness and Capacity.** Last but not least, a clear commitment to open access for data and publications from all partners is needed. Managers, researchers and information managers must understand that sharing will enhance our capacity to

---

[29] http://www.opencalais.com/
[30] http://ring.ciard.net/

create knowledge for agricultural development. They also must understand the instruments that have been created by international standard bodies to achieve this.

Fig. 4 provides an intuitive view of the data flow from individual, local repositories into web based applications, through web services that consume linked data.

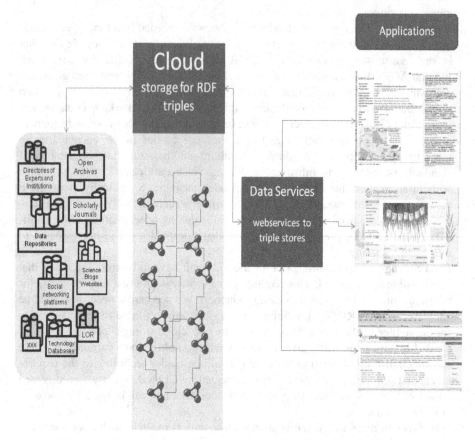

**Fig. 4.** Data flow from local repositories to data accessible through the web

This is a program that can be implemented gradually. FAO is already working towards the implementation of some of these points, with activities supported by both its own budget and also external funding. For example, recently the EU has funded the agINFRA project[31], that deals with the implementation of an infrastructure for the agricultural domain. This fact proves that the CIARD view may be presented in a convincing way to donors and interested stakeholders. In the following section we go on presenting in details FAO´s contribution to the CIARD view.

---

[31] http://aginfra.eu/

# 5     How FAO Is Contributing to the CIARD View

FAO is contributing to two of the main components of the CIARD view. FAO is active in the area of "Vocabularies" (point 2 above), because it maintains the AGROVOC thesaurus. We have introduced AGROVOC in Section 3 above, together with the Linked Data project. AGROVOC is central to FAO's effort within CIARD, because it triggered the development of VocBench, a tool for vocabulary editing, and because it is used in other applications such as AgroTagger, an agriculture-specific tagger, that we introduce below. Moreover, FAO is planning on turning AGRIS[32], originally a bibliographic repository, into a pilot service based on an infrastructure dedicated to interoperability.

VocBench[33] is a web-based tool originally developed to manage the AGROVOC thesaurus. Nowadays, many ontology editors have been developed over the years, including Protégé[34], the NeOn Toolkit[35], Altova Semantic Works[36], and TopBraid Composer[37]. But a few years back, very few editors were available and usually they did not support collaborative work, a formalized workflow with user roles and editing rights also by languages, or UTF-8 – and none all these features together. This is the reason why FAO started the development of its own web-based, fully multilingual vocabulary editor supporting collaboration structured into an explicit workflow. Moreover, in order to allow individuals and organizations to contribute to AGROVOC while maintaining the information about the provenance of their authorship, VocBench applies a fine-grained mechanism of track changes. These features have made the interest around VocBench grow, so that its community of users has grown beyond the one originally envisaged. VocBench is now used to maintain the FAO Biotechnology Glossary[38] and much of the bibliographic metadata used by FAO. Currently in version 1.3, VocBench supports the export of data into RDF and it will soon support RDF natively.

Much of the available data was born before the Linked Data era, so it will need to be converted to formats suitable to that style of publication. This is certainly the case for both relational data and for unstructured textual documents. In particular, in the case of unstructured textual documents, information will have to be extracted from documents and then turned into sets of RDF triples. The information extraction process may involve the extraction of the author and title of the document at hand, or the events discussed in it, or specific topics such as "salmon" or "crops" and the like. For the case of agricultural related topics, IIT Kanpur[39], FAO  and MIMOS Berhad[40] have developed Agrotagger, a web-based tool to tag documents according to

---

[32] http://agris.fao.org/
[33] http://aims.fao.org/tools/vocbench-2
[34] http://protege.stanford.edu/
[35] http://neon-toolkit.org/wiki/Main_Page
[36] http://www.altova.com/semanticworks.html
[37] http://www.topquadrant.com/products/TB_Composer.html
[38] http://www.fao.org/biotech/biotech-glossary/en/
[39] http://www.iitk.ac.in/
[40] http://www.mimos.my/

keywords taken from AGROVOC. Currently in beta version, AgroTagger is available for trial through a web interface[41], where one may select a document and get in return the AGROVOC keywords suggested by AgroTagger, completed with their URIs. The set of suggested keywords can also be downloaded as an XML document. AgroTagger may also be accessed via web services so that one may submit document in bulk, for use within other applications.

Finally, we want to mention the Routemap to Information Nodes and Gateways (RING)[42], a registry of tools, services and data sources developed by the Global Forum on Agricultural Research (GFAR)[43]. The idea behind the RING is to make available from a single access point the data and services relevant to agricultural information professionals. Currently implemented as a Drupal[44] website, it aims at offering extra services on top of the role of registry of data and tools. For example, it could not only provide links to RDF repositories, but also provide view/navigation functionalities to the data. Or it could include harvesters for all data and metadata providers registered in it. It could also become a repository of code to the benefit of programmers wishing to implement services embodying the philosophy shared by the CIARD.

## 6    Conclusions and Future Work

The AIMS team at FAO has concentrated its efforts for years on providing semantic standards to facilitate cooperation and data exchange. Now, thanks to the community grown around the CIARD initiative, a more comprehensive view about information sharing in agriculture is taking shape. All aspects involved, technical, human and political, are being analyzed. In this paper we concentrated on the technical level, and placed the main products of the AIMS team within the infrastructure envisaged by the CIARD view. In the coming years, we will continue working towards the completion of the CIARD view, and publish as much as possible of the FAO data as Linked Data. Our first step will be to open up the Agris bibliographic repository to the web. The connection will be provided by AGROVOC, which is used to index all documents in Agris: a search in Agris will then provide access to data residing in all data repositories linked through AGROVOC. It is for applications like that the Linked Data Given seems to show all its potential.

We are convinced that the landscape of data production will remain heterogeneous for many years, which makes important that tools to convert existing data into RDF and Linked Data formats be widely available to the community. Also, we encourage the development and sharing of applications that exploit the data produced. In this sense, we expect that interesting results will come from the recently started, EU funded project agINFRA.

---

[41] http://kt.mimos.my/AgroTagger/
[42] http://ring.ciard.net/
[43] http://www.egfar.org/egfar/
[44] An open source content management system. See: http://drupal.org/

**Acknowledgments.** Several people and groups in FAO, GFAR, CIARD, IIT Kanpur, and MIMOS Berhad have contributed to the ideas, tools and data sets mentioned in this paper. Among them, special our team in FAO Rome. Special thanks to MIMOS Berhad which founded the participation of Johannes Keizer in the Knowledge Technology Week in 2011, which took place in Malaysia.

# References

[1]  Bizer, C., Heath, T., Berners-Lee, T.: Linked Data - The Story So Far. International Journal on Semantic Web and Information Systems (IJSWIS), Special Issue on Linked Data 5(3), 1–22 (2009)

[2]  Bizer, C., Cyganiak, R., Heath, T.: How to Publish Linked Data on the Web (July 27, 2007), http://www4.wiwiss.fu-berlin.de/bizer/pub/LinkedDataTutorial/

[3]  Dadzie, A.-S., Rowe, M.: Approaches to Visaulising Linked Data: A Survey. Semantic Web (1) (2011)

[4]  Caraccciolo, C., Morshed, A., Stellato, A., Johannsen, G., Jaques, Y., Keizer, J.: Thesaurus Maintenance, Alignment and Publication as Linked Data: the AGROVOC use case. In: Metadata and Semantic Research 5th International Conference Proceedings, Izmir, Turkey (2011)

[5]  Coherence in Information for Agricultural Research and Development, Manifesto (2010), http://www.ciard.net/sites/default/files/CIARD%20Manifesto%20EN_0.pdf

# E-Learning Based on Metadata, Ontologies and Competence-Based Knowledge Space Theory

Dietrich Albert[1,2], Cord Hockemeyer[2], Michael D. Kickmeier-Rust[1]
Alexander Nussbaumer[1], and Christina M. Steiner[1]

[1] Graz University of Technology, Knowledge Management Institute,
Cognitive Science Section,
Brückenkopfgasse 1/VI, 8020 Graz, Austria
[2] University of Graz, Department of Psychology
Universitätsplatz 2 / III, 8010 Graz, Austria
{dietrich.albert,michael.kickmeier-rust,
alexander.nussbaumer,christina.steiner}@tugraz.at
{dietrich.albert,cord.hockemeyer}@uni-graz.at

**Abstract.** The 21st century is challenging the future educational systems with 'twitch-speed' societal and technological changes. The pace of (technological) innovations forces future education to fulfill the need of empowering people of all societal, cultural, and age groups the acquire competences and skills in real-time for demands and tasks we cannot even imagine at the moment. To realize that, we do need smart novel educational technologies that can support the learners on an individual basis and accompany them during a lifelong personal learning and development history. This paper gives some brief insights in approaches to adaptive education based on sound psycho-pedagogical foundations and current technologies.

**Keywords:** Competence-based Knowledge Space Theory, e-learning systems, ontology, game-based learning, self-regulated learning, adaptivity.

## 1 Introduction

Creative, smart, personalized, focused, and lifelong – these are some characteristics of future education. Much more than today, future learning will be an individual process, right in contrast to today's one-fits-all frontal classroom education. This individual process will be enabled and supported by smart technologies that guide the learners, support them, motivate them, and offer them new and rich possibilities of acquiring new understanding and new skills. Smart technologies will assess and understand the learners, their needs, preferences, strength and weaknesses, their goals and visions and those technologies will achieve that understanding in a subtle, unobtrusive way.

These are, to some extent, visions and also hopes for future education; these are the needs for the future educational landscape. The way there is so important because 21st century education, although being a big buzz word, needs to successfully address the demands of the new millennium which is accompanied by substantial technological

D. Lukose, A.R. Ahmad, and A. Suliman (Eds.): KTW 2011, CCIS 295, pp. 24–36, 2012.

evolutions; we became a highly diverse, globalized, complex, real-time media-,knowledge-, information-, and learning-society. Since the 1990s, the progress of media and technology was breathtaking; during these one or two decades, we were facing the rise of a serious and broad use of computers at home (although the development started earlier, of course), the rise of the internet and how it revolutionized our society, becoming a "collective unconscious" (in the words of Carl Gustav Jung). We faced the spread of mobile phones and their evolution from telephones to omnipresent computer and communication devices; we saw spread of mp3, twitch speed computer games and TV shows. We saw how our world got closer by changing the bridges over continents and oceans from 56k wires to hyper speed fiber glass networks. Some say, this rapid and pervasive technological revolution will have greater impact on society than the transition from an oral to a print culture.

The following paper summarizes approaches to smart, adaptive educational technologies based on sound theoretical foundations of cognitive psychology (in particular Competence-based Knowledge Space Theory) and psycho-pedagogical/didactic research (e.g., in the fields of self-regulated learning or game-based learning) and gives some brief insights in novel educational technologies based on metadata and ontologies.

## 2    Knowledge Space Theory

Knowledge Space Theory (KST) is a mathematical-psychological theory for representing domain and learner knowledge (Doignon&Falmagne, 1985, 1999; Falmagne & Doignon, 2011). In KST a *knowledge domain* is identified with a set $Q$ of problems. The subset of problems that a person is able to solve represents the *knowledge state* of this individual. Among the problems of a domain mutual dependencies will exist, such that not all potential knowledge states (i.e. subsets of problems) will actually occur. These dependencies are captured by a so-called *prerequisite relation* (also referred to as precedence relation), which restricts the number of possible knowledge states. Two problems $a$ and $b$ are in a prerequisite relation whenever the correct solution of problem $a$ is a prerequisite for the mastery of problem $b$. Illustrated in a Hasse diagram (see Fig. 1), ascending sequences of line segments indicate a prerequisite relationship. The collection of knowledge states corresponding to a prerequisite relation is called a *knowledge structure*. In a knowledge structure a range of different *learning paths* from the naïve knowledge state to the expert knowledge state are possible (see Fig. 1). A knowledge structure enables *adaptive assessment* procedures for efficiently identifying the current knowledge state of an individual (see e.g., Doignon & Falmagne, 1999; Hockemeyer, 2002). Through defining individual starting and goal states for a learner, meaningful learning sequences with reasonable choices for navigation and appropriate levels of challenge can be realized for each learner.

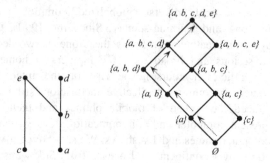

**Fig. 1.** Example of a prerequisite relation and the induced knowledge structure. Dashed arrows show a possible learning path.

## Application: ALEKS

The commercial ALEKS (Adaptive Learning with Knowledge Spaces) system is a fully automated, multi-lingual, adaptive tutor that grounds on KST (Canfield, 2001). The system provides individualized learning including explanations, practice, and feedback on learning progressfor various disciplines, ranging from maths and natural science to social sciences. ALEKS adaptively and accurately assesseswhich concepts a learner already knows, what he/she is ready to learn next, which previously learned material should be addressed for review, and continuously updates a precise map of the learner's knowledge state (Falmagne, Doignon, Cosyn, & Thiery, 2004; Hardy, 2004).

## 3    Component and Demand Analysis

Opposed to the mere behavioral approach of Doignon and Falmagne, the group of Albert (see Albert & Lukas, 1999) investigated the latent cognitive structures and processes underlying the observable behavior. In this section, we focus on the analysis of problem components and demands, which was introduced into knowledge space research by Albert and Held (1994). In a first step, components of test problems in a certain area and attributes for these components occurring in the problem set are searched for. In elementary probability theory, for example, Held (1999) found for problems on drawing balls from an urn as one component the way of drawing with three attributes, (1) drawing one ball, (2) drawing multiple balls with replacement, and (3) drawing multiple balls without replacement.

In a second step, the demands (i.e., required skills) posed on the learner by the different attribute values are investigated. In the example above, Held identified four demands related to the attributes for the component way of drawing, (a) knowing the Laplace rule for computing probabilities, (b) being able to determine the number of convenient events if only one ball is drawn, (c) knowing that for stochastically independent events the probability of their intersection can be computed as product of the single probabilities, and (d) knowing that drawing without replacement reduces the number of balls in the urn. For the three attributes and the four demands specified above, Held found the assignment shown in Table 1.

Table 1. Demand assignments for attributes of way of drawing

| Attribute | Demands |
|-----------|-----------|
| 1 | a, b |
| 2 | a, b, c |
| 3 | a, b, c, d |

Based on the demand assignment, the attributes for each component can be ordered, e.g. using the set inclusion principle. In this case, we would obtain an ordering $1 \le 2 \le 3$. Regarding set of test problems for which multiple components have been identified, the test problems can be characterized through attribute vectors and can be ordered according to a component-wise ordering of these attribute vectors.

Held (1999; Albert and Held, 1994) has shown that this method can also be used to complete a set of test problems by constructing problems for those attribute vectors for which there do not yet exist any problems.

**Application: RATH**

The results of Held's structuring process have been applied in the adaptive tutoring system RATH (Hockemeyer, Held, & Albert, 1998). For each of the demands identified by Held (1999) in the field of elementary probability theory, they developed a lesson teaching the respective skill. They derived a demand structure from the demand assignment following the set inclusion principle again; a demand $a$ is considered a prerequisite of demand $b$ if all test problems which require $b$ do also require $a$ (Albert & Hockemeyer, 2002). The RATH system then provides navigation guidance through a combined structure of lessons and test problems.

## 4    Competence-Performance Approach

The competence-performance-approach was developed by Korossy (1997, 1999) as an extension to the original, rather behavioral KST by Doignon and Falmagne (1985, 1999). He distinguishes between observable performances, i.e. test item solving behavior, and their underlying competencies (in other approaches also denoted as skills). This is done by mapping each item to the subset of competencies required for solving this item and, vice versa, by mapping each subset of competencies to the subset of items which can be solved by a person who has all (and only) the competencies of the given subset. From these mappings, prerequisite structures on the sets of competencies and of performances (i.e. items) can be derived through the set inclusion principle: an item $a$ is a prerequisite of item $b$ (in the sense of the aforementioned surmise relation) if the set of competencies assigned to $a$ is a subset of the set assigned to $b$.

Fig. 2 shows an example of three items $A$, $B$, and $C$ to which subsets of competencies $x$, $y$, and $z$ are assigned (denoted by the arrows). The set inclusion relation on the competence subsets (denoted by dashed lines) induces a surmise relation between the items (denoted by straight lines).

Based on this background, Hockemeyer (2003; Hockemeyer, Conlan, Wade, & Albert, 2003) extended Korossy's competence–performance approach to the concept of competence learning, which is summarized below. The basic idea is to specify, for each learning object, separate subsets of required and taught competencies. Given two mappings $r$ and $t$, which assign to each learning object these subsets of competencies one can derive a surmise mapping $\sigma_L$ on the set of competencies. For each competence $c$, its set $\sigma_L(c)$ of clauses contains the sets $r(l) \cup t(l)$ of all learning objects $l$ teaching competence $c$, i.e. all learning objects $l$ for which $c \, \varepsilon \, t(l)$ holds. This surmise mapping $\sigma_L$ can then easily be closed under transitivity etc. using well-documented efficient procedures (Dowling & Hockemeyer, 1998). Through this surmise mapping $\sigma_L$ (and its closure) on the set of competencies, a competence space, i.e. the set of all possible competence states, is well–defined.

The main difference between competence learning structures and Korossy's competence–performance approach lies first of all in the separation of taught and required competencies (in the case of test items, competencies which are actually to be tested by the item  and other required competencies would be separated, instead). The advantage of this separation is that authors (or metadata creators) do not have to specify those prerequisites, which can be derived through transitivity. This is especially important if the objects are to be used in different contexts where different sets of (especially low–level) competencies might be used.

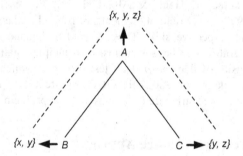

**Fig. 2.** Competence assignment and induced prerequisite relationships

## Application: APeLS

The concept learning structures were applied in the prototypical adaptive learning system APeLS (Conlan, Hockemeyer, Wade, & Albert, 2002). It can easily merge content from different sources to build an adaptive course. The only requirement is that the individual learning objects carry metadata information on required and taught competencies according to the competence learning structure approach given that the metadata author use the same competence terminology (Albert, Hockemeyer, Conlan, & Wade, 2001).

# 5     Ontology-Based Skill Approach

In Technology-Enhanced Learning (TEL) systems there is often a need to use ontologies or concept maps to represent learning or subject domains. Concept maps are directed graphs consisting of concepts (nodes) and semantic relations between concepts (edges). All pairs of connected concepts form a set of propositions that represents the semantic or declarative knowledge of the respective subject domain (Steiner, Albert, & Heller, 2007). Semantic technologies open up the possibility to use concept maps in TEL systems for several purposes, such as the definition of learning goals.

In order to incorporate ontologies and concept maps into KST and its competence-based extension, structure and elements of concept maps are used to build prerequisite structures on skills and assessment problems. Different approaches have been identified how competence and knowledge structures can be derived from concept maps (Steiner & Albert, 2008). In order to build prerequisite relations between assessment problems, a technique based on set inclusions of related concepts is proposed. Using a concept map, the problems are analyzed regarding the concepts (or propositions) of the concept map needed to solve the respective problems. In a second step the assigned sets of concepts (or propositions) can be compared whether one set is a sub-set of another set. In this case, a problem A is prerequisite for problem B, if the concept set of problem A is a sub-set of the concept set of problem B.

Building structures on skills based on concept map information can be done in a similar way. Skills can be defined through concepts (or propositions) from concept maps. The same set inclusion strategy is used to build prerequisite relations between skills. It is important to mention that the definition of skills not only takes into account declarative knowledge (concepts), but also procedural knowledge. Action verbs are used to indicate in which way conceptual knowledge is applied. Through the application of a component-attribute approach (Albert & Held, 1994), structures on concepts and on action verbs can be combined to derive prerequisites on skills defined as pairs consisting of an action and a concept component (Heller, Steiner, Hockemeyer, & Albert, 2006).

There are several advantages of the described approach to use ontologies and concept maps as a basis for CbKST. Existing curricula can be opened up to CbKST procedures, and in the future information from a growing Semantic Web can be taken into account. Personalized learning strategies can be grounded on conceptual knowledge that might be pre-existing or automatically generated.

### Application: iClass

In the iClass project a semantic structure in Web Ontology Language (OWL) has been created to capture CbKST and concept map information and to relate their elements accordingly (Görgün et al., 2005). In this way a curriculum can be expressed as a knowledge map, which in turn can be used for creating personalized learning paths and efficient assessment procedures. In order to create such knowledge maps, a software tool has been created which allows for defining the entities of the map and their relationship in a graphical manner (Steiner, Nussbaumer, & Albert, 2009).

# 6    Self-Regulated Learning (SRL)

In the last decade CbKST has successfully been used in TEL systems for adapting learning content and system behavior to the learners' competence and knowledge state. In addition to adaptivity research in the TEL area there is the research stream on self-regulated learning (SRL), which argues for giving more control to the learner (Zimmermann, 2002). While adaptive systems guide the learner through the learning experience, the SRL approach considers learning guided by meta-cognition, strategic action, and motivation of the learner. Though these two approaches appear contradicting at first sight, they can be harmonized and complement each other.

**Application: iClass, ROLE**

While SRL might run the risk of overloading the learner with control and choice, adaptive learning realized through CbKST might risk overriding learners' involvement and will. In the context of the European iClass (Intelligent Distributed Cognitive-Based Open Learning System for Schools) project, research and development has aimed to find a balance between 'the system decides everything' and 'the learner decides everything'(Steiner, Nussbaumer, & Albert, 2009).

In an attempt to harmonize the approach of SRL with the tradition of adaptation and personalization, a comprehensive pedagogical model has been defined. The key pedagogical process of this model is self-regulated personalized learning (SRPL), which aims to realize personalization embedded in self-regulation. SRPL therefore stands for providing learners with the opportunity to self-regulate their learning process and supporting them in this self-personalization through adaptation technologies. SRPL thus also builds upon the typical cyclical phases of an SRL process, characterized by planning (forethought), learning (performance) and reflection. In this way, the learner has the freedom to control the own learning behavior, but is also guided by adaption strategies.

For the technical realization of this model, several tools have been developed which support the phases of the SRPL cycle. A planning tool allows the learner for creating a learning goal and learning plan based on graphically depicted domain and user model. Unlike as for adaptive systems, these models are not hidden from the learner, but used as structures that guide the learner in her self-regulated learning process. An assessment tool performing adaptive knowledge assessment also gives insight in the actual performance. Another tool supports the learner to self-evaluate her competence state based on domain model information. The knowledge presenter tool graphically depicts the knowledge and competence state on a graphical map and reveals the competence gap between learning goal and available competences.

In the ROLE project (http://www.role-project.eu/) learners are empowered to build their own learning environment consisting of learning tools coming for a huge tool repository. For supporting SRL in this environment, the SRPL model has been extended. Self-regulatory competences have been defined and are used as a basis for the user profile. The learner is primarily supported to build a personalized learning environment, which can be used for SRL. Recommendation strategies based on self-regulatory competences support the learner for this activity (Fruhmann, Nussbaumer, &Albert, 2009). Furthermore, recommendation tools guide the learner through the SRL process using the individually compiled learning environment.

# 7    Game-Based Learning (GBL)

A medium that increasingly got in the focus of research on smart educational technologies is utilizing the strength and potentials of modern computer games for educational purposes. The genre of learning games has become a major field of psycho-pedagogical research and quite certainly such games will play an important and accepted role in the future educational landscape.

A key topic of research is seen in equipping the games with an educational artificial intelligence (AI) that allows adapting to the individual needs of learners. Especially in the genre of educational computer games, a psycho-pedagogically sound personalization is a key factor since the games' success heavily depends on fragile and highly individual constructs like motivation, immersion, and flow experience.

**Application: ELEKTRA, 80Days**

We commenced transposing the ideas of CbKST and CbKST-based educational technologies to the genre of educational computer games in the context of the European ELEKTRA project (www.elektra-project.org), which was no trivial attempt since it requires subtle, unobtrusive, strongly embedded mechanisms for assessing psycho-pedagogical states (e.g., learning progress or motivation) and for adaptations accordingly. ELEKTRA was a multi-disciplinary research and development project, running from 2006 to 2008, funded by the European Commission.

Within the project a methodology for successful design of educational games has been established and a game demonstrator was developed based on a state-of-the-art 3D adventure game, teaching optics according to national (i.e., French, Belgian, and German) curricula. More importantly, ELEKTRA addressed research questions concerning data model design as basis for adaptivity and resource description enabling interoperability of systems as well as the data model itself (Kickmeier-Rust & Albert, 2008). In the course of the project, an approach to adaptivity, that is, micro adaptivity (Albert, Hockemeyer et al., 2007), was developed that allows assessing learning performance and cognitive states in a non-invasive way by interpreting the learners' behavior within the game and by responding on the conclusions drawn from their behavior. Attuned to the assessed competencies (or lack of competencies), meaningful feedback, for example hints, suggestions, reminders, critical questions, or praise, can be triggered, without destroying the gaming experience.

The work of ELEKTRA was continued in a successor project named 80Days (www.eightydays.eu); inspired by Jules Verne's novel "Around the world in eighty days", the project was running from April 2008 until September 2010. During the project, the consortium could make significant progress by elaborating a joint formal model of cognitive assessment of learning progress on a probabilistic and non-invasive level, the provision of suitable support and interventions, and open interactive adaptive storytelling (cf. Augustin, Hockemeyer, Kickmeier-Rust, & Albert 2011; Kickmeier-Rust & Albert, 2010; Kickmeier-Rust, Mattheiss, Steiner, and Albert, 2011). From a technical point of view, an accurate analysis of learning and game design requirements has been carried out and the results have constituted the starting point for the study on system architectures and software modules that best could have fulfilled the requirements. Research in the area of open, interactive storytelling achieved a technical realization of the developed formal model in form of a story engine, which implements the psycho-pedagogical model and which drives and adapts the game. Overall, psycho-pedagogical and technical efforts lead to a compelling demonstrator game teaching geography (Fig. 3). Significantly, this demonstrator also represents the substantial steps towards achieving a multi-adaptive system that not only adapts discrete elements of the game towards educational purposes, but also adapts the story to accommodate larger educational objectives.

The demonstrator game (Fig.3) is teaching geography for a target audience of 12 to 14 year olds and follows European curricula; therefore an adventure game was realized within which the learner takes the role of an Earth kid. The game starts when a UFO is landing in the backyard and an alien named Feon is contacting the player. Feon is an alien scout who has to collect information about Earth. The player assists the alien to explore the planet and to create a report about the Earth and its geographical features. This is accomplished by the player by means of flying to different destinations on Earth, exploring them, and collecting and acquiring geographical knowledge.

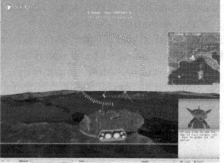

**Fig. 3.** Screenshots from the prototype games developed in the ELEKTRA project (left) and its successor 80Days (right)

The demonstrator game was subject to broad evaluation activities. The evaluation work has been geared towards its objectives of defining an evaluation framework and of implementing an array of evaluative activities. In close collaboration of different disciplines, the game design concepts were validated in schools in England and Austria. Multi-method approaches have been applied to analyze the empirical data thus collected.

Empirical findings yielded beneficial effect of playing the game, as evident and an overall satisfying usability and user experience (e.g., Kickmeier-Rust, Hillemann, & Albert, 2011). Implications for the future development of the game prototypes and the design of evaluative activities have been drawn. In particular, the theoretical knowledge and practical experience thus gained will contribute to advancing the research area of evaluating usability and user experience in digital educational games.

The achievements of both projects feed back into the work of further research initiatives – in the field of game-based learning and beyond.

## 8     Current Trends in Technology-Enhanced Learning

The work on Competence-based Knowledge Space Theory (CbKST) and its applications as presented in this paper constitutes a continuous elaboration and evolution of this theoretical framework towards new directions in the field of technology-enhanced learning (see also Conlan, O'Keeffe, Hampson, & Heller, 2006; Pilato, Pirrone, & Rizzo, 2008). This is an ongoing process of taking up current and new trends in education, in general, and technology-enhanced learning, in particular, and is reflected by the work inother projects, like MedCAP (http://www.medcap.eu/), TARGET (http://www.reachyourtarget.org/), ImREAL (http://www.imreal-project.eu/), NEXT-TELL (http://www.next-tell.eu/), and CULTURA (http://www.cultura-strep.eu/).

The development of skill- and competence-oriented, highly adaptive, context-sensitive systems for personalized learning clearly continues in current projects, for example by researchinginnovative forms of multi-dimensional personalization, by augmenting virtual learning environments and learning experiences, or by investigating appraisal methodsand new forms of detailed certification on different levels (individual, group, school, company, etc.) for different stakeholders. Adaptive, context-sensitive support for teachers and groups of students will foster distributed cognitions and competences. The fact that learning takes place every time and everywhere is increasingly acknowledged by the creation of systems capturing and supporting such informal and social learning situations.The developments on self-regulated learning and learner empowerment are progressing towards the vision of enabling every learner to create his or her own transparent learning environmentthat suits his or her actual needs in an ideal manner, and which can be adapted and changed throughout the learner's lifelong learning path, also depending on the context. Transparency means experiencing the own learning process and progress and the learning mechanisms - so that learners and teachers turn into experts of their own learning and teaching. Totally new curricula are evolving in the educational landscape, regarding learning and performance strategies and information literacy: the acquisition of competences in creativity, innovation, complex problem solving and decision making is getting more and more important in basic and higher education, as well as at the workplace. These developments towards explicitly considering and accessing new areas of competence are increasingly taken up in technology-enhanced learning in order to provide ICT support in formal and informal learning settings for pupils and adult learners.

For appropriately demonstrating the potential of technology-enhanced learning and for further evolving research and development in response to the current and future trends, inter-disciplinary cooperation is necessary across different disciplines of computer science, mathematics, psychology, and pedagogy. Moreover, the users in question and their tasks need to be focused early and continually in these research activities to ensure the success and uptake of the new technologies.

**Acknowledgments.** The research and development reported in this paper was financially supported by the European Commission within their Framework Programmes 3 to 7 through the grants ERBCHBICT941599 (FP3-HCM), IST-1999-10051 (FP5-IST; EASEL), 507922 (FP6-IST; iClass), 027986 (FP6-IST; ELEKTRA), 231396 (FP7-ICT; ROLE), 215918 (FP7-ICT; 80Days)

# References

Albert, D., Held, T.: Establishing knowledge spaces by systematical problem construction. In: Albert, D. (ed.) Knowledge Structures, pp. 78–112. Springer, New York (1994)

Albert, D., Hockemeyer, C.: Applying demand analysis of a set of test problems for developing an adaptive course. In: Proceedings of the International Conference on Computers in Education ICCE 2002, pp. 69–70. IEEE Computer Society, Los Alamitos (2002)

Albert, D., Hockemeyer, C., Conlan, O., Wade, V.: Reusing adaptive learning resources. In: Lee, C.-H., et al. (eds.) Proceedings of the International Conference on Computers in Education ICCE/SchoolNet 2001, vol. 1, pp. 205–210 (2001)

Albert, D., Hockemeyer, C., Kickmeier-Rust, M.D., Peirce, N., Conlan, C.: Microadaptivity within complex learning situations – a personalized approach based on competence structures and problem spaces. In: Hirashima, T., Hoppe, H.U., Shwu-Ching Young, S. (eds.) Supporting Learning Flow through Integrative Technologies. Frontiers in Artificial Intelligence and Applications, vol. 162, IOS Press (2007) ISBN 978-1-58603-797-0; Albert, D., Lukas, J. (eds.): Knowledge Spaces: Theories, Empirical Research, Applications. Lawrence Erlbaum Associates, Mahwah (1999)

Augustin, T., Hockemeyer, C., Kickmeier-Rust, M.D., Albert, D.: Individualized Skill Assessment in Digital Learning Games: Basic Definitionsand Mathematical Formalism. IEEE Transactions on Learning Technologies 4(2), 138–148 (2011)

Canfield, W.: ALEKS: A Web-based intelligent tutoring system. Mathematics and Computer Education 35, 152–158 (2001)

Conlan, O., Hockemeyer, C., Wade, V., Albert, D.: Metadata driven approaches to facilitate adaptivity in personalized eLearning systems. The Journal of Information and Systems in Education 1, 38–44 (2002)

Conlan, O., O'Keeffe, I., Hampson, C., Heller, J.: Using knowledge space theory to support learner modeling and personalization. In: Reeves, T., Yamashita, S. (eds.) Proceedings of

World Conference on E-Learning in Corporate, Government, Healthcare, and Higher Education 2006, pp. 1912–1919. AACE, Chesapeake (2006)

Doignon, J., Falmagne, J.: Spaces for the assessment of knowledge. International Journal of Man-Machine Studies 23, 175–196 (1985)

Doignon, J., Falmagne, J.: Knowledge Spaces. Springer, Berlin (1999)

Dowling, C.E., Hockemeyer, C.: Computing the intersection of knowledge spaces using only their basis. In: Dowling, C.E., Roberts, F.S., Theuns, P. (eds.) Recent Progress in Mathematical Psychology, pp. 133–141. Lawrence Erlbaum Associates Ltd., Mahwah (1998)

Falmagne, J.-C., Doignon, J.-P.: Learning spaces. Interdisciplinary applied mathematics. Springer, Berlin (2011)

Falmagne, J.-C., Doignon, J.-P., Cosyn, E., Thiery, N.: The assessment of knowledge in theory and practice (2004), http://www.aleks.com/about/Science_Behind_ALEKS.pdf (retrieved August 23, 2011)

Fruhmann, K., Nussbaumer, A., Albert, D.: A Psycho-Pedagogical Framework for Self-Regulated Learning in a Responsive Open Learning Environment. In: Hambach, S., Martens, A., Tavangarian, D., Urban, B. (eds.) Proceedings of the International Conference eLearning Baltics Science (eLBa Science 2010). Fraunhofer (2010)

Görgün, I., Türker, A., Ozan, Y., Heller, J.: Learner Modeling to Facilitate Personalized E-Learning Experience. In: Kinshuk, Sampson, D.G., Isaías, P.T. (eds.) Cognition and Exploratory Learning in Digital Age (CELDA 2005), pp. 231–237. IADIS (2005)

Hardy, M.E.: Use and evaluation of the ALEKS interactive tutoring system. Journal of Computing Sciences in Colleges 19(4), 342–347 (2004)

Held, T.: An integrated approach for constructing, coding, and structuring a body of word problems. In: Albert, D., Lukas, J. (eds.) Knowledge Spaces: Theories, Empirical Research, Applications, pp. 67–102. Lawrence Erlbaum Associates, Mahwah (1999)

Heller, J., Steiner, C., Hockemeyer, C., Albert, D.: Competence-Based Knowledge Structures for Personalised Learning. International Journal on E-Learning 5(1), 75–88 (2006)

Hockemeyer, C.: A Comparison of Non–Deterministic Procedures for the Adaptive Assessment of Knowledge. Psychologische Beiträge 44, 495–503 (2002)

Hockemeyer, C.: Competence based adaptive e-learning in dynamic domains. In: Hesse, F.W., Tamura, Y. (eds.) The Joint Workshop of Cognition and Learning through Media-Communication for Advanced E-Learning (JWCL), pp. 79-82, Berlin (2003)

Hockemeyer, C., Conlan, O., Wade, V., Albert, D.: Applying competence prerequisite structures for eLearning and skill management. Journal of Universal Computer Science 9, 1428–1436 (2003)

Hockemeyer, C., Held, T., Albert, D.: RATH - a relational adaptive tutoring hypertext WWW–environment based on knowledge space theory. In: Alvegård, C. (ed.) CALISCE 1998: Proceedings of the Fourth International Conference on Computer Aided Learning in Science and Engineering, pp. 417–423. Chalmers University of Technology, Göteborg (1998)

Kickmeier-Rust, M.D., Albert, D.: The ELEKTRA ontology model: A learner-centered approach to resource description. In: Leung, H., Li, F., Lau, R., Li, Q. (eds.) ICWL 2007. LNCS, vol. 4823, pp. 78–89. Springer, Heidelberg (2008)

Kickmeier-Rust, M.D., Albert, D.: Micro adaptivity: Protecting immersion in didactically adaptive digital educational games. Journal of Computer Assisted Learning 26, 95–105 (2010)

Kickmeier-Rust, M.D., Hillemann, E., Albert, D.: Tracking the UFO's Paths: Using Eye-Tracking for the Evaluation of Serious Games. In: Shumaker, R. (ed.) Virtual and Mixed Reality, Part I, HCII 2011. LNCS, vol. 6773, pp. 315–324. Springer, Heidelberg (2011)

Kickmeier-Rust, M.D., Mattheiss, E., Steiner, C.M., Albert, D.: A Psycho-Pedagogical Framework for Multi-Adaptive Educational Games. International Journal of Game-Based Learning 1(1), 45–58 (2011)

Korossy, K.: Extending the theory of knowledge spaces: A competence-performance approach. Zeitschrift für Psychologie 205, 53–82 (1997)

Korossy, K.: Modeling knowledge as competence and performance. In: Albert, D., Lukas, J. (eds.) Knowledge Spaces: Theories, Empirical Research, Applications, pp. 103–132. Lawrence Erlbaum Associates, Mahwah (1999)

Pilato, G., Pirrone, R., Rizzo, R.: A KST-based system for student tutoring. Applied Artificial Intelligence 22, 283–308 (2008)

Steiner, C.M., Albert, D.: Personalising Learning through Prerequisite Structures Derived from Concept Maps. In: Leung, H., Li, F., Lau, R., Li, Q. (eds.) ICWL 2007. LNCS, vol. 4823, pp. 43–54. Springer, Heidelberg (2008)

Steiner, C., Albert, D., Heller, J.: Concept Mapping as a Means to Build E-Learning. In: Buzzetto-More, N.A. (ed.) Advanced Principles of Effective e-Learning, pp. 59–111. Informing Science Press, Santa Rosa (2007)

Steiner, C., Nussbaumer, A., Albert, D.: Supporting Self-Regulated Personalised Learning through Competence-Based Knowledge Space Theory. Policy Futures in Education 7(6), 645–661 (2009)

Zimmerman, B.J.: Becoming a Self-Regulated Learner: An Overview. Theory Into Practice 41(2), 64–70 (2002)

# Linked Library Data:
# Offering a Backbone for the Semantic Web

Joachim Neubert and Klaus Tochtermann

ZBW German National Library of Economics – Leibniz Centre for Economics
Neuer Jungfernstieg 21, 20354 Hamburg, Germany
{j.neubert,k.tochtermann}@zbw.eu

**Abstract.** Since the publication of Tim Berner-Lees "Linked Data – Design Issues" [1] in 2006, the number of linked datasets in the Semantic Web has exploded. However, coverage and quality of the datasets are seen as issues, since many of them are the outcome of time-limited academic projects and are not curated on a regular basis. Yet, in the domain of cultural heritage institutions, large and high quality datasets have been built over decades, not only about publications but also about the personal und corporate creators and about the subjects of publications. Libraries and information centres have started to publish such datasets as Linked Open Data (LOD). The paper will introduce types of such datasets (and services built on them), will present some examples and explore their possible role in the linked data universe.

## 1    Introduction

Libraries and Information Centres gather and organize information, often for centuries. Their cataloging rules and data formats seem arcane to every outsider. Recently however, some of them opened up to the Semantic Web and especially to Linked (Open) Data – for their own purposes and for inter-library exchange and re-use of data, but also for making their results available to the general public.

In May 2010, within the World Wide Web Consortium (W3C) a "Library Linked Data Incubator Group"[1] was formed in order to better coordinate such efforts and suggest further action. An open wiki[2] allows access to the work of the group. The publicly accessible datasets were collected in the CKAN data registry[3]. Mainstream library organizations such as the International Federation of Library Associations and Institutions (IFLA) and the American Library Association (ALA) are currently forming special interest groups on Library Linked Data. Focussed conferences such as

---

[1]  http://www.w3.org/2005/Incubator/lld/
[2]  http://www.w3.org/2005/Incubator/lld/wiki/Main_Page
[3]  http://ckan.net/group/lld

D. Lukose, A.R. Ahmad, and A. Suliman (Eds.): KTW 2011, CCIS 295, pp. 37–45, 2012.

the LOD-LAM Summit[4] in San Francisco or the SWIB conference[5] in Germany organize a vivid exchange of ideas [2].

In the remainder of this paper, typical datasets from the library community will be presented. Authority files, as discussed in the first section, are used to disambiguate especially personal and corporate bodies names. Thesauri and classifications are used to organize knowledge mostly in confined domains, as shown in the second section. As demonstrated in the third section, bibliographic datasets not only give us data about publications, but also link together data about people, organizations and subjects. Conclusions about the applicability of library linked data in the broader semantic web are drawn in the fourth, final section.

## 2    Agents Identified – Personal and Corporate Bodies Name Authority Files

Especially large scientific libraries have a constant need for identifying persons and – to a lesser extent – institutions which are creators, editors or subjects of works: They have to tell apart persons with the same name, or to track different names back to one single person. To this end, over decades so-called authority files have been built, typically under the curation of national libraries, often as a collaborative effort of many scientific libraries. Rules apply for additional properties which are necessary to identify a person, such as date of birth and/or death, profession or affiliation to an organsiation. Different spellings of the name – possibly in different scripts – are recorded, and publications by the person are referenced.

The rules for the individualization of entries differ from country to country. For the German "Personennormdatei" (PND, personal names authority file) for example, dates of birth and death are used, the field of activity, the title of a work by the person, the profession or occupation, the designation, country or location, relations to other persons or an affiliation to an organisation. Interpersonal relationships, such as child/parent/sibling/spouse of, are also denoted [3]. Normally, a minimum of two identifying properties (other than the name itself) is required for building a valid authority record. Since this information is carefully checked by professional staff, the resulting data is normally of high quality.

The German National Library (DNB) has published this data in its "Linked Data Service"[4]. And like many other national institutions it feeds its personal and corporate bodies name authority files into the Virtual International Authority File (VIAF)[6][5]. National libraries or union catalogs of Australia, the Czech Republic, France, Hungary, Israel, Italy, Portugal, Spain, Sweden, Switzerland, the Library of Congress, the Vatican Library, the Bibliotheca Alexandrina and the Getty Research Institute add their data on a regular basis; Canada, Poland and Russia do this with a test status. VIAF merges the records from different sources to clusters under own URIs, which "expands the concept of universal bibliographic control by (1) allowing

---

[4]  Linked Open Data in Libraries Archives and Museums (2011),
     http://lod-lam.net/summit/
[5]  Semantic Web in Libraries (since 2009), http://swib.org
[6]  http://viaf.org/

national and regional variations in authorized form to coexist; and (2) supporting needs for variations in preferred language, script and spelling"[7]. VIAF offers access to RDF representations and an autosuggest lookup service for the clusters. An alternate, more artifical-intelligence-inspired approach is taken by the (currently not publicly accessible) ONKI People service [6].

An example for a VIAF personal name authority entity is given in Fig. 1. It shows how a person (the "real world entity" Anton Chekhov) is described by VIAF itself and by the aggregated authority files (e.g., of the German and Russian National Library). The collected foaf:name entries sum up to 165 (!) variants in different spellings and scripts. As may be noticed, VIAF itself takes no choice of a preferred name. The different libraries however do so.

```
<http://viaf.org/viaf/95216565>
  rdaGr2:dateOfBirth "1860" ;
  rdaGr2:dateOfDeath "1904" ;
  a rdaEnt:Person, foaf:Person ;
  owl:sameAs <http://d-nb.info/gnd/118638289>, <http://dbpedia.org/resource/Anton_Chekhov>,
            <http://libris.kb.se/resource/auth/201439> ;
  foaf:name "Bogemskiĭ, A., 1860-1904", "Cecov, A. 1860-1904", "Cecov, Anton, 1860-1904", "Cekhava, Ențana, 1860-
            1904", "Cekhovha, Ayāṇtan, 1860-1904", "Cekoff, Antonio 1860-1904", "Chechov, Anton 1860-1904",
            ... .

<http://viaf.org/viaf/sourceID/DNB%7C118638289#skos:Concept>
  a skos:Concept ;
  skos:altLabel "Bogemskiĭ, A., 1860-1904", "Cekhava, Ențana, 1860-1904", "Chehov, Anton, 1860-1904", ... ;
  skos:inScheme <http://viaf.org/authorityScheme/DNB> ;
  skos:prefLabel "Čechov, Anton P. 1860-1904" ;
  foaf:focus <http://viaf.org/viaf/95216565> .

<http://viaf.org/viaf/sourceID/RSL%7Cnafpn-000082167#skos:Concept>
  a skos:Concept ;
  skos:altLabel "Chekhov, Anton, 1860-1904", "Chéjov, Antón 1860-1904", "Csehov 1860-1904", ... ;
  skos:inScheme <http://viaf.org/authorityScheme/RSL> ;
  skos:prefLabel "Чехов, Антон Павлович, 1860-1904" ;
  foaf:focus <http://viaf.org/viaf/95216565> .
```

**Fig. 1.** VIAF data about the Russian writer Anton Chekhov (heavily shortened)

Together, libraries' authority files form a large fund of interlinked identities for persons and organisations. As table 1 shows, it outnumbers in this respect by far the most used dataset on the Semantic Web, DBpedia: 364,000 persons in DBpedia as compared to 10 million in VIAF. The VIAF entries cover persons (and in parallel organisations) who created works, but also persons – living as well as historical– who are or were subjects of works. In conjunction with different national cultures, this leads to a very broad coverage of "publicly known" persons and organisations.

The authorities become even more valuable when interlinked with already existing Linked Data hubs such as DBpedia/Wikipedia. To this aim, a project of DNB and the German Wikipedia initiated the crowdsourced enrichment of Wikipedia pages with

---

[7] http://www.oclc.org/research/activities/viaf/

PND ids [7], which resulted in up to now 55,000 DBpedia–PND links. Projects such as "Linked History"[8] (University Leipzig) already make use of these links.

The potential of interlinking through authorities as shared identities for web resources is demonstrated by the PND-BEACON project. Though not in any Semantic Web format, it connects more than 50 datasets[9] by use of identifiers from the personal name authority file as a common key.

**Table 1.** Numbers of individual persons and corporate bodies identified by selected sources

| | Persons | Organisations / Corporate Bodies |
|---|---|---|
| DBpedia[10] | 364,000 | 148,000 |
| Library of Congress Authorities[11] | 3,800,000 | 900,000 |
| German National Library Authority Files[12] | 1,797,911 | 1,262,404 |
| VIAF[13] | 10 million | 3.25 million |

## 3     Structured Knowledge Organisation – Thesauri and Classifications

In contrast to the comparatively simple and well definable entities of persons and organisations, things get much more differentiated and fuzzy when it comes to general subjects. Libraries deal with the thematic scope of works by assigning subject headings or, more sophisticated, descriptors from a thesaurus or classes from a classification. These instruments form knowledge organization systems in that their concepts normally are well defined and separated against each other. Classifications form a strict mono-hierarchy of concepts about the world as a whole (Universal or Dewey Decimal Classification, UDC and DDC respectively) or about a domain of specific knowledge (e.g., the classification of the Journal of Economic Literature, JEL). Thesauri include much richer properties and relationships, especially preferred and alternate labels, possibly multilingual, editorial and scope notes, and (poly-) hierarchical as well as associative relations.

---

[8]  http://aksw.org/Projects/LinkedHistory?v=oi6
[9]  http://ckan.net/package/pndbeacon
[10] http://wiki.dbpedia.org/Ontology (as of 2011-01-11)
[11] http://authorities.loc.gov/help/contents.htm. The LoC name authority is currently available in RDF only through VIAF.
[12] http://www.slideshare.net/ah641054/linkedrdadata-in-der-praxis (as of 2010-11-30). 58,307 entities, which represent geographical entities, have been subtracted (Email by Alexander Haffner (DNB), 2011-03-22)
[13] Clusters of merged personal and corporate name records from 18 participating libraries, http://outgoing.typepad.com/outgoing/2010/11/corporate-viaf.html (as of 2010-11-23) and email by Jeff Young (OCLC), 2011-03-21

The SKOS standard[14], developed by W3C, was "designed to provide a low-cost migration path for porting existing [knowledge] organization systems to the Semantic Web" and to provide "a lightweight, intuitive conceptual modeling language for developing and sharing new KOSs. ... SKOS can also be seen as a bridging technology, providing the missing link between the rigorous logical formalism of ontology languages such as OWL and the chaotic, informal and weakly-structured world of Web-based collaboration tools, as exemplified by social tagging applications." [8]

An example of a SKOS concept, as a human-readable XHTML page with embedded RDFa data prepared for use in the Semantic Web, and in its Turtle representation, is given in Fig. 1 and 2 (taken from STW Thesaurus for Economics[15]).

More and more thesauri are published using SKOS, in the fields of public administration[16], social sciences[17], environmental information[18], medical subjects[19],

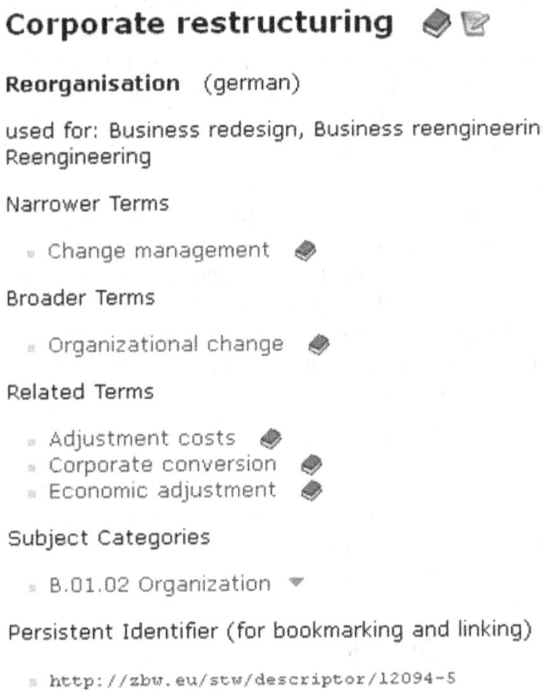

## Corporate restructuring

**Reorganisation** (german)

used for: Business redesign, Business reengineering, Reengineering

Narrower Terms

- Change management

Broader Terms

- Organizational change

Related Terms

- Adjustment costs
- Corporate conversion
- Economic adjustment

Subject Categories

- B.01.02 Organization ▼

Persistent Identifier (for bookmarking and linking)

- http://zbw.eu/stw/descriptor/12094-5

**Fig. 2.** STW Thesaurus for Economics concept, XHTML+RDFa representation

---

[14] http://www.w3.org/TR/skos-reference/
[15] http://zbw.eu/stw
[16] Eurovoc (the EU's multilingual thesaurus), http://eurovoc.europa.eu/, and UK ESD standards, http://standards.esd.org.uk/
[17] Thesaurus for the Social Sciences (TheSoz), http://www.gesis.org/en/services/tools-standards/social-science-thesaurus/
[18] General Multilingual Environmental Thesaurus, http://eionet.europa.eu/gemet
[19] Medical subject Headings (MeSH),
http://neurocommons.org/page/Bundles/mesh/mesh-skos

```
<stw/descriptor/12094-5>
   gbv:gvkppn "091386640"^^xsd:string ;
   a skos:Concept, zbwext:Descriptor ;
   rdfs:isDefinedBy <stw/descriptor/12094-5/about> ;
   rdfs:seeAlso <econis/search/descriptor/Corporate%20restructuring>, <econis/search/descriptor/Reorganisation> ;
   skos:altLabel "Business redesign"@en, "Business reengineering"@en, "Reengineering"@en,
           "Reorganisationsprozess"@de, "Reorganisationsprozeß"@de, "Unternehmensrestrukturierung"@de ;
   skos:broader <stw/descriptor/12105-5>, <stw/thsys/70562> ;
   skos:closeMatch <http://dbpedia.org/resource/Restructuring> ;
   skos:inScheme <stw> ;
   skos:narrower <stw/descriptor/24661-1> ;
   skos:prefLabel "Corporate restructuring"@en, "Reorganisation"@de ;
   skos:related <stw/descriptor/12196-4>, <stw/descriptor/12697-3>, <stw/descriptor/19254-2> ;
   skos:scopeNote "Für betriebliche Anpassungsmaßnahmen der Organisationsstruktur."@de ;
   zbwext:indexedItem <econis/search/descriptor/Corporate%20restructuring>,
           <econis/search/descriptor/Reorganisation> .
```

**Fig. 3.** STW Thesaurus for Economics concept, Turtle representation (extract)

agriculture and food[20], astronomy[21] and many more. The Library of Congress Subject Headings[22] try to cover the complete domain of human knowledge, similar to the German (SWD)[23] and French (RAMEAU)[24] subject heading files. The UDC is working on a "SKOSification", for the DDC [9] the first levels of the classification are already published[25].

These thesauri and classifications are a valuable source for well defined concepts. Although it is not straightforward, as Hyvönen et al. [10] assess, to convert such vocabularies to ontologies – since their understanding often implies implicit human knowledge – they can inform work on general or specific ontologies. A rough outline for a "thesaurus-to-ontology transformation" is given in the cited work.

In the field of subject thesauri, as well as in that of authority files, more and more mappings are published. This includes the results of experiments in multi-lingual matching, e.g., between LCSH, Rameau and SWD concepts [11], as well as domain centric mappings. The Agrovoc thesaurus of FAO for example was mapped to Eurovoc, NALT, GEMET, LCSH, and STW.[26] This adds inter-scheme relationships for concepts to the inner-scheme relationships described above, and adds value through often multi-lingual labels. The resulting concept network can be exploited for retrieval applications as well as for ontology building.

## 4    Tied Together – Publications

Publications – working papers, journal articles, books – and their archiving and provision are the libraries' main business. The collected metadata about publications is essential for

---

[20] Agrovoc, http://aims.fao.org/website/Download/sub;
   NALT, http://agclass.nal.usda.gov/
[21] International Virtual Observatory Alliance astronomy vocabularies
   http://www.ivoa.net/Documents/latest/Vocabularies.html
[22] http://id.loc.gov/
[23] http://www.d-nb.de/eng/hilfe/service/linked_data_service.htm
[24] http://stitch.cs.vu.nl/rameau
[25] http://dewey.info/
[26] http://www.taxobank.org/content/agrovoc-thesaurus

the administration of their holdings, for their internal workflows and for providing access to their patrons. Seen from outside, publications can be viewed as small nodes or linking hubs, structuring the landscape of science. Publications tie together subjects (ideally represented as links to formally expressed concepts), people (as authors or editors), organizations (involved as publishers or via affiliation of authors), both ideally represented as links to authorities, and – more and more frequently – machine readable contents, which are open to text analysis via natural language processing (NLP) tools. This offers opportunities for interrelating data, for example about co-authorship networks or about main foci of research for different countries.

Different libraries have started to release data about publications in Linked Data formats: The National Library of Sweden [12] published the complete catalog as RDF (dcterms, skos, foaf, bibo ontology), including links to DBpedia and the Library of Congress Subject Headings. The Hungarian National Library [13] did the same, and the French National Library presented plans to follow [14]. The British Library released parts of its catalog as open data and also prepares for LOD [15]. In Germany, the Library Service Center of North-Rhine Westphalia [16] and some university libraries published their catalogs, too. The "Open Library", as an "open, editable library catalog, building towards a web page for every book ever published"[27], makes about 20 million book description in RDF available, more than a million with searchable full text. It combines different editions of a work and links out to OCLC's WorldCat and the Library of Congress.

Since all of this happened recently, few of the opportunities offered by this large amount of RDF structured bibliographic data have been exploited yet. Issues remain: For example, merging all the data will not be an easy task, because common identifiers are lacking (ISBN numbers apply only to a fraction of the data, and they are not reliably unique) and the mappings to RDF structures differ widely. However, we expect to see much more use in the future.

## 5     Conclusions

The linked open datasets published by libraries and information centres create opportunities for the Semantic Web as a whole. As Hannemann and Kett [17] put it: "Library data tends to be of very high quality, being collected, revised and maintained by trained professionals. As such, it has the potential to become a much-needed backbone of trust for the growing semantic web." The curation of large datasets and terminologies is expensive, especially if strict and transparent policies are to be applied, because this means personnel expenditures over a long time. Therefore, it is difficult to achieve by academic projects. Publicly funded cultural heritage institutions on the other hand have long experience in data curation. They have a high score in long-term stability, reliability and independence from commercial interests. The Linked Data community could benefit from the offering of their autoritative and trusted data sets, using it for linking hubs in the web of data.

---

[27] http://openlibrary.org/

# References

1. Berners-Lee, T.: Linked Data - Design Issues (2006),
   http://www.w3.org/DesignIssues/LinkedData.html
2. Borst, T., Fingerle, B., Neubert, J., Seiler, A.: How do Libraries Find their Way onto the Semantic Web (2010), http://liber.library.uu.nl/publish/articles/000482/index.html
3. Deutsche Nationalbibliothek ed: PND-Praxisregel zu RAK-WB § 311. Individualisierung von Personennamen beim Katalogisieren mit der Personennamendatei, PND (2010), http://www.dnb.de/standardisierung/pdf/praxisregel_individualisierung_311.pdf
4. German National Library: The Linked Data Service of the German National Library. Version 3.0 (2011), http://files.d-nb.de/pdf/linked_data_e.pdf
5. Tillett, B.B., Harper, C.: Library of Congress controlled vocabularies, the Virtual International Authority File, and their application to the Semantic Web. Presented at the World Library and Information Congress: 73rd IFLA General Conference and Council, Durban, South Africa (May 22, 2007)
6. Kurki, J., Hyvönen, E.: Authority Control of People and Organizations on the Semantic Web. In: Proceedings of the International Conferences on Digital Libraries and the Semantic Web 2009 (ICSD 2009), Trento, Italy (2009)
7. Danowski, P.: Library 2.0 and User-Generated Content. What can the users do for us? Presented at the World Library and Information Congress: 73rd IFLA General Conference and Council, Durban, South Africa (August 2007)
8. Isaac, A., Summers, E.: SKOS Simple Knowledge Organization System Primer. W3C Working Group Note (2009), http://www.w3.org/TR/2009/NOTE-skos-primer-20090818/
9. Panzer, M., Zeng, M.L.: Modeling classification systems in SKOS: some challenges and best-practice recommendations. In: Proceedings of the 2009 International Conference on Dublin Core and Metadata Applications, pp. 3–14. Dublin Core Metadata Initiative, Seoul (2009)
10. Hyvönen, E., Viljanen, K., Mäkelä, E., Kauppinen, T., Ruotsalo, T., Valkeapää, O., Seppälä, K., Suominen, O., Alm, O., Lindroos, R., Känsälä, T., Henriksson, R., Frosterus, M., Tuominen, J., Sinkkilä, R., Kurki, J.: Elements of a National Semantic Web Infrastructure - Case Study Finland on the Semantic Web (Invited paper). In: Proceedings of the First International Semantic Computing Conference (IEEE ICSC 2007), Irvine, California (2007)
11. Wang, S., Isaac, A., Schopman, B., Schlobach, S., Van Der Meij, L.: Matching multilingual subject vocabularies. In: Proceedings of the 13th European Conference on Research and Advanced Technology for Digital Libraries, pp. 125–137. Springer, Heidelberg (2009)
12. Malmsten, M.: Making a Library Catalogue Part of the Semantic Web. In: Proc. Int'l Conf. on Dublin Core and Metadata Applications (2008)
13. Horváth, Á.: Linked data at the National Széchényi Library - road to the publication. In: SWIB 2010, Köln (2010)
14. Wenz, R.: data.bnf.fr: describing library resources through an information hub. In: SWIB 2010, Köln (2010).

15. Wilson, N.: Linked Open Data Prototyping at The British Library. UK Library Metadata Services: Future Directions, London (2010)
16. Ostrowski, F., Pohl, A.: Linked Data - und warum wir uns im hbz-Verbund damit beschäftigen. BIT Online 13, 259–268 (2010)
17. Hannemann, J., Kett, J.: Linked Data for Libraries. Presented at the World Library and Information Congress: 76th IFLA General Conference and Assembly, Gothenburg, Sweden (August 15, 2010)

# The CRM of Tomorrow with Semantic Technology

Sheng-Chuan Wu

Franz Inc.,
2201 Broadway, Suite 715, Oakland, California 94612 USA
scw@franz.com

**Abstract.** Customer Support or Relation Management (CRM) used to be a post-sale after-thought, an obligation stemming from product sales and a cost of doing business. No longer! In today's abundance of product/service choices, CRM must be an integral part of business to maintain brand loyalty, which is becoming critical for any business to survive. What if you can anticipate what you can do for your customers before they know it, predict what your customers may like or dislike and take steps to address potential problems before your customers switch vendors, or target individual marketing campaign to very specific and appreciative group of customers instead of spamming. How would this improve a business bottom-line and change its operation? This paper discusses an Intelligent Decision Automation platform for such a CRM system of tomorrow. It was built by Amdocs and Franz using semantic technology, machine learning and scalable java middleware. The Semantic platform consists of a number of elements: an Event Collector, a Decision Engine, the AllegroGraph triple store, a Bayesian Belief Network and a Rule Workbench. Combined, this pipeline of technology implements an event-condition-action framework to drive business process in real time.

**Keywords:** Customer Relation Management, Semantic Technology, Machine Learning, Rule, Bayesian Belief Network, AllegroGraph RDF Database.

## 1  Introduction

Today's market can be characterized by over-abundance of product offerings. Because of automation and the vast manufacturing capacity in Asia, one can find almost endless choices from many suppliers for any product with good quality and low price. Good products alone no longer suffice. This is particularly true for electronic devices and computing hardware. It becomes much harder to stand out in the market today comparing with just 20 years ago. Furthermore, in some sectors such as telecom, it has become a zero-sum game. For a vendor to gain a new customer, a competitor must lose one. Therefore, to stand out and survive, a business today must:

1) Resolve individual customer problems effectively and efficiently;
2) Anticipate potential issues with individual customers and address them before the customers are even aware of or before they become a serious problem;

D. Lukose, A.R. Ahmad, and A. Suliman (Eds.): KTW 2011, CCIS 295, pp. 46–51, 2012.

3) Offer added value to existing products or services, customized for individual customers rather than just spamming;

4) Cultivate product and brand loyalty among the customer base; and

5) Gain additional business from current customers for the long term.

## 2 Current Situation Analysis

Unfortunately, today most businesses fail to do all these five things well. How many times have you been frustrated by your long and ineffective interaction with the call center staff (Figure 1)? How often have you been bombarded by spam from your vendors trying to get more money out of you by pushing additional, unwanted products or services? According to a study, a call-center agent on average must go through 68 screens before getting the right data to address a customer's problem, and the customer often needs to call more than once to resolve the issue. Another example is the continuous spam using scare tactics from anti-virus software vendors (built into most PC's) pushing users to pay to upgrade.

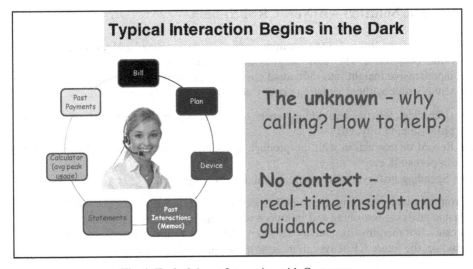

**Fig. 1.** Typical Agent Interaction with Customers

What can businesses do to remedy the situation? Traditional Business Intelligence (BI) tools can help a business understand the pattern and trend of *all* its customers, but tell nothing about *each* individual customer (Figure 2). That's why BI cannot provide personalized services to individual customers. Most current CRM (Customer Relation Management) systems, while intended for managing a company's interactions with customers, clients and sales prospects, are mainly used to organize, automate, and synchronize business processes, principally for sales, marketing and technical support activities. There is little analysis and reasoning capability to proactively serve the customer base individually for their long-term loyalty.

BI can tell you ALL about the average customer

but NOTHING about the individual ones

**Fig. 2.** The Shortcoming of BI Tools

## 3    The Solution – A New CRM System

Specifically, the future CRM system must provide personalized and proactive services for the customers. The only way to achieve such is to develop a critical and comprehensive insight into individual customers. A business must understand, to the extent possible without violating privacy concern:

- User characteristics individually such as gender, age, preference, educational level, profession, family, etc.
- Record of interaction with the product, the service, the company and its sales and support staff, etc.
- Spending history, credit worthiness and payment pattern.

From this understanding, a detailed model of individual users can be established to enable analysis, reasoning and inference to proactively provide better user services, to create customer loyalty and to discover additional future business opportunity. In essence, the future CRM system must be an integral part of any business.

## 4    Technical Challenges

Technically, such a CRM system must support multiple data sources such as sales, operation, support, accounting, etc, entailing many legacy data and yet-to-come new data. There are many challenges to integrate such diverse data together to form a complete view of individual customers, including:

- Heterogeneity of data distributed over multiple departments, functions and locations;
- Data stored with different database schema; and
- Information encoded with inconsistent data semantics.

Figure 3 below is a snap shot of a typical telecom IT system, encompassing many different database systems with applications written in nearly every programming language.

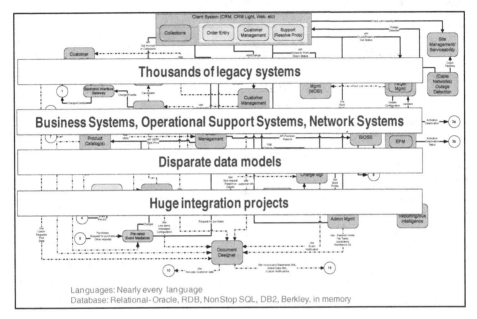

**Fig. 3.** Complexity of a Typical Telecomm IT System

Furthermore, the user model must be conducive to supporting semantic reasoning, rule-based inference, machine learning (supervised and unsupervised) and complex intelligent scenarios. And the CRM system architecture must be flexible enough to meet increasing and changing demands. Traditional data warehouse approach to data integration is totally inadequate and too rigid (both architecturally and schema-wise) to be useful for the future CRM system.

# 5    A Semantic Technology Based CRM System

A semantic model, employing URI (Universal Resource Identifier) and RDF (Resource Description Framework), uniquely represents data or more accurately concepts within the organization and globally in a schema-less RDF format (i.e., subject-predicate-object). It totally avoids the ambiguity and semantic inconsistence in most relational models that has plagued many data integration effort, while preserving maximum flexibility to meet future demand. Such a model can also connect easily to external metadata (also encoded in RDF and URI) from fast-growing information sources such as LOD (Linked Open Data cloud).

The RDF data model is also very natural for semantic reasoning (e.g., RDFS++ reasoning), machine learning and rule-base inference. Such analysis can be applied directly on individual customers to achieve personalized and proactive services. That's why Amdocs (the world's largest billing service provider for Telco's) chooses Semantic Technology for its next generation CRM system (AIDA) for its Telecom clients such as Bell Canada, Verizon and AT&T [1].

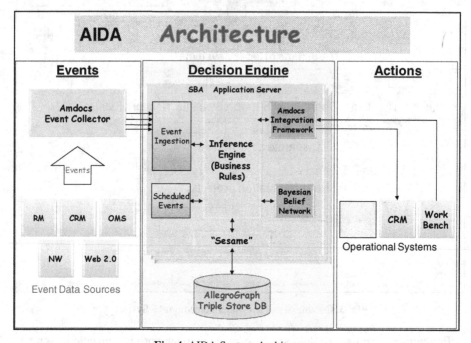

**Fig. 4.** AIDA System Architecture

The AIDA system consists of three main functional blocks (Figure 4):

### 1)  Event Collector
Here the system automatically extracts critical user events from very diverse sources with disparate data models, cleanses the data, and transforms (maps) them into semantic model (i.e., RDF and URI). For a telecom company with 80 million subscribers, it could generate 1 trillion RDF event triples every two months, an extreme scalability requirement for the AIDA architecture.

### 2)  Decision Engine
After the event triples are ingested into the triple database, a rule-based inference engine automatically classifies data of interest. For example, a customer may be classified as high credit risk or high retention risk. This classification process is run in the background controlled by a scheduler. Results of classification are stored back into the triple database as new properties of the entity being classified. Additionally, a Bayesian Belief Network (BBN) model that has been trained with historical data provides predictive analysis, forecasting potential consequence based on current observations (events). Both the rule-based inference and predictive analysis operate directly on the user model in the RDF triple database. This is the core of this new CRM system.

### 3)   Action Support

This block consists of two main functions, rule definition by business analysts and interface with existing CRM system in operation. A workbench is provided for business analysts to define business rules (e.g., what constitutes a customer with high-retention risk) by drag and drop. The workbench translates user defined rules into standard rule languages such as Prolog and Jess, which will be compiled and run by the system later. Additionally, all the inference and reasoning results are fed back into existing CRM system to help support staff to serve the customers correctly.

AIDA is currently being deployed in production at the call center of a major telecom company. According to an Amdocs study in a pilot deployment, AIDA provides significant operating cost savings by:

- Increases First Call Resolution by 15%
- Reduce average handling time by 30%
- Decrease training cost by 25%

This translates to a whopping 40% direct cost saving, besides the intangible but potentially even more significant benefits – good will and brand loyalty from the customers.

## 6    Summary

In summary, for a business to excel or even just to survive today, it must develop a comprehensive insight into its customer base and proactively and effectively serve them. Traditional BI tools and CRM systems cannot meet such requirements. Only a new CRM system based on a comprehensive user model with reasoning and inference capability can achieve such goals. With Semantic Technology, your business can realize such a future CRM system TODAY.

## Reference

1. Guinn, B., Aasman, J.: Semantic Real Time Intelligent Decision Automation. In: Proceedings of Semantic Technologies for Intelligence, Defence, and Security Conference (STIDS 2010), pp. 125–128 (2010)

# Identifying Metabolic Pathway within Microarray Gene Expression Data Using Combination of Probabilistic Models

Abdul Hakim Mohamed Salleh and Mohd Saberi Mohamad

Artificial Intelligence and Bioinformatics Research Group,
Faculty of Computer Science and Information System, Universiti Teknologi Malaysia,
81310 UTM Skudai, Johor, Malaysia
abdhakim.utm@gmail.com, saberi@utm.my

**Abstract.** Extracting metabolic pathway that dictates a specific biological response is currently one of the important disciplines in metabolic system biology research. Previous methods have successfully identified those pathways but without concerning the genetic effect and relationship of the genes, the underlying structure is not precisely represented and cannot be justified to be significant biologically. In this article, probabilistic models capable of identifying the significant pathways through metabolic networks that are related to a specific biological response are implemented. This article utilized combination of two probabilistic models, using ranking, clustering and classification techniques to address limitations of previous methods with the annotation to Kyoto Encyclopedia of Genes and Genomes (KEGG) to ensure the pathways are biologically plausible.

**Keywords:** Metabolic pathway, biological response, probabilistic models, annotation.

## 1    Introduction

A metabolic pathway which comprise of coordinated sequence of biochemical reactions is a small segment of the overall metabolic network that contribute to a specific biological function. However, a complete metabolic network is so huge and highly complex that the key pathways contributing to the responses are usually hidden. Therefore, an appropriate and effective model to extract and identify the pathways which at the same time takes account of the biological interactions between the components is required so that the real underlying structure of the system can be precisely obtained.

Many of the approaches that have been done before can successfully identify a pathway within the metabolic networks but none of them can clearly justify that the pathway extracted has a significant contribution in a certain metabolic response since none are considering the genetic interactions within the components level. Models such as network expansion [1] and Flux Balance Analysis (FBA) [2] only focused on

D. Lukose, A.R. Ahmad, and A. Suliman (Eds.): KTW 2011, CCIS 295, pp. 52–61, 2012.

chemical properties of metabolic network and do not directly consider the genetic component in the network.

Numerous amount of research incorporate the genetic factors that contribute to the function of metabolic networks as proposed by Karp et al. (2010) [3] and Mlecnik et al. (2005) [4], but they can only identify groups of specified genes are important although only some genes within this known groups are contributing to the observe response. Other research such as Gene Set Enrichment Analysis (GSEA) [5] do not incorporate the known networked structure of genes but instead rely on structure of simple test statistics. Probabilistic network models such as Markov Random Field [6] and Mixture Model on Graph [7] on the other hand able to confirm that the features to be logically connected within the metabolic network but an assumption has to be made that is the gene expression is discretely distributed. This may not correctly describe the underlying structure and mechanisms of the system.

This article discuss about the implementation based on combination of probabilistic models which has similar concept with GSEA but additionally takes account of the network structure [8]. With the use of pathway annotation from Kyoto Encyclopedia of Genes and Genomes (KEGG) this approach can overcome the limitations mentioned before and produce biologically plausible results. First, pathway ranking method [9] is applied to extract a number of pathways with maximum correlation through metabolic network. Then 3M Markov mixture model [10] is used to identify the functional components within the extracted pathways and finally Hierarchical Mixture of Experts, HME3M model [11] utilized as the classification model to identify set of pathways related to a particular response label.

The techniques are implemented on GSE121 dataset, the observation of genetic differences between obese patients that are divided into insulin resistance and insulin sensitive. This article extend the findings by calculating the *p-value* for the best HME3M component and annotating the gene set to enzyme accession number from KEGG. The outcomes of the methods are represented as directed graph pathway comprises of the relations between reaction, compounds, genes and also enzymes involved in that particular pathway.

## 2    Methods

This research is conducted by implementing the framework of model developed by Hancock et al. (2010) [8] with the extension of finding enzymes involved in particular pathway. The first step is defining pathway to precisely identify the location of each gene denotes a specific function, by the fact that same gene can be found in multiple location with different biological functions within the metabolic network. This step will define specific location of each gene using node and edge annotations extracted from KEGG database [12]. In pathway definition, each gene is defined as node in the network and annotated by its gene code ($G$), reaction ($R$) and KEGG pathway membership ($P$) as in (1).

$$\text{nodes}: = (G,R,P) \text{ ; edges}: = (C_F, C_M, C_T, P).$$ (1)

In addition, the edges that connect the nodes will be identified as first substrate compound ($C_F$), the product compound of first reaction ($C_M$), final product compound ($C_T$) and ($P$) the final KEGG pathway membership of $C_T$. Then, using annotation in equation (1), genetic pathway will be defined through metabolic network to be an extending connected sequence of genes, $g$, starting from specified start ($s$) and end compound ($t$) as shown in equation (2).

$$s \cdots \xrightarrow[\text{label}_{k-1}]{f(g_{k-1}, g_k)} g_k \xrightarrow[\text{label}_k]{f(g_k, g_{k+1})} g_{k+1} \xrightarrow[\text{label}_{k+1}]{f(g_{k+1}, g_{k+2})} \cdots t \tag{2}$$

Each of the edges will also be evaluated by the functions $f(g_k, g_{k+1})$ which measure the strength of relationship between $g_k$ and $g_{k+1}$ where label $k$ is the edge annotation in equation (1).

## 2.1    Pathway Ranking

This second step is to find the pathway of maximum correlation trough metabolic network. This particular technique will identify $K$ number of shortest and loop-less path within the weighted network [9] which is a non-parametric ranking procedure using Empirical Cumulative Distribution Function (ECDF) over all edge weights in the network.

The ranking procedure will usually tend to biased towards shorter path consisting same genes due to high levels of redundancy in metabolic network. To overcome this problem two parameter are set. First, a parameter to control number of minimum genes in a pathway to remove small and insignificant pathways from pathway set. Secondly, as the result of redundancy, there will also be chains of reactions involving similar or identical genes therefore the second parameter is the user specified penalty $p$ which control over the diversity of genes selection. An assigned of edge correlation, $f(g_k, g_{k+1})$ for all same gene edges will be used to specify penalty value.

## 2.2    Pathway Clustering

The goal for this important step is to identify set of pathways that produce the specific response and directly can be used to classify a particular response label. This research will utilize a pathway classifier based on the 3M Markov Mixture Model (3M) [10] which will provide the basic framework for the model. The 3M model will be used to identify M functional components by mixture of first order Markov chains as shown in equation (3). This method achieved competitive performance in terms of prediction accuracies with combination of two types of data sets, pathway graph and microarray gene expression data.

$$p(x) = \sum_{m=1}^{M} \pi_m p(s|\theta_{1m}) \prod_{k=2}^{K} p(g_k, label_k | g_{k-1}; \theta_{km}) \tag{3}$$

The $\pi_m$ is the probability of each components, transition probabilities $\theta_{km}$ defines each components, $p(s_i|\theta_{1m})$ is the start compound probability of $s_i$ and $p(g_k, label_k|g_{k-1}; \theta_{km})$ is the probability of path travers on edge labelk. The result of this 3M is M components defined by $\theta_m = \{ \theta_{sm}, [\theta_{2m},..., \theta_{tm},..., \theta_{Tm}]\}$. The $\theta_m$ is probabilities of each gene clustered within each component and indicate the importance of the genes.

## 2.3 Pathway Classification

For pathway classification, an extension to the previous 3M model, HME3M [11] will be used which incorporate Hierarchical Mixture of Experts (HME) that enables it to create a classification model from 3M model directly. In order to do so, additional term, $p(y|X, \beta_m)$ which is a classification model will be added to the equation (3) into equation (4).

$$p(y|X)= \sum_{m=1}^{M} \pi_m p(y|X,\beta_m)\prod_{k=2}^{K} p(g_k, label_k|g_{k-1}; \theta_{km}) \qquad (4)$$

$y$ is a binary response variable and $X$ is a binary matrix where the columns represent genes and the rows represent a pathway and value of 1 indicates that the particular gene is included within specific path.

The parameters $\pi_m$, $\theta_{km}$ and $\beta_m$ are estimated simultaneously with an EM algorithm [11]. The additional term p(y|X,βm) which takes the binary pathway matrix $X$ weighted by the EM component probabilities as input and returns the output as the posterior probabilities for classification of the response variable $y$. To ensure a scalable and interpretable solution, HME3M uses a penalized logistic regression for each component classifier. The goal of HME3M is to identify a set of pathways that can be used to classify a particular response label, $y_l \in y$.

By using set of genes that involved in the particular pathway, p-values for each pathway are calculated using the hypergeometric distribution. If the whole genome has a total of ($m$) genes, of which ($t$) are involved in the pathway under investigation, and the set of genes submitted for analysis has a total of ($n$) genes, of which ($r$) are involved in the same pathway, then the p-value can be calculated to evaluate enrichment significance for that pathway by equation (5):

$$p = 1 - \sum_{x=0}^{r-1} \frac{\binom{t}{x}\binom{m-t}{n-x}}{\binom{m}{n}} \qquad (5)$$

## 2.4 Pathway Visualization

The most important HME3M pathway is visualize in nodes and edge representation by connected pathways, genes, compounds and reactions. One of the enhancements made to this visualization technique is by incorporating the enzyme information that involved in the particular pathway based on set of genes that made up the pathway using the EC (Enzyme Commission) accession as well as the KO (KEGG Orthology) which both are annotated from KEGG database.

## 3    Results and Discussion

The dataset used is obtained from Gene Expression Omnibus (GEO) (GSE121) derived from an experiment of global transcript profiling to identify differentially expressed muscle genes in insulin resistance which is the prime causes of Type II diabetis-melitus [13].

Here this experiment presents the minimum path analysis of the HME3M [8]. The result shown in figures below are the key component for insulin resistant as identified by HME3M in terms of connected pathways (Figure 1), genes (Figure 2) and compounds (Figure 3) involved in that particular pathways. The edge thickness indicates the importance of that edge to the network and pathway with higher probability. This experiment is only focusing on insulin resistant as it is the key factor that contributes to Type II diabetes.

This experiment is conducted by using number of minimum path to be extracted of 5 paths. From Figure 1 it can be concluded that there are 2 main pathway components to the insulin resistance biological response that is the purine metabolism as the primary driver as well as pyrimidine metabolism which also serve as the shortest path. Another significant path includes glutathione metabolism, alanine, aspartate and glutamate metabolism and also arginine and proline metabolism. These observations may cause by the ability of this model to classify genes into the correct pathway map and calculate the *p-value* to estimate membership as in Table 1. With the combination of probabilistic models, this method able to extract probable pathways that are biologically significant based on the annotation to the pathway database.

It is clear from set of compounds that made up the pathway, the highest path probability would be the transition and conversion from C00002 (ATP) through C00046 (RNA), C00075 (UTP), C00063 (CTP), C00044 (GTP) and C01261 (GppppG) in Figure 3. This particular pathway result in production in ATP which is the significant signaling molecule in diabetes and insulin secretion as describe in Koster et al., 2005 [14]. In addition, the production of ATP that are occurring from C01260 (ApppppA), C06197 (ApppA), C06198 (UppppU) or converted to C00575 (cAMP), C00020 (AMP) and C00008 (ADP) and then back to ATP by using is supported by previous researches to have impact on insulin resistance. Verspohl and Johannwille (1998) prove that ApppppA and ApppA play important part in insulin secretion which may relate to diabetes [15] as well as production of GLP-1 by C00575 (cAMP) and nucleoside diphosphate kinase (NDK) enzyme in ADP to ATP conversion known factor in insulin secretion and Type II diabetes [16].

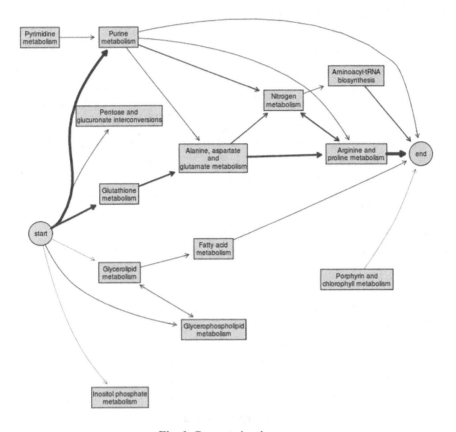

**Fig. 1.** Connected pathways

From Figure 2 we can clearly see there is a gene with the accession number 318 which code for nudix (nucleoside diphosphate linked moiety X) –type motif 2 also known as NUDT2. This gene encodes a member of nucleotide pyrophosphatases which can asymmetrically hydrolyzes Ap4A to yield AMP and ATP and responsible for maintaining intracellular level of dinucleotide Ap4A.

This research extend the findings of this experiment by using the set of genes involve in this particular pathway from HME3M classifier to calculate *p-value* for each related pathways to measure the gene membership in the pathway (Table 1). Here the top 15 pathways correspond to the set of genes are presented in the table.

The gene ratio indicates the number of genes that are the members of the pathway from the number of genes produced by HME3M. Besides providing calculation for *p-value* this research also provides the FDR-corrected *q-values* (if applicable) for reducing the false positive discovery rate.

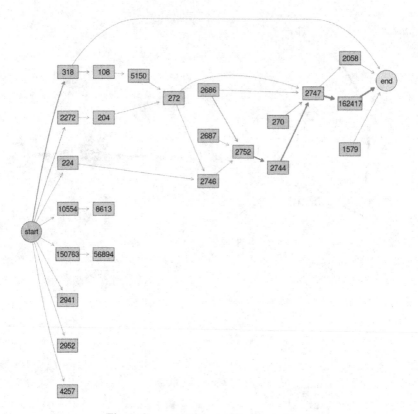

**Fig. 2.** Connected genes of insulin resistant

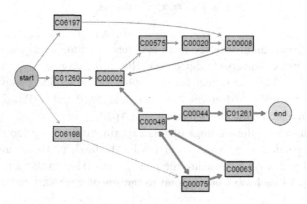

**Fig. 3.** Connected compound of insulin resistant

From the table it is obvious that purine metabolism pathway has the lowest *p-value* with the highest gene ratio indicating the significant of the pathway with the gene set produce by HME3M component. The pathways are considered to be highly statistically significant if having $p < 0.01$.

**Table 1.** Gene ratio, background ratio, *p-value* and *q-value* for each pathway

| Path | Pathway Name | Gene Ratio | Background Ratio | pvalue | qvalue |
|------|--------------|------------|------------------|--------|--------|
| 00230 | Purine metabolism | 53/221 | 161/25668 | 0.0000 | 0.0000 |
| 00330 | Arginine and proline metabolism | 35/221 | 79/25668 | 0.0000 | 0.0000 |
| 00565 | Ether lipid metabolism | 14/221 | 35/25668 | 0.0000 | 0.0000 |
| 00250 | Alanine, aspartate and glutamate metabolism | 21/221 | 58/25668 | 0.0000 | 0.0000 |
| 00591 | Linoleic acid metabolism | 14/221 | 29/25668 | 0.0000 | 0.0000 |
| 00240 | Pyrimidine metabolism | 32/221 | 99/25668 | 0.0000 | 0.0000 |
| 04370 | VEGF signaling pathway | 15/221 | 76/25668 | 1.11E-16 | 7.51E-16 |
| 00340 | Histidine metabolism | 11/221 | 29/25668 | 4.44E-16 | 2.84E-15 |
| 04664 | Fc epsilon RI signaling pathway | 14/221 | 79/25668 | 6.22E-15 | 3.76E-14 |
| 04270 | Vascular smooth muscle contraction | 16/221 | 126/25668 | 1.63E-14 | 9.12E-14 |
| 00592 | alpha-Linolenic acid metabolism | 9/221 | 19/25668 | 1.89E-14 | 1.03E-13 |
| 00620 | Pyruvate metabolism | 11/221 | 41/25668 | 3.79E-14 | 1.98E-13 |
| 00260 | Glycine, serine and threonine metabolism | 10/221 | 31/25668 | 6.91E-14 | 3.45E-13 |
| 04912 | GnRH signaling pathway | 14/221 | 101/25668 | 2.15E-13 | 1.03E-12 |

From the set of genes this research also extends the findings to identify the enzymes involved in the particular pathway. In order for researchers to gain benefits from this extension, they should have a prior knowledge in the study of enzymes involved in a particular pathway. Figure 4 shows the enzyme involve using undirected graph with the correlation to every members. EC: 3.6.1.5 for example is ATP diphosphohydrolase which responsible for the formation of AMP and phosphate using ATP and water as substrate as well as its role as modulator of extracellular nucleotide signaling and also contribute to changes in metabolism [17].

Figure 4 shows the corresponding enzymes that may contribute to insulin resistant. Some of the enzymes that potentially related to insulin resistant are for example EC: 1.7.1.7 is GMP reductase that has a role of producing NADPH, guaosine 5' phosphate. EC: 2.7.1.73 is inosine kinase which has the role of converting ATP to ADP and the other way around which gives an impact on insulin resistance as mention before as well as EC: 3.6.1.8 (ATP diphosphatase) which also involved in ATP conversion to AMP. The AMP-activated protein kinase plays important part in lipid and glucose metabolism where it promotes glucose uptake into muscle and suppressed glucose output from liver via insulin independent mechanism [18].

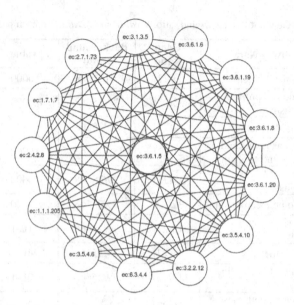

**Fig. 4.** The related enzymes that contribute to diabetes and insulin resistant

## 4    Conclusion

In this article, we describe an experiment of identifying and analyzing biologically significant pathway using gene expression dataset within global metabolic network. The key aspect of this research is that it takes into account for analysis of the sub networks, compound, reaction and interaction which allows a better picture of metabolic response without neglecting the underlying structure and mechanisms of metabolic network. The method discussed in this research has shown its effectiveness in extracting biologically significant pathway by using a combine approach with pathway ranking, clustering and classification technique by using two algorithms as the core structure that is the 3M and HME3M.

**Acknowledgement.** We also would like to thank Universiti Teknologi Malaysia for supporting this research by UTM GUP research grant (Vot number: Q.J130000.7107.01H29).

## References

1. Handorf, T., Ebenhoh, O., Heinrich, R.: Expanding metabolic networks: scopes of compounds, robustness, and evolution. J. Mol. Evol. 61(4), 498–512 (2005)
2. Smolke, C.D.: The Metabolic Engineering Handbook: Tools and Applications. CRC Press, Boca Raton (2010)

3. Karp, P.D., Paley, S.M., Krummenacker, M., Latendresse, M., Dale, J.M., Lee, T.J., Kaipa, P., Gilham, F., Spaulding, A., Popescu, L., Altman, T., Paulsen, I., Keseler, I.M., Caspi, R.: Pathway tools version 13.0: integrated software for pathway/genome informatics and systems biology. Brief Bioinform. 11(1), 40–79 (2010)

4. Mlecnik, B., Scheideler, M., Hackl, H., Hartler, J., Sanchez-Cabo, F., Trajanoski, Z.: PathwayExplorer: web service for visualizing high-throughput expression data on biological pathways. Nucleic Acids Research 33(1), 633–637 (2005)

5. Subramanian, A., Tamayo, P., Mootha, V.K., Mukherjee, S., Ebert, B.L., Gillette, M.A., Paulovich, A., Pomeroy, S.L., Golub, T.R., Lander, E.S., Mesirov, J.P.: Gene set enrichment analysis: A knowledge-based approach for interpreting genome-wide expression profiles. PNAS 102(43), 15545–15550 (2005)

6. Wei, Z., Li, H.: A markov random field model for network-based analysis of genomic data. Bioinformatics 23(12), 1537–1544 (2007)

7. Sanguinetti, G., Noirel, J., Wright, P.C.: Mmg: a probabilistic tool to identify submodules of metabolic pathways. Bioinformatics 24(8), 1078–1084 (2008)

8. Hancock, T., Takigawa, I., Mamitsuka, H.: Mining metabolic pathways through gene expression. Gene Expression 26(17), 2128–2135 (2010)

9. Takigawa, I., Mamitsuka, H.: Probabilistic path ranking based on adjacent pairwise coexpression for metabolic transcripts analysis. Bioinformatics 24(2), 250–257 (2008)

10. Mamitsuka, H., Okuno, Y., Yamaguchi, A.: Mining biologically active patterns in metabolic pathways using microarray expression profiles. SIGKDD Explorations 5(2), 113–121 (2003)

11. Hancock, T., Mamitsuka, H.: A Markov classification model for metabolic pathways. In: Workshop on Algorithms in Bioinformatics (WABI), pp. 30–40 (2009)

12. Kanehisa, M., Goto, S.: KEGG: Kyoto Encyclopedia of Genes and Genomes. Nucleic Acids Res. 28, 27–30 (2000)

13. Yang, X., Pratley, R.E., Tokraks, S., Bogardus, C., Permana, P.A.: Microarray profiling of skeletal muscle tissues from equally obese, non-diabetic insulin-sensitive and insulin-resistant pima indians. Diabetologia 45, 1584–1593 (2002)

14. Koster, J.C., Permutt, M.A., Nichols, C.G.: Diabetes and insulin secretion: the ATP-sensitive k+ channel (k ATP) connection. Diabetes 54(11), 3065–3072 (2005)

15. Rusing, D., Verspohl, E.J.: Influence of diadenosine tetraphosphate (ap4a) on lipid metabolism. Cell Biochem. Funct. 22(5), 333–338 (2004)

16. Yu, Z., Jin, T.: New insights into the role of camp in the production and function of the incretin hormone glucagon-like peptide-1 (glp-1). Cell Signal 22(1), 1–8 (2010)

17. Enjyoji, K., Kotani, K., Thukral, C., Blumel, B., Sun, X., Wu, Y., Imai, M., Friedman, D., Csizmadia, E., Bleibel, W., Kahn, B.B., Robson, S.C.: Deletion of Cd39/Entpd1 Results in Hepatic Insulin Resistance. Diabetes 57, 2311–2320 (2007)

18. Hegarty, B.D., Turner, N., Cooney, G.J., Kraegen, E.W.: Insulin resistance and fuel homeostasis: the role of AMP-activated protein kinase. Acta Physiol. (Oxf) 196(1), 129–145 (2009)

# A Transparent Fuzzy Rule-Based Clinical Decision Support System for Heart Disease Diagnosis

Adel Lahsasna[1], Raja Noor Ainon[1],
Roziati Zainuddin[1], and Awang M. Bulgiba[2]

[1] Faculty of Computer Science and Information Technology,
University of Malaya, 50603 Kuala Lumpur, Malaysia
lahsasna@perdana.um.edu.my, {ainon,roziati}@um.edu.my
[2] Julius Centre University of Malaya and CRYSTAL, Faculty of Science,
University of Malaya, 50603 Kuala Lumpur, Malaysia
awang@um.edu.my

**Abstract.** Heart disease (HD) is a serious disease and its diagnosis at early stage remains a challenging task. A well-designed clinical decision support system (CDSS), however, that provides accurate and understandable decisions would effectively help the physician in making an early and appropriate diagnosis. In this study, a CDSS for HD diagnosis is proposed based on a genetic-fuzzy approach that considers both the transparency and accuracy of the system. Multi-objective genetic algorithm is applied to search for a small number of transparent fuzzy rules with high classification accuracy. The final fuzzy rules are formatted to be structured, informative and readable decisions that can be easily checked and understood by the physician. Furthermore, an Ensemble Classifier Strategy (ECS) is presented in order to enhance the diagnosis ability of our CDSS by supporting its decision, in the uncertain cases, by other well-known classifiers. The results show that the proposed method is able to offer humanly understandable rules with performance comparable to other benchmark classification methods.

**Keywords:** heart disease, fuzzy system, transparency, medical diagnosis.

## 1 Introduction

Heart disease (HD) is the leading cause of death worldwide [1] and its diagnosis is a quite challenging task as the physician needs to examine a combination of symptoms that may overlap with each other causes [2]. In order to deal with this problem in a more efficient way, many systems have been proposed to serve as a clinical decision support system (CDSS) to help the physician to decide whether HD is likely to exist or not.

Soft computing techniques are widely used to solve real-world problems including medical diagnosis problems [3]. Their ability to handle uncertain information, upon which medical diagnosis is based, is the main motivation behind their use

D. Lukose, A.R. Ahmad, and A. Suliman (Eds.): KTW 2011, CCIS 295, pp. 62–71, 2012.
© Springer-Verlag Berlin Heidelberg 2012

as CDSS [3]. Artificial Neural Networks (ANNs) (see for example [4] and [5]) and Support Vector Machines (SVMs) ([6],[7]) are the commonly used methods for developing CDSS for HD diagnosis. ANNs and SVMs were successfully used to build highly accurate CDSS for HD diagnosis; they have been however, criticized for their lack of transparency, that is, the user is prevented from knowing the reasons behind their decisions and how particular decision has been made [8]. This drawback is critical in the case of CDSS as the physician needs to know about the logic behind any decision in order for him to consider it or not and using such black box CDSS has little benefit in real-life cases.

Unlike black-box methods, fuzzy rule-based systems offer a convenient format to express a peace of knowledge in the form of linguistic fuzzy rules [9]. Although, in the expert-driven (manual) approach, the transparency feature of FRBS is maintained during the rules generation process, it is usually lost during the learning process in the data-driven (automatic) approach [9].

The literature shows that the automatic approach has dominated the fuzzy-based CDSS for HD diagnosis (see for example [10] and [11]). Their only objective, however, is to develop higher diagnosis systems without considering the transparency feature. As a result, this kind of fuzzy-based CDSS is acting as black-box system and then has the same drawbacks as ANNs and SVMs. It is important then to maintain the transparency during the learning process as its importance in CDSS is equivalent to that of the accuracy [9].

To preserve the transparency in the data-driven approach, many techniques have been proposed ([9], [12]-[14]). Specifically, genetic algorithm has been demonstrated a good ability to deal with this problem by applying constrains on the optimization process ([12]-[14]). In fact, the design process has to tackle some drawbacks of genetic algorithm such as the computational cost needed to find the optimum solution especially when the number of attributes, as in our case, is relatively high. So it is desirable to apply feature selection method at the preprocessing stage to reduce the number of features and as a result the time needed for the optimization process. In addition, it is better to provide more information about the rules by extending the format of rules proposed in [14] to include also, in addition to the degree of confidence or certainty, the degree of support of each rule.

The aim of this study is to propose a FRBS that serves as a CDSS for HD diagnosis. To achieve this aim, FRBS is designed in a way to fulfill two main criteria, namely, the transparency and accuracy. These criteria are maintained and balanced during the rule generation and optimization phases using multi-objectives genetic algorithm. In addition, the transparency measure is further enhanced by extending the rule format to include more information about both its degree of certainty and support to help the physician to evaluate the strength of the rules. Finally, by introducing Ensemble Classifier Strategy (ECS) method, the accuracy of the CDSS is clearly boosted specially in the uncertainly cases.

The rest of the paper is organized as follows. In Section 2 a brief description of the data set used in this study is provided while section 3 details the proposed methodology. The results obtained and their discussions are given in section 4 while conclusion is drawn in section 5.

## 2  Data Set

The heart disease data set used in this study was supplied by Robert Detrano, M.D., Ph.D of the V.A. Medical Center, Long Beach, CA [15]. It is now publicly available at the UCI Repository of Machine Learning Databases. The original data has 76 attributes, but most of the studies use only 13 attributes that represent the clinical and noninvasive test results of 303 patients undergoing angiography. The total number of cases considered is 270 -after removing the cases with missing values-, out of which 150 are identified as patients without HD while the other 120 cases are diagnosed as patient with HD.

## 3  Methodology

The proposed methodology composed of five main steps:

### 3.1  Feature Selection

The literature shows that the most effective feature selection methods are the Sequential Floating Forward Selection (SFFS) [16] and the genetic algorithms (GAs) [17]. while many studies ([16], [17]) suggest the superiority of SFFS over the sequential search algorithms , there is no clear cut case study of which of the two methods -SFFS and GAs - is better than the other [17]. In our study, SFFS is selected as it is much faster than the GAs method.

### 3.2  Fuzzy Rule-Based System Generation

**Extending the format of fuzzy rules** Formally, medical diagnosis problem can be solved by finding a suitable classifier or mathematical function $f$ that maps a set of symptoms $x^n$ to a diagnosis class label $c_j$. This function can be written as follow:

$$f : x^n \rightarrow c_j$$

where $x^n = \{x_1, x_2, , x_n\}$ is a set of symptoms, and $cj \in \{c_1, c_2\}$ is the class label. In fuzzy set theory, the classifier $f$ is a set of fuzzy rules where the $k^{th}$ rule has the following format:

$$R_k : if \, x_1 \, is \, A_1^k \, and \, x_2 \, is \, A_2^k \, and \, ... \, and \, x_n \, is \, A_n^k \, then \, Y \, is \, c_j \qquad (1)$$

where $A_j^k$ are fuzzy sets of the input variables $(x_1, x_2, , x_n)$ represented by linguistic values such as low, moderate and high while $c_j$ is the class label of the class variable $Y$ (which in this case is either $Y = c_1$ for present class or $Y = c_2$ for absent class).

Some studies ([12],[14]) used an extended format of (1) that includes the degree of certainty or confidence $r_k$ of the $k^{th}$ rule where $r_k \in [0\,1]$. The value $r_k$ represents also the certainty's degree of the decision made by this rule. This feature allows the physician to know about the degree of confidence of the decision made by the fuzzy classifier or more precisely by the winner rule and whether to consider it or not. This format of fuzzy rule-based system can be written as follows:

$$R_k : if\ x_1\ is\ A_1^k\ and\ x_2\ is\ A_2^k\ and\ ...\ and\ x_n\ is\ A_n^k\ then\ Y\ is\ c_j\ with\ certainty\ r_k \tag{2}$$

where $r_k$ is the degree of certainty of the $k^{th}$ rule. Despite (2) gives more information about the k-th rule than (1), but it does not allow the decision maker to know the degree of importance of the rule and precisely how many times $R_k$ rule holds true in the training data. This concept is known in data mining field as *support* and it is an important criterion to assess the rules. Using the degree of certainty of a rule alone might be misleading in decision making, as it does not give information about how many cases can be covered by this rule and then it does not reflect the degree of the rule's importance. In our study, the format of fuzzy rules is extended to include the support value of the rule as follows:

$$R_k : if\ x_1\ is\ A_1^k\ and\ x_2\ is\ A_2^k\ and\ ...\ and\ x_n\ is\ A_n^k\ then$$
$$Y\ is\ c_j\ with\ certainty\ r_k\ and\ support\ s_k \tag{3}$$

where $s_k$ is the support value of the $k^{th}$ rule.

**Fuzzy Rule Generation Procedure.** Let $T_i$ be the number of linguistic values associated with the input variable $x_i$. One of these linguistic values is used as the fuzzy antecedent $A_i^k$ for the input variable $x_i$ in each rule $R_k$. In addition "*dont care*" is an additional linguistic value that represents the irrelevant fuzzy antecedents that can be deleted without affecting the fuzzy systems performance. Thus, the number $N$ of possible combinations of the antecedent part in this case is:

$$N = (T_1 + 1) * (T_2 + 1) * (T_3 + 1) * ... * (T_n + 1) \tag{4}$$

Using this method, all the possible antecedents of fuzzy rules can be generated, but when the number of inputs is high (high-dimensional data set), the number of rules generated will be exponentially increased. Since we are applying feature selection procedure at the pre-processing stage, the number of inputs will be adequate to apply this approach. After defining the antecedents of fuzzy rules, the consequent class of the rule $R_k$, its degree of certainty can calculated as in [18]. Support $s_k$ can be calculated with the following formula:

$$s_k = \frac{NCC_{R_k}}{m_j}, R_k \in FRBS_{c_j} \tag{5}$$

where $NCC_{R_j}$ is the number of training patterns correctly classified by the rule $R_k$, $m_j$ is the number of training patterns belonging to the class $c_j$ and $FRBS_{c_j}$ is the set of fuzzy rules associated with the class $c_j$.

**Fuzzy Reasoning Method.** The fuzzy reasoning adopted in this study is based on a single rule winner [18].

## 3.3   Fuzzy Rule-Based System Optimization

After generating all possible rules, the optimization procedure is applied to search for a subset of a small number of rules with the highest classification ability. The first objective represents the accuracy of FRBS while transparency (a subset with small number of rules) defines the second objective. These two modeling objectives of our FRBS can be written as follows:

$$Maximize f_{acc}(FRBS), Minimize f_{rule}(FRBS) \tag{6}$$

Where $f_{acc}(FRBS)$ is the accuracy of FRBS measured by the rate of classification accuracy, $f_{rule}$ (FRBS) is the number of fuzzy rules of FRBS. To optimize simultaneously the above mentioned objectives, controlled NSGA-II is utilized, and the results of the optimization are Pareto-front solutions that represent a number of different FRBSs solutions with their corresponding 2-tuple of values $t_{FRBS}$ where $t_{FRBS}$ = (accuracy of FRBS, rules' number of FRBS). The selection between these solutions usually depends on user preference and the type of problem under investigation. In our case, and since the accuracy in medical diagnosis is critical, the FRBS with the highest accuracy is selected. In the following subsection, a brief description of the multi-objective genetic algorithm used is introduced.

**Multi-objective Genetic Algorithms.** Genetic algorithms are heuristic techniques inspired by natural evolution for searching for optimum solution. Multi-objective genetic algorithms (MOGAs) are classes of genetic algorithms which are mainly applied for optimization problems that have multiple and even conflicting objectives [21]. There are a number of MOGAs proposed in the literature ([19]), of which NSGA-II algorithm is the most used multi-objective genetic algorithms to this kind of problem. In the current study, an enhanced version of NSGA-II called Controlled elitist genetic algorithm [20] is used. It allows for better convergence comparing with original NSGA-II. The following subsection describes the chromosome representation used in our implementation.

**Chromosome Representation.** Let $N$ be the total number of fuzzy rules generated and $S$ is the subset of rules selected, where $S \leq N$. The chromosome is coded as binary string of length $N$ and each binary bit represents one rule. When the value of $q^{th}$ bit is set to 1, the $q^{th}$ rule is selected while in the other case ($q^{th}$ bit is set to 0), the $q^{th}$ rule is not considered. Figure 1 shows an example of chromosome coding of rule selection procedure with total number of rules $N = 10$, and number of selected rules $S = 5$. The rules number 1,5,6,7 and 10 are selected to form a candidate fuzzy rule-based system while rules number 2,3,4,8 and 9 are ignored.

## 3.4   Enhancing the Classification Accuracy of the Fuzzy Rule-Based System by an Ensemble Classifiers Strategy (ECS)

After producing the fuzzy rule-based system, we propose an approach to enhance the performance of the FRBS by supporting its decision (classification) by an

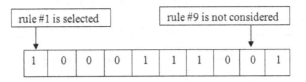

**Fig. 1.** Chromosome coding of rule selection procedure with $N = 10$, $S = 5$

ensemble of classifiers in the case where the classification decision is made by a rule (winner rule) which its degree of certainty is below a threshold $\alpha$ where $\alpha$ is defined by the user, in this study $\alpha = 0.5$. The decision made by the ensemble of classifiers is based on the majority vote. In addition to our FRBS, six other popular classifiers are selected, out of which 3 are non-fuzzy based methods and the others are fuzzy-based methods. The non-fuzzy based methods are: Artificial Neural Networks (ANNs), the popular decision tree algorithm C4.5 [21] and statistical method Linear Discriminant Analysis (LDA) while the fuzzy based methods are: FH-GBML [14], SLAVE [12], and GP-FCS [13].

### 3.5 Fuzzy Rule-Based System Evaluation

**Transparency Measures.** Most of the researchers agreed on some measures related to the transparency concept [9] in the fuzzy systems such as the number of fuzzy rules, the number of antecedent conditions in each rule and more important the using of linguistic terms to describe the antecedent conditions [14]. In addition to the above mentioned measures, we provide an additional measure which is the degree of certainty and support for each rule.

**Performance Measures.** To evaluate the classification accuracy, we used two well-known metrics, namely, the Percentage of Correctly Classified testing patterns (PCC) and the area under the Receiver operating characteristic (ROC) curve or AUC for short. In addition, our results are compared with benchmark methods. All PCC and AUC values are estimated using 10-fold cross-validation method.

## 4  Results and Discussions

### 4.1  Feature Selection Procedure

In this step, SFFS method was applied to select the subset of features that can produce the best classification model. The results of this step are shown in Table 1. As can be noted from the table, SFFS selects the subset that includes the following four features: sex, chest pain type, Major Vessels and Thal. In addition, while the number of features decreases from 13 to only 4, the classification accuracy of the model is still maintained at almost 83% which reflects the ability of SFFS method in handling feature selection problem.

**Table 1.** Results of feature selection procedure

|                          | All features | Selected features                                        |
|--------------------------|--------------|----------------------------------------------------------|
| ♯ of features            | 13           | 4                                                        |
| Description              | See [16]     | 2 (sex),3 (Chest pain type) ,12 (Major Vessels) ,13 (Thal) |
| Misclassification Error  | 17.07%       | 17.04%                                                   |

## 4.2 Fuzzy Optimization

**Transparency Criterion.** Table 2 shows the number of rules, fuzzy sets and number of sets per rule for different folds. As can be seen from table2, fuzzy rules selected from 300 rules -the initial number of generated rules- have relatively small number with an average of 8 rules per fold. The difference in the number of rules between folds is due to the changing of training data from a fold to another. Another improvement is in the number of fuzzy sets selected per rule which decreases from 4 to 2.5 per rule. In addition, as can be seen in figure2-which depicts the fuzzy rules of fold2-, the fuzzy rules use understandable description of the antecedent conditions in the form of if-then rules and they include in addition to their class labels, their degree of certainty and support. The linguistic description helps the physician to understand the relations between the factors or symptoms and the diagnosis outcome while degree of certainty degree helps to know how well this relation is accurate. This kind of reasoning is matching with the human thinking which allows partial truth to exist. Rules support at the other hand allows knowing to what extent this relation can be generalized.

**Table 2.** The number of fuzzy rules and fuzzy sets selected in the optimization step

| Subsets                   | S1  | S2  | S3  | S4  | S5  | S6  | S7  | S8  | S9  | S10 | Average |
|---------------------------|-----|-----|-----|-----|-----|-----|-----|-----|-----|-----|---------|
| ♯ of rules                | 7   | 8   | 8   | 10  | 8   | 10  | 6   | 7   | 7   | 9   | **8**   |
| ♯ of initial fuzzy sets   | 28  | 32  | 32  | 40  | 32  | 40  | 24  | 28  | 28  | 36  | **32**  |
| ♯ of selected fuzzy sets  | 17  | 22  | 20  | 27  | 20  | 25  | 15  | 15  | 17  | 21  | **19.9**|
| ♯ of fuzzy sets/rule      | 2.4 | 2.8 | 2.5 | 2.7 | 2.5 | 2.5 | 2.5 | 2.1 | 2.4 | 2.5 | **2.5** |

**Classification Accuracy Criterion.** Table 3 shows the PCC and AUC values of our FRBS before applying ensemble classifiers strategy (ECS) and it is named in the table as B-FRBS and after using this strategy (A-FRBS). In addition, it shows also the PCC and AUC values of a number of benchmark methods.

As can be seen from Table3, the initial results show that the ensemble classifiers strategy applied to support the classification of rules which their certainty value are less than 0.5 has improved the accuracy from 81.85% to 84.44% and from 0.812 to 0.839 for PCC and AUC respectively. This accuracy is comparable or even better than the other the classifiers. This result also indicates that more reliable classification decision can be obtained when combined with other classifiers known by their precision ability. This conclusion is shared by West et al.[22]

---

**Disclaimer:** The fuzzy rules listed here should not be used in clinical diagnosis without consulting experienced physicians.

(1)   IF ♯ of MajorVessels is (0) THEN HD is: (Absent) with Certainty: 0.28 and Support: 0.337

(2)   IF ChestPainType is (asymptomatic) AND Thal is (reversable defect) THEN HD is: (Present) with Certainty: 0.80 and Support: 0.231

(3)   IF Sex is (Male) AND ChestPainType is (non-anginal pain) AND ♯ of MajorVessels is (0) AND Thal is (normal) THEN HD is: (Absent) with Certainty: 0.72 and Support: 0.124

(4)   IF Sex is (Male) AND ChestPainType is (asymptomatic) THEN HD is: (Present) with Certainty: 0.27 and Support: 0.074

(5)   IF Sex is (Male) AND ChestPainType is (asymptomatic) AND ♯ of MajorVessels is (1) AND Thal is (reversable defect) THEN HD is: (Present) with Certainty: 0.74 and Support: 0.037

(6)   IF ChestPainType is (non-anginal pain) AND of ♯ of MajorVessels is (1) AND Thal is (reversable defect) THEN HD is: (Present) with Certainty: 0.74 and Support: 0.037

(7)   IF Sex is (Female) AND ChestPainType is (typical angina) AND ♯ of MajorVessels is (0) AND Thal is (normal) THEN HD is: (Absent) with Certainty: 1.0 and Support: 0.012

(8)   IF Sex is (Male) AND ♯ of MajorVessels is (2) AND Thal is (normal) THEN HD is: (Present) with Certainty: 0.2 and Support: 0.004

---

**Fig. 2.** Linguistic fuzzy rules of subset 2

who suggested that the accuracy of a set of classifiers is generally better than the best single classifier. In fact, it is common in medical practice to have more than one opinion in cases where the degrees certainty of the diagnosis is weak. Thus, these classifiers are playing the role of experts who provide their opinions about complicated cases that have low degree of certainty.

**Sensitivity and Specificity.** We notice from Table 3 that the specificity value for FRBS is clearly higher than the sensitivity value. This remark is also true for all classifiers (expect for MLP algorithm) and especially for fuzzy systems such as SLAVE, HF-GBML and B-FRBS where the difference between the specificity and the sensitivity is very significant, for example the specificity in B-FRBS is 0.873 while the sensitivity is 0.750. Another remark is that the non-fuzzy classifiers generally have better sensitivity values than the fuzzy systems and then exhibits better ability in detecting patients with HD while fuzzy-based systems achieve relatively better results in specificity which means better accuracy in detecting patients without HD. These two remarks show that the role of the two sets of classifiers can be complementary and can be used in ensemble classifier strategy to improve the quality of the FRBS. The results achieved by A-FRBS (after applying ECS) shows that the improvement in sensitivity (from 0.750 to 0.792) is more significant than the improvement in specificity (from 0.873 to 0.887). Thus, ECS has generally improved the ability of FRBS for HD diagnosis and specifically for detecting the patient with HD which is desirable feature for any HD diagnosis system.

**Table 3.** PCC, AUC, sensitivity and Specificity Values of FRBS before (B-FRBS) and after (A-FRBS) applying Ensemble classifier strategy (ECS) and a number of benchmark classifier algorithms using 10-fold cross-validation method

|  | Methods | PCC | AUC | Sensitivity | Specificity |
|---|---|---|---|---|---|
| Training data | C4.5 | 0.863 | 0.858 | 0.810 | 0.906 |
|  | ANNs | 0.868 | 0.837 | 0.831 | 0.843 |
|  | HF-GBML | 0.862 | 0.857 | 0.807 | 0.907 |
|  | SLAVE | 0.848 | 0.840 | 0.767 | 0.913 |
|  | GP-FCS | 0.795 | 0.817 | 0.751 | 0.882 |
|  | LDA | 0.839 | 0.838 | 0.826 | 0.850 |
|  | B-FRBS | 0.853 | 0.846 | 0.788 | 0.905 |
|  | A-FRBS | 0.860 | 0.854 | 0.802 | 0.907 |
| Testing data | C4.5 | 0.833 | 0.828 | 0.783 | 0.873 |
|  | ANNs | 0.814 | 0.819 | 0.825 | 0.813 |
|  | HF-GBML | 0.823 | 0.815 | 0.738 | 0.878 |
|  | SLAVE | 0.803 | 0.793 | 0.692 | 0.893 |
|  | GP-FCS | 0.781 | 0.770 | 0.700 | 0.840 |
|  | LDA | 0.825 | 0.823 | 0.800 | 0.847 |
|  | B-FRBS | 0.819 | 0.812 | 0.750 | 0.873 |
|  | A-FRBS | 0.845 | 0.839 | 0.792 | 0.887 |

## 5   Conclusions

In the present paper, a fuzzy based-CDSS for HD diagnosis is proposed to help the physician in the diagnosis decision process by providing a set of decision rules. These rules describe the relationship between the symptoms and the diagnosis outcome in an accurate and understandable way. In addition, the format of fuzzy rules is extended to provide additional information about the degree of certainty and support of each decision rule so that the physician can decide whether to take it into consideration or not. This feature specifically makes the proposed CDSS a more suitable tool for HD diagnosis as it will not only enable the physician to accurately diagnosis the HD but also understand the reason behind the system decision which may reveals unexpected findings and knowledge about the disease.

## References

1. Thom, T., Haase, N., Rosamond, W., Howard, V.J., Rumsfeld, J., Manolio, T., et al.: Heart disease and stroke statistics 2006 update: A report from the American Heart Association Statistics Committee and Stroke Statistics Subcommittee. Circulation 113, e85–e151 (2006)
2. Reddy, K.S.: Cardiovascular diseases in the developing countries: Dimensions, determinants, dynamics and directions for public health action. Public Health Nutr. 5, 231–237 (2002)
3. Zadeh, L.: Soft Computing and Fuzzy Logic. Computer Journal of IEEE Software 11, 48–56 (1994)
4. Lapuerta, P., Azen, S.P., Labree, L.: Use of neural networks in predicting the risk of coronary artery disease. Comput. Biomed. Res. 28, 38–52 (1995)
5. Babaoglu, I., Baykan, O.K., Aygul, N., Ozdemir, K., Bayrak, M.: Assessment of exercise stress testing with artificial neural network in determining coronary artery disease and predicting lesion localization. Expert Systems with Applications 36, 2562–2566 (2009)

6. Comak, E., Arslan, A., Turkoglu, I.: A Decision Support System Based on Support Vector Machines for Diagnosis of The Heart Valve Diseases. Computers in Biology and Medicine 37, 21–27 (2007)

7. Babaoglu, I., Findik, O., Bayrak, M.: Effects of principle component analysis on assessment of coronary artery diseases using support vector machine. Expert Systems with Applications 37, 2182–2185 (2010)

8. Ainon, R.N., Bulgiba, A.M., Lahsasna, A.: AMI Screening Using Linguistic Fuzzy Rules. Journal of Medical Systems (Published online May 2, 2010), doi:10.1007/s10916-010-9491-2

9. Nauck, D.: Data Analysis with Neuro Fuzzy Methods (Habilitation thesis. Otto-von-Guericke University of Magdeburg, Faculty of Computer Science, Magdeburg, Germany (2000)

10. Jain, R., Mazumdar, J., Moran, W.: Application of fuzzy classification system to coronary artery disease and breast cancer. Australasian Phys. Eng. Sci. Med. 21, 141–147 (1998)

11. Polat, K., Gunes, S., Tosun, S.: Diagnosis of heart disease using artificial immune recognition system and fuzzy weighted pre-processing. Pattern Recognition 39, 2186–2193 (2006)

12. Gonzalez, A., Perez, R.: SLAVE: a genetic learning system based on an iterative approach. IEEE Trans. Fuzzy Systems 7, 176–191 (1999)

13. Snchez, L., Couso, I., Corrales, J.A.: Combining GP Operators With SA Search To Evolve Fuzzy Rule Based Classifiers. Information Sciences 136, 175–192 (2001)

14. Ishibuchi, H., Yamamoto, T., Nakashima, T.: Hybridization of Fuzzy GBML Approaches for Pattern Classification Problems. IEEE Trans. on Systems, Man, and Cybernetics- Part B: Cybernetics 35, 359–365 (2005)

15. Detrano, R., Janosi, A., Steinbrunn, W., Pfisterer, M., Schmid, J., Sandhu, S., Guppy, K., Lee, S., Froelicher, V.: International application of a new probability algorithm for the diagnosis of coronary artery disease. American Journal of Cardiology 64, 304–310 (1989)

16. Pudil, P., Novovicova, J., Kittler, J.: Floating Search Methods in Feature Selection. Pattern Recognition Letters 15, 1119–1125 (1994)

17. Oh, I.S., Lee, J.S., Moon, B.R.: Hybrid genetic algorithms for feature selection. IEEE Transactions on Pattern Analysis and Machine Intelligence 26, 1424–1437 (2004)

18. Ishibuchi, H., Nozaki, K., Tanaka, H.: Distributed Representation of Fuzzy Rules and Its Application to Pattern Classification. Fuzzy Sets and Systems 52, 21–32 (1992)

19. Coello, C.A.C.: A comprehensive survey of evolutionary based multi-objective optimization techniques. Knowl. Inf. Syst. 1, 269–308 (1999)

20. Deb, K., Goel, T.: Controlled Elitist Non-dominated Sorting Genetic Algorithms for Better Convergence. In: Zitzler, E., Deb, K., Thiele, L., Coello Coello, C.A., Corne, D. (eds.) EMO 2001. LNCS, vol. 1993, pp. 67–81. Springer, Heidelberg (2001)

21. Quinlan, J.R.: C4.5: Programs for Machine Learning. Morgan Kaufmann Publishers (1993)

22. West, D., Mangiameli, P., Rampal, R., West, V.: Ensemble strategies for a medical diagnostic decision support system: a breast cancer diagnosis application. Eur. J. Oper. Res. 162, 532–551 (2005)

# Knowledge Modeling
# for Personalized Travel Recommender

Atifah Khalid, Suriani Rapa'ee, Norlidza Mohd Yassin, and Dickson Lukose

MIMOS Berhad,
Technology Park Malaysia,
Kuala Lumpur, Malaysia
{atifah,suriani,mwliza,dickson.lukose}@mimos.my

**Abstract.** Looking for information for travel or holiday destinations can be troublesome when the information gathered is often too much and when this happens, a traveler has to spend a lot of time to filter or digest the information before a sensible travel plan can be prepared. A traveler normally needs more than one piece of travel information to decide on destinations to visit and a collection of such information is normally extracted from several information sources such as travel guides, forums, blogs, social applications and websites of accommodations, transportations and eating places. In this paper, we present a mobile personalized travel recommender that is known as Travel Advisor. It is designed to provide a context-aware support for itinerary planning for both individual and group travelers according to the context, static profile, and dynamic profile of the travelers. It leverages on the use of semantic knowledge representations to make semantically meaningful recommendations to meet the need of travelers. This paper also describes the challenges associated with personalized travel recommendations.

**Keywords:** Semantic Knowledge Representation, Context-Aware Computing, Personalization, Travel Plan, Personalized Traveler.

## 1 Introduction

Prior to a trip, a traveler normally needs to plan for the trip in advance. This will involve arranging for accommodations, transports and places to visit throughout the travel period. The challenge in making a travel plan is that a traveler may not be familiar with the travel destinations, hence he/she may not be able to make informed decisions to suit interests. Having a system that can assist in the decision making process can be very useful. This system is capable of making recommendations to the traveler based on his/her personal preferences.

Recommending travel information to an individual is quite simple since the information is tailored only to a single individual. However, it becomes more complicated to recommend itineraries for group travelers since the recommendations for travel information should be based on summative preferences of all the travelers

D. Lukose, A.R. Ahmad, and A. Suliman (Eds.): KTW 2011, CCIS 295, pp. 72–81, 2012.

involved. Such situation often leads to conflicts because the decisions on places to visit or travel route may weigh differently for different individuals. For instance, visiting a theme park is highly preferred by a child-traveler but for a senior-traveler, a historical place is a better option. In balancing such requirements, the system should be able to come out with a proposed plan that can fulfill the needs of all travelers in the group. This is another challenge that the paper is aiming to resolve.

This paper is organized as follows. Section 2 presents the related work, and subsequently, the user profile aspects are explained in Section 3. In Section 4, the recommendation approach for planning, creation and demand travel information features for individual traveler are discussed. Section 5 covers recommendation approach for Group Travelers. Section 6 illustrates the semantic representation for travel knowledge-base and how it is used. Finally, in the last section we conclude with qualitative feature comparisons and potential future work.

## 2      Related Works

In the tourism domain, efforts have been made to address the problems faced by travelers in obtaining information on more suitable recommendations for places of interest or accommodations through their mobile phones. For instance, Speta [7] has created a system that provides context aware personalized guides and personalized services based on user interest. It gives recommendations on places of interest, accommodation, matched with context information such as GPS locations, date and time, and weather conditions. The filtering is done on the basis of user predetermined preferences, static information imported from social network sites, and collaborative filtering. Another work by PSiS [13] provides features such as Web portal planning and re-planning, Web portal recommendation and mobile travel companion known as Traveler. The recommendations are based on contextual information such as location (obtained through GPS), weather, phone status, date and time. Traveler [10] is a computer based software agent for travel agencies. It recommends tours and holiday packages through e-mails to customers. It combines collaborative filtering with content-based recommendations and demographic information to model the preferences of users. Meanwhile, ITS [18] focuses on the ontology based modeling techniques are adopted to design a personalized route planning system. It recommends the best route to the travelers based on user preferences (for both mobile devices and computers). User models are built based on preferences (predefined), demographics, contact information, travelers' name and travelers' context information such as traveler location. In addition, the context model is built based on tourist attractions, facilities, traffic conditions and weather forecast. Finally, Murshid [5] provides guidance to tourists travelling to UAE country. It gives special event notifications, weather forecasting, currency exchange rate, language translations and travel services. Users provides information on their interest during registration and the system will push recommendations solely based on the user's current locations, without considering the different types of users, who may demand different services.

# 3    User Profile

The user profile consists of three main components which are static, dynamic profile and context. The static profile of a user is basically the user's background and static preference information. Such information can easily be acquired during user registration process which is normally based on user's manual input and are fixed in nature. The user details such as name, date of birth, gender, ethnicity, religion are considered static profile. The static profile in our terminology is defined as GEAR [8].

The dynamic profile information of a user is considered implicit facts about the user which are learnt by the system. In the process of learning, it will generate a pattern in determining the user behavior based on past travel activities and experiences. User behavior can be modeled according to interest, tendency and habits model. The tendency model incorporates the scope of preference and interest model to summarize the tendency of a user to perform a particular act [9].

Almost any information available at the time of an interaction can be seen as context information. Context information can be spatial information, temporal information, environmental information, social situation, nearby resources and activity. The context profile stores information on the user's current context, e.g. location and date or time. This profile is required when context data are expected by the system, such as when a recommendation is to be generated based on the user's current location and date or time. Meanwhile all other resources of travel information are accessible by the use of referencing to the travel ontology which is defined in Section 6.

# 4    Recommendation for Itinerary, Accommodation, Transportation and Restaurant

A variety of techniques have been used to make recommendations in the travel domain that includes content, collaborative, knowledge-based and many others [15]. Travel Advisor makes a recommendation based on the following four techniques i) content based which recommends the similar types of activities from traveller's past historical data, ii) static profile which includes traveller's gender, ethnicity, age and religion (GEAR) [8], iii) dynamic profile of a user that is user's behavioural pattern learnt based on the tendency model [9] that translate the learnt activity travel pattern into different tendency measures (such as never, sometimes, and always) and iv) knowledge base which supports the inference of traveller needs and recommendation item. The Travel Advisor uses two types of recommendation media which are travel recommendation during itinerary planning (Web based) and search on demand for travel recommendation. These two recommendation media will be explained next.

## 4.1    Travel Recommendation for Itinerary Planner (Web Based)

Each traveller is required to provide information on destination to visit, date, and duration of visit, when creating a travel itinerary [17]. Once user provides the necessary travel information, the Travel Advisor will then recommend Places of

Interest (POI), accommodations, transportations and eating places based on the result of filtered travel information by using the four filtering technique stated in Section 4. Fig. 1 below depicts the process of recommendation in Travel Advisor. First, Travel Advisor queries the list of items to recommend to the traveler from the Travel knowledge-base (Ontology) based on location. Secondly, the static and dynamic profiles data will be retrieved from the Profile Database. Then, the item list is filtered using static and dynamic profile to get a list of recommended items that match the traveler's preferences. In the case where a user's dynamic profile is not available, the list is only filtered against the static profile. Finally, a recommendation list is deduced using the Tendency Model [9] to rank the highest preferred item.

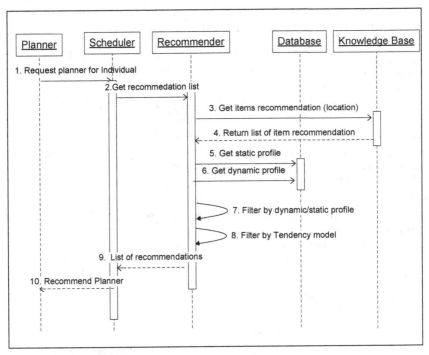

**Fig. 1.** Travel Advisor Interaction Diagram – Get Recommendation

The mechanism of generating a travel itinerary includes the process of arranging the sequence of activities and events during a trip. This includes routine travel activities such as checking in to a hotel, lunch, dinner and travelling time. It also considers the opening hours of the POI and time spent to tour the places. When such itinerary is generated, it is first presented to the traveller for approval. Additionally, the system can suggest more than one itinerary for traveller to opt and approve the choices. The example of a created itinerary is shown in Fig. 2a.

## 4.2    Search on Demand for Travel Recommendation

Even though a travel-plan or an itinerary has already been created, some unexpected events may happen during the trip, which requires new plans to be prepared. In such a scenario, the traveler may have to decide at that instance what to do next. For example, the place of interest that a traveler is supposed to visit is closed due to unexpected event. This information is made known to the traveler only when he or she arrives at the destination. Thus the traveler has to instantly make a new decision as to where to visit next. The traveler is able to request further recommendation on nearby POIs (Fig. 2b). Additional information may be requested in the recommendation list such as advertisements, more information on recommended items, map of nearby hotspots and transport will also be pushed to the traveller. The recommendation process is the same as described in Section 4.1.

Fig. 2a. Itinerary Created

Fig. 2b. POI Recommendation

# 5    Travel Group Recommendation

In Section 5, this paper will discuss the creation of itinerary planning for recommending places of interest based on travel group. The Travel Advisor employs the win-win strategy that tries to accommodate everyone in the group rather than majority wins. It applies location, GEAR model [8], convergence list and the group preferences as the basis for filtering criteria to give the best recommendation for all during planning stage for group of travelers. Travel group normally consists of combination of different individual background as depicted in Fig. 3 that shows sample of three different travel groups. Knowing the different combination of demographic factors for each group as shown in Fig. 3 stated in the GEAR box, the recommendation then is returned based on the GEAR subgroup list convergence [17].

Fig. 4 depicts the interaction diagram for itinerary planning for group travel until the itinerary is created. The diagram shows that when a request comes in for group itineraryplanner, the GEAR subgroup percentages are calculated and the knowledge base is inferred on the POI based on the location and the GEAR subgroups identified. The list is then amalgamated and any overlapping of POI of at least 2 GEAR subgroups is selected. POI having more than one overlapping will be ranked higher and will be at the top of the recommendation list. The planner is then matched against any available group preferences.

The group preferences are historical data that capture the preferences or groups' feedback on past travel experiences. The list is further filtered according to opening hours/day before the planner is generated and sent to the requestor. The outcome of the recommendations is as depicted in Fig. 3 as stated in the Recommendation box for each group. However, for on-demand travel recommendation, information pushed will be tailored according to both individual personal tendency model [9] and GEAR model [8] as mentioned in Section 4.2 regardless of the grouping breakdown as this involves personal needs and does not need to satisfy group request.

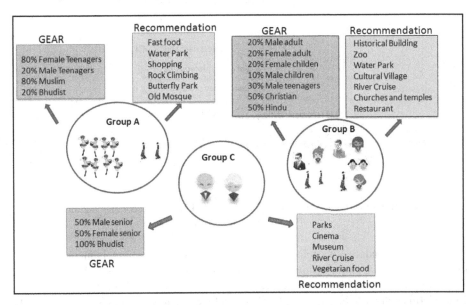

**Fig. 3.** Example of Different Travel Group and Recommendation Result

## 6    Semantic Knowledge Representation

The travel domain Ontology is used to represent the overall model of the concepts and properties of the travel knowledge representation in Travel Advisor. The already defined concept such as Person which has been created by the Friend of Friend

Project (http://www.foaf-project.org) is adopted or referred in this Ontology. FOAF has created the vocabulary to describe the people. In this case it is used to represent a person who is the traveler. The traveler's Gender, Ethnicity, Age and Religious belief (GEAR) [8] is defined as a subclass of the concept Person. Besides Person, the concept Location which has been developed before in the previous work is reused to represent the location and in this case the location will not only represent the town and city but include the physical place such as room and building.

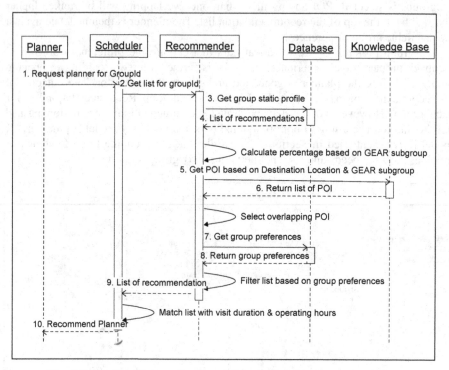

**Fig. 4.** Interaction Diagram for Group Itinerary Planning Recommendation

The recommendation system uses ontology for inference and content filtering purposes during the planning and on-demand travel information search. In Fig. 5, the Transportation concept link to the other major concepts such as **Accommodation** and **PlaceOfInterest** by the property *isAvailableAt*. One example of Ontology inferencing can be seen from the main concepts where **PlaceOfInterest, Accomodation** and **Transportation** have a relationship to the **CostType** by the property *hasCost*. Hence, knowledge on accommodation, transport or place of interest according to the traveler's budget can be recommended. Another example, in recommending POI to traveler, the knowledgebase can be inferred using the relationship of the concept **PlaceOfInterest** with **PlaceType** which have the property isPreferBy of **foaf:Person** of GEAR[8] concept. One example of content filtering can be shown when POI is recommended based on the relationship *isLocatedIn* between concepts **Location** and **PlaceOfInterest** when the traveler's location is detected.

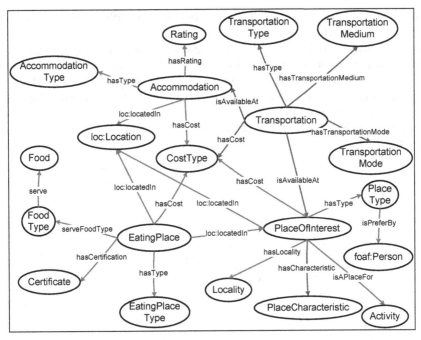

**Fig. 5.** Travel Ontology Model

## 7    Conclusion

We have described the travel advisor in the tourism domain which combines context-based, GEAR model [8], and content-based user tendency model [9] filtering to recommend places of interest, accommodations, transportation, and food venues for individual traveler as well as group of travelers. Additionally this hybrid approach is enhanced further by the use of knowledge base technique which uses the travel domain knowledge for better interpretation of the recommendation. Our work differs from other similar works done in the travel domain and aims to give better recommendations for traveller on the move in planning and during the travelling period. We have shown how the challenges in creating an itinerary can be accommodated according to context, GEAR, and dynamic profiles and also accommodating the variety of needs coming from travel groups. Before we can conclude, we present the qualitative feature comparisons outlined in Fig. 6 and Fig. 7 (the spider chart and table) in terms of functionalities of the related research work done on the other travel recommender and the travel advisor.

In summary, the work of Speta [7] and ITS [18] is quite similar to our work in terms of the comparative features. The major differences is that we provide the group itinerary planner, travel advisors which also includes recommendation of advertisements against personal profile during the travel activity, and lastly the use of learned dynamic profiles to indicate user preferences. In a nutshell, our Travel

Advisor addresses some issues that are absent in other approaches, which we believed are useful especially in a tourism context.

In the future, we would like to improve further on the planning stage to include the use of agent negotiations in our system to make it more robust. Secondly, we want to explore the automatic way of consuming and refreshing the Travel Ontology for better maintenance on the use of Ontology as knowledge repository for travel resources. Lastly, we want to improve on the user experience such as making use of the traveler feedback after each actual travel has taken place through social networking.

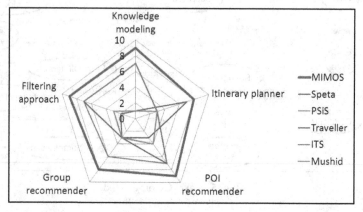

**Fig. 6.** Spider Chart

| Features | MIMOS | Speta | PSIS | Traveler | ITS | Murshid |
|---|---|---|---|---|---|---|
| **1. Knowledge modeling** <br> 1-5    No semantic representation <br> 6-10  Semantic representation | 9 | 7 | 1 | 1 | 7 | 1 |
| **2. Itinerary planner** <br> 1-3 Manual <br> 4-7 Semi-auto <br> 8-10 Automatic | 8 | 3 | 3 | 7 | 3 | 3 |
| **3. POI recommender** <br> 1-3  Location & time-based <br> 4-7  Location ,time & static profile <br> 8-10 Location, time, static, & dynamic profile | 9 | 7 | 4 | 3 | 7 | 4 |
| **4. Group recommender** <br> 1-3  Manual <br> 4-7  Semi-auto <br> 8-10 Auto | 8 | 3 | 3 | 3 | 6 | 3 |
| **5. Personalization Type** <br> 1-3 Context, Demographic & Collaborative. <br> 4-7 Content, Context, Demographic & Collaborative <br> 8-10 Dynamic Profile, Content, Context, Demographic & Collaborative | 9 | 7 | 3 | 2 | 7 | 3 |

**Fig. 7.** Qualitative Feature Comparison

# References

1. Abowd, G., Atkeson, C., Hong, J., Long, S., Kooper, R., Pinkerton, M.: Cyberguide: a mobile context-aware tour guide. ACM Wireless Networks 5(3), 421–433 (1997)
2. Cheverst, K., Davies, N., Michell, K., Friday, A., Efstratiou, C.: Developing a Context-aware Electronic Tourist Guide: Some Issues and Experiences. CHI Letters 2(1), 17–24 (2000)
3. Setten, M.V., Pokraev, S., Koolwaaij, J.: Context-Aware Recommendations in the Mobile Tourist Application COMPASS. In: De Bra, P.M.E., Nejdl, W. (eds.) AH 2004. LNCS, vol. 3137, pp. 235–244. Springer, Heidelberg (2004)
4. Krösche, J., Baldzer, J., Boll, S.: MobiDENK-Mobile Multimedia in Monument Conservation. IEEE Multi-Media 11(2), 72–77 (2004)
5. Echtibi, A., Zemerly, J., Berri. J.: Murshid: A Mobile Tourist Companion. In: Proceedings of the 1st International Workshop on Context-Aware Middleware and Services (2009)
6. Digital Marketing Strategy Alert, http://www.smartinsights.com/digital-marketing-strategy-alerts/mobile-marketing-statistics (last Visited: April 2011)
7. Garcia-Crespo, A., et al.: SPETA: Social pervasive e-Tourism advisor. Telematic and Informatics 26, 306–315 (2009)
8. Zaini, N., Yassin, N.: MIMOS Patent Pending; Gender, Ethnicity, Age, Religion (GEAR) Ontology as Personalization Filter (PI 20072006)
9. Zaini, N., Omar, H., Yassin, N.: MIMOS Patent Pending; Context Based Behavioral Modelling (PI 20092240)
10. Schiaffino, S., Amandi, A.: Building an expert travel agent as a software agent. Expert System with Application 36, 1291–1299 (2009)
11. Zografos, G.K., Androutsopoulos, K.N., Nelson, J.D.: Identifying Travelers' Information Needs and Services for an Integrated International Real Time Journey Planning System. In: 13th International IEEE Annual Conference on Intelligent Transportation Systems (2010)
12. Anacleto, R., Figueiredo, L., Luz, N., Almeida, A., Novais, P.: Recommendation and Planning through Mobile Devices in Tourism Context. In: Novais, P., Preuveneers, D., Corchado, J.M. (eds.) ISAmI 2011. AISC, vol. 92, pp. 133–140. Springer, Heidelberg (2011)
13. Cardoso, J., Sheth, A.: Developing an owl ontology for e-Tourism. In: Cardosa, J., Sheth, A. (eds.) Semantic Web Services, Process and Application. Springer, New York (2006)
14. Montaner, M., Lopez, B., de la Rosa, J.L.: A Taxanomy or Recommender Agents on the Internet. Artificial Intelligence Review 19(4), 285–330 (2003)
15. Maswera, T., Edwards, J., Dawson, R.: Recommendations of e-commerce systems in tourism industry of sub-Saharan Africa. Telematics and Informatics 26, 12–19 (2009)
16. Zaini, N., Omar, H., Yassin, N.: MIMOS Patent Pending; A Method of Providing Travel Recommendations (PI 20092324)
17. Niaraki, A.S., Kim, K.: Ontology based personalized route planning system using a multi-criteria decision making approach. Expert System with Application 36, 2250–2259 (2009)

# Formulating Agent's Emotions
# in a Normative Environment

Azhana Ahmad, Mohd Sharifuddin Ahmad,
Mohd Zaliman Mohd Yusoff, and Moamin Ahmed

Universiti Tenaga Nasional,
Jalan IKRAM-UNITEN, 43009 Kajang, Selangor, Malaysia
{azhana,sharif,zaliman}@uniten.edu.my, moamin84@gmail.com

**Abstract.** In this paper, we propose emotion as another additional component to the belief-desire-intention (BDI) agent architecture, which could enhance the reasoning process for agents to deliberate changes that occur in a normative framework. We postulate that emotions triggered in agents could influence the agents' decisions on the appraisal process, which subsequently enhance their performance. We discuss how emotions are deliberated in an agent's reasoning process that involves its beliefs, desires, and intentions in time-constrained situations.

**Keywords:** Norms, Normative Environment, Software agent, Multi-agent Systems, Emotions.

## 1   Introduction

In an organization, cooperation is one of the important issues to be resolved. The decision-making process requires many variables to be considered especially when there are competing personal and organizational goals to be achieved. Members of the organization are imposed with obligations (and prohibitions) in performing their jobs. Usually, obligations (and prohibitions) come with complete information in terms of tasks, durations, rewards, and penalties. Events associated with these obligations (and prohibitions) could come in the form of personal and organizational goals, which could either be mandatory or discretionary.

An agent's environment changes with the tasks of achieving its normative goals, which changes its belief due to changes in its environment. In our earlier work [1], we proposed the Obligation-Prohibition-Recommended-Neutrality-Disliked (or OP-RND) normative framework, in which agents are influenced by the normative periods (RND) based on the reward and penalty imposed in these periods. We used the BDI model to compute agents' reasoning in their beliefs, which are perceptions from the environment, and desires, which are the possible options of action available to the agent. Their intentions, which are based on desires and beliefs, lead to actions. We discovered that without any constraints, agents were able to comply with the norms.

The rest of the paper is organized as follows: Section 2 discusses the related work in this research. Section 3 presents the case for our study. Section 4 deliberates the

D. Lukose, A.R. Ahmad, and A. Suliman (Eds.): KTW 2011, CCIS 295, pp. 82–92, 2012.
© Springer-Verlag Berlin Heidelberg 2012

formalization of emotions in our framework. In Section 5, we compare the results of simulations between the non-emotional (rational) normative and emotional normative models. Finally, Section 6 concludes the paper.

## 2     Related Work in Emotions

Emotions occur when an agent's cognitive, physiological and motor component dissociates in serving separate functions as a consequence of a situation-event appraised as highly relevant for the agent [13]. Researches in emotion-based BDI architecture have shown some results in which emotion is represented as an extended component in the architecture. Pereira et al. [11] introduces the emotional-BDI architecture, as an extended version of the classic BDI architecture with the addition of three new components – the Sensing and Perception Manager, the Effective Capabilities and Effective Resources revision function and the Emotional State Manager.

Jiang et al. [5] introduce the EBDI architecture by adding the influence of primary and secondary emotions into the decision making process of a traditional BDI architecture by adding possible belief candidates directly from communication and contemplation. Jun et al. [7] introduce a new architecture for emotional agent based on the traditional BDI agent model. The emotional agent architecture includes components of belief, desire, intention, plan, rational reasoning machine, emotional knowledge base and emotional reasoning machine. To treat emotions, the architecture extends the traditional BDI agent model by adding an emotional knowledge base and an emotional reasoning machine. The emotional reasoning machine implements the logical reasoning relevant with emotions. The above deliberation shows that every architecture uses data perceived from the environment to reflect belief, desire and intention of the agent and the agent adapts to the environment based on the result from different approaches.

Our research also draws inspiration from the important concepts and theories of emotion. Cognitive appraisal theories (CATs), for example, explain human emotions as a result of the subjective evaluation of events that occur in the environment. Emotions allow flexibility both in event interpretation and in response choice [12]. In social environment, emotions inform people about social events and prepare their bodies and minds for actions; coordinate social interactions; help individuals define their identities and play their roles within groups; and mark or strengthen boundaries between groups [8]. Theorists assume that emotions solve problems relating to social relationships in the context of on-going interactions [9]. If we look at a specific type of emotion, for example embarrassment, such emotion could result in one's loss of self-esteem. One will thus be concerned for other's evaluation of oneself [4]. Such event prevents one from violating one's culture or norms because emotions have become a mechanism that enables people to maintain and attend to the responsibility to the society.

The relationship between CATs and BDI is very clear. Emotions result from a BDI agent's evaluation of the environment, which consists of the agent's goals, needs, belief, and desires. Such evaluation is the agent's cognitive appraisal on the environment. The influence of emotions could be measured in terms of the agent's performance. In this paper, we exploit the OCC theory [10] to determine the type of affective reaction in a normative environment. We propose that emotions are triggered when unexpected events occur, and the agent needs to use its resources and efforts to complete the tasks to achieve the normative goal. We postulate that the ratio of cooperation and flexibility increases when agents use emotions to decide on which course of action to take [3]. A domain is specified as a process of document submission within a stipulated time, which is the normative goal. We use $G_{Nm}$ to represent the normative goal and $G_P$ to represent the personal goal. We propose two types of personal goal: *mandatory* and *discretionary*. A mandatory personal goal, $G_{PM}$, is immutable and an agent must mandatorily attempt to achieve the goal stipulated by all its requirements. A discretionary personal goal, $G_{PD}$, is subject to changes and an agent can discretionarily decide to achieve the goal, put it on hold, or disregard it completely. The agent's performance is measured on their emotional states in different environments.

Researches have been conducted on extending the computational study of social norms with a systematic model of emotions. Some researchers argue that emotions sanction norms as exemplified by Alexander et al. [2] on embarrassment. They claim that people will feel embarrass if they violate social norms (for example, wearing jeans in formal dinner). Such phenomenon shows that emotions would help to sustain the social norms and the compliance of norms would improve agent's performance in certain processes. They present an emotion-based view on the scenario by Conte et al. (1995), and show that the computational study of social norms can profit by modeling emotions among agents in artificial societies, and propose the TABASCO architecture for the development of appraisal-based agents.

Based on these findings, we conclude that the emotional computation model could be a viable attempt to enhance collaborative interactions in multi-agent systems, which is the motivation for our work.

## 2.1    The Ortony, Clore, and Collins (OCC) Theory

The OCC model is based on the types of emotions and has three branches that relate to emotions arising from aspects of objects, events, and agents. The theory imposes some structure on the unlimited number of possible emotion-eliciting situations, with the assumption that emotions arise because of the way in which the situations that initiate them are construed by the experiencer. The theory emphasizes that emotion is capable of accommodating the fact that there are significant individual and cultural differences in the experience of emotions. The theory proposes that the structures of individual emotions are always along the lines that *if* an individual conceptualizes a situation in a certain kind of way, *then* the potential for a particular type of emotion exists.

We use this theory to lay the foundation for a computationally tractable model of emotions, i.e., to use emotions for agents to reason about events occurring in their environment. In this paper, we focus on the reactions to events since our domain relates to event occurrences to achieve the goal of document submission. We choose the Well-Being category of emotions resulting from *"Consequences of Events"* and focusing on *"Consequences for Self"* and *"Prospects Irrelevant"* (see Figure 1).

Conception of an event is defined as construal to things that happen. The affective reaction for a Lecturer is desirability (consist of desirable and undesirable) to the events of normative and personal goals. These are the criteria we consider to evaluate the effect of agents' emotions in our domain.

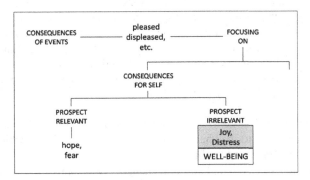

**Fig. 1.** The Well-Being Emotions

## 3     Case Study

We initiate a case study involving the examination paper preparation and moderation process (EPMP) of our faculty, which attempts to determine the actions and emotions of a Lecturer in executing the process of preparing and submitting the examination paper to the Examination Committee [14]. We use the OCC theory to classify a Lecturer's emotion in evaluating events that occur while he/she is trying to attend to the normative goal (i.e. preparing and submitting examination paper) and at the same time attending other personal goals.

The Lecturer analyzes the situation and recalls similar experience from which an emotion emerges on "unexpectedness" of event. Such situation could incur a loss in opportunity to prepare the examination paper on that day. At the same time, he/she is likely to have a feeling of having the opportunity to work on the normative goal, and may use all efforts or resources if any other personal goals occur on that day. The situation positively or negatively affects the Lecturer's belief which triggers an emotion as a reaction to the situation. For example, if the Lecturer is in distress, the event must be construed as undesirable. Since construing the environment is a cognitive process, the elicited emotion embodies the cognitive representations that result from such construal [2].

## 4     Formalizing Emotions in a Normative Framework

In this paper, we formalize the elicited emotions of Lecturers in the EPMP process when the events change and influence the Lecturer's action in submitting the document. However, we only consider the positive and negative categories of emotions represented by *joy* and *distress* respectively from the Well-being emotions category of the OCC theory (see Figure 1). We assert that the emotion of *joy* is under the positive emotions and *distress* under the negative emotions. We standardize such type of emotions for the purpose of discussion, but later in the simulation, we use *joy* and *distress* respectively.

We exploit our normative OP-RND framework in which agents have an obligation to the authority to submit the document in a time-constrained situation. The emotion triggers on event occurrences whenever the agents lose or gain one time slot between t[1] to t[14] to achieve the normative goal. The agents reason on a plan at time t[0] and revise their beliefs motivated by their elicited emotion within the duration. Consequently, we formalize the elicitation of emotions in terms of belief, action, plan, and goal based on the OCC theory of emotions.

**Definition 4.1:** Elicited Emotion, $E_{Em}$, is a consequence of cognitive evaluation and a mechanism of adaptation to the changes of events, at specific time t, which contribute to a change in agent behavior to achieve goals.

The fact that an agent's elicited emotion changes with changes to events, requires a method to update that change in emotion. Such change is implemented by an *emotion update function (euf)*. If an agent's emotion changes, the emotion update function updates the emotion and determines the agent actions for that instance:

$$E_{Em} = euf(\text{Event, t[i]})$$

i.e., the new elicited emotion is the update function of the occurrences of goals in the normative environment at time t[i].

In the EPMP domain, events are represented by the occurrence of goals within the duration. An Event Generator generates random goals: $G_{Nm}$, $G_{PM}$, and $G_{PD}$ at t[i], where i = 1 . . 14 and replaces the current goal with either $G_{Nm}$, $G_{PM}$, or $G_{PD}$ at t[i].

Emotions are triggered when the following events occur:

- $G_{PM}$ **changes to** $G_{Nm}$ – *positive* emotion such as *joy* for getting one extra day to achieve the goal $G_{Nm}$, and *pride* for the ability to submit the document early in the duration.
- $G_{Nm}$ **changes to** $G_{PM}$ – *negative* emotion such as *distress* for losing one day to achieve the goal $G_{Nm}$, and *shame* for the inability to submit the document earlier in the duration.
- $G_{PM}$ **changes to** $G_{PM}$ – *neutral* emotion or no changes to emotion.
- $G_{Nm}$ **changes to** $G_{Nm}$ – *neutral* emotion or no changes to emotion.
- $G_{Nm}$ **changes to** $G_{PD}$ – *neutral* emotion or no changes to emotion, because the agents can still sacrifice the $G_{PD}$ to execute the original plan set at t[0].

## 4.1    Modeling Emotions

The emotions are based on the OCC's event-based group of emotions that consists of *joy* and *distress*, representing the positive and negative emotions, respectively. The Well-Being emotions depend on the value of desirability and undesirability to the event. An agent is likely to see $G_{Nm}$ in every slot, but the event generator generates random goals depending on the percentage of goals set for each type.

The effort of document submission depends on event occurrences. If $G_{Nm}$ changes to $G_{PM}$, a negative emotion is triggered, thus influencing the plan to submit. Therefore, agents need to look for other resources to catch up with lost time within the set duration, e.g. work harder to submit on time. The agent is always influenced by the degree of its belief to decide on which action to take, either to proceed or re-plan the action.

### 4.1.1    Agent's Belief, Desire, Intention and Emotion

In non-emotional environments, an agent's belief represents its knowledge about the environment. In an environment of emotions-endowed agents, the elicited emotion contributes significantly to an agent's belief revision function. If there are changes in the state of the environment (i.e., when $G_{Nm}$ is changed to $G_{PM}$), the agent updates its belief. The possibilities of emotion to trigger depend on such changes of event. These changes are additions to the agent's belief on the consequence of submitting in different normative periods and the associated reward or penalty.

**Definition 4.2:** The normative belief, $B_{Nm}$, is a belief revision function of a normative autonomous agent endowed with emotion. In such environment, there is an elicited emotion to trigger, $E_{Em}$ with changes in the belief on the reward and penalty (evaluated points, $P_E$).

$$B_{Nm} : brf(Env_{Nm} \cup P_E \cup E_{Em}) \rightarrow \delta_S$$

where $\delta_S$ = submission date at t[i], i = 1...14, and $P_E$ and $E_{Em}$ are the agent's belief candidates. The expression means that the agent's normative belief is its *belief revision function (brf)* of the environment, the evaluated points, and the elicited emotion, which implies the document's submission date.

The agent's desire $D$, is represented by the tasks which the agent could perform based on its belief. We model the agent's desire as a set of pre-compiled tasks, which the agent could perform. However, due to the dynamics of the environment the ensuing elicited emotion influences the agent's desire. As mentioned earlier, changes of event determine the positive or negative elicited emotion, which influences the agent's desirability.

Based on the EPMP, the agent could perform any of the following tasks, $\tau_i$, depending of the current state of the environment, its belief, and the elicited emotions:

- $\tau_1$: If the elicited emotion is positive, the plan to submit the document is executed as per original plan.
- $\tau_2$: If the elicited emotion is negative, the agent re-evaluates the plan such as increasing its diligence until a new plan is found and executed.

The tasks which the agent selected based on its belief, emotions, and desire represent its intention, $I$. We use the *filter* function to formulate the selection of task based on the elicited emotion, $E_{Em}$, the evaluation of the reward and penalty, $P_E$, and the set of pre-compiled tasks, $T$.

**Definition 4.3**: Intention, *I*, is a function of the environment, the evaluated points, the set of pre-compiled tasks, the elicited emotion, and the evaluation of reward and penalty, which implies the selected task, $\tau_i$ from T.

$$I : filter(Env_{Nm} E_{Em}, P_E, T) \rightarrow \tau_i, i \geq 1.$$

Figure 2 below shows the relationship between emotions and the agent's belief, desire, and intention (BDI).

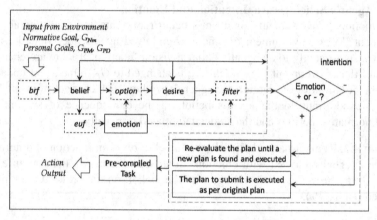

**Fig. 2.** Emotions and Agent's BDI

### 4.1.2    Optimization Functions

In the OP-RND framework, we introduce two optimization functions, i.e. Sacrifice and Diligence that enable an agent to disregard or postpone their discretionary personal goals and work harder by increasing its effort to achieve the normative goal. For example, when the agent revises its belief, its elicited emotion contributes to its intention. It uses its diligence as an alternative way when a negative emotion is triggered. Thus, if $G_{Nm}$ changes to $G_{PM}$, negative emotion is triggered and the agent believes on increasing its effort on the task as an alternative way to achieve the normative goal in the Recommended period [14]. Thus, we define diligence function, $D_X$ under a normative period X as:

$$D_X : filter(Env_{Nm}, G_{PM}, E_{negative}) \rightarrow \tau_i$$

i.e., the agent performs the Diligence function based on the environment, $Env_{Nm}$, the evaluated points, $P_E$, the number of $G_{PM}$, and the elicited emotion, $E_{negative}$, which implies a selected task from its pre-compiled tasks. When Diligence is invoked, the agent increases its effort to perform the task quicker to get the reward and avoid the penalty. We redefine the agent's intention to include the Diligence function as follows:

$$I : filter(T, D_X, E_{Em}) \rightarrow \tau_i, i \geq 1.$$

**Definition 4.4:** An agent's intention, $I$, is a function of the set of pre-compiled tasks, $T$ its diligence, $D_X$, with the influence of elicited emotion, $E_{Em}$ which implies a selected task, $\tau_i$ from $T$.

## 4.2    BDI with Emotions and Optimization functions

The execution of document submission by an agent is influenced by the elicited emotion after an event is generated. Due to the time-constrained environment, the action is based on the types of emotion, which is either positive or negative. If it is positive, the agent maintains its diligence and executes the original plan set at t[0], but if the emotion is negative, it optimizes its resources and efforts to accomplish the normative goal. One such approach is to boost its remaining diligence computed from the previous time slots. For example, suppose originally at t[0] the agent has four $G_{Nm}$ slots from which it plans to contribute 25% of its diligence to achieve the goal. But if the event generator changes the goal type to $G_{PM}$ on the fourth slot, it re-evaluates its plan and optimizes its effort by boosting its diligence to 50% at the third slot, i.e. t[$1^{st}$]=25%, t[$2^{nd}$]=25%, and t[$3^{rd}$]=50%, thus achieving the normative goal.

Every decision depends upon the output of the event generator that generates the probability of the agent working to achieve the normative or its personal goals. If there is no change to the goal, its emotion remains neutral and no changes are made to its decision.

# 5    Simulation and Analysis of Results

We simulate the following two modes using 300 agents and present the results in graphical form. The two modes are identified as (i) Mode A - The Rational Normative Agent, and (ii) Mode B - The Emotional Normative Agent.

## 5.1    Mode A - The Rational Normative Agent

This simulation is run to observe the performance of rational agents when given the capability to disregard or postpone personal discretionary goals, $G_{PD}$, and increase their efforts in $G_{Nm}$ slots due to the time constraints of personal mandatory goals, $G_{PM}$ in the dynamic environment. Two possible scenarios emerged due to the dynamic nature of the environment: constraints are relaxed when $G_{Nm}$ or $G_{PD}$ replaces $G_{PM}$, and conversely, constraints are tightened when $G_{Nm}$ or $G_{PD}$ are replaced by $G_{PM}$. The agents' performance depends on these two possible scenarios. Intuitively, performance improves in the first scenario but degrades in the second.

Figure 3 shows the simulation results of the Rational Normative Agent mode in the dynamic environment. The results show the distribution of submissions between t[1] and t[15]. About 74.7% of agents managed to submit their documents within the Recommended period, 20.7% within the Neutrality period, 3% within the Disliked period and the rest (1.6%) fail to submit the document within the duration.

## 5.2    Mode B - The Emotional Normative Agent

This simulation is conducted to observe the performance of emotion-endowed agents when given the capability to disregard or postpone personal discretionary goals, $G_{PD}$, and increase its effort due to the time-constraint effects of $G_{PM}$ in the dynamic environment. The simulation attempts to tease out the effect of agents' emotions on their performance in a comparable environment.

Figure 4 shows the simulation results of the Emotional Normative Agent mode with agents' elicited emotions. The results show the distribution of submissions between t[1] and t[15] indicating that the emotional agents' performance is better than that of the rational agents. About 83% of the agents managed to submit their documents within the Recommended period, 16% within the Neutrality period, 0.6% within the Disliked period and only one agent (0.3%) fail to submit the document within the duration.

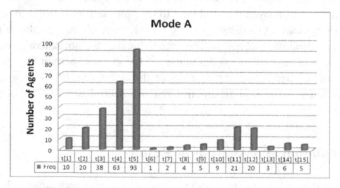

**Fig. 3.** Simulation for Rational Normative Agents

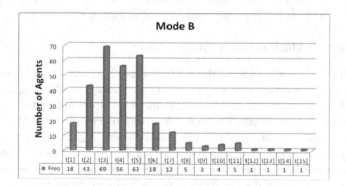

**Fig. 4.** Simulation for Emotional Normative Agents

The normative emotional, or the OP-RND-E, framework presents a norms-based BDI agent architecture that offers an alternative approach to enhanced agent performance in situations when agents are imposed with multiple goals in time-constrained environments. The results show that with elicited emotion, agents submit much earlier than the rational agents, demonstrating a better performance. We have not compared our framework's performance with the other E-BDI frameworks' performances because our emotion model is embedded in a dedicated normative framework. Hence any comparison would be inconclusive.

# 6    Conclusions

We propose that emotions motivate an agent to plan for better actions in achieving the normative goal. In this instance, emotions become a mechanism for the agent to evaluate the environment and decide accordingly when there are changes to its belief. It optimizes its efforts to achieve the normative goal if there is a possibility of failure to achieve its goal within a favorable normative period.

We equipped the agents with human-like traits that enable them to prioritize their goals and the ability to maximize their efforts when faced with overwhelming constraints and extreme urgencies. Consequently, the agents were able to decide on using these traits to improve their performance and maintain compliance with the norms. Elicited emotion plays the motivating role for agents to comply with the norms and submit earlier than the rational agents. The results show that in severe time-constrained, multiple goals situations, agents require an architecture that enables them to seek ways and means to achieve their normative goal.

In our future work, we will investigate on emotions for different roles, and other factors that influence emotions, such as emotions' intensity, global and local variables, and complex requirements as extended issues in the computational study of emotions.

# References

1. Ahmad, A., Ahmad, M.S., Mohd Yusoff, M.Z., Mustapha, A.: A Novel Framework for Normative Agent-based Systems. MJCAI (2009)
2. Alexander, S., Paolo, P.: Introducing Emotions into the Computational Study of Social Norms: A first Evaluation. Journal of Artificial Societies and Social Simulation (JASSS) 4(1) (2001)
3. Bazzan, A.L.C., Adamatti, D.F., Bordini, R.H.: Extending the Computational Study of Social Norms with a Systematic Model of Emotions. In: Bittencourt, G., Ramalho, G.L. (eds.) SBIA 2002. LNCS (LNAI), vol. 2507, pp. 108–117. Springer, Heidelberg (2002)
4. Broekens, J., DeGroot, D.: Scalable Computational Models of Emotion for Virtual Characters. In: 5th Game-On International Conference: Computer Games: Artificial Intelligence, Design and Education, CGAIDE 2004, Reading, UK, pp. 208–215 (2004)
5. Jiang, H., Vidal, M.J.: From Rational and to Emotional Agents. In: Proceedings of the AAAI Workshop on Cognitive Modeling and Agent-based Social Simulation (2006)
6. Julia, F., Christian, V.S., Daniel, M.: Emotion-based Norm Enforcement and Maintenance in Multi-Agent Systems: Foundations and Petri Net Modeling, AAMAS 2006 (2006)
7. Hu, J., Guan, C.: An Architecture for Emotional Agent. IEEE (2006)
8. Keltner, D., Haidt, J.: Social functions of emotions at four levels of analysis. Cognition and Emotion 13(5), 505–521 (1999)

9. Keltner, D., Buswell, B.N.: Embarrassment: Its distinct form and appeasement functions. Psychological Bulletin 122, 250–270 (1997)
10. Ortony, A., Clore, G., Collins, A.: The cognitive structure of emotion. Cambridge U. Press (1988)
11. Pereira, D., Oliveira, E., Moreira, N.: Towards an Architecture for Emotional BDI Agents, Technical Report DCC-2005-09, Universidade do Porto (2005)
12. Scherer, K.R.: Appraisal theories. In: Dalgleish, T., Power, M. (eds.) Handbook of Cognition and Emotion, pp. 637–663. Wiley, Chichester (1999)
13. Scheutz, M.: Agents with or without emotions? In: American Association for Artificial Intelligence (2002)
14. Yusoff, M.Z.M., Ahmed, M., Ahmad, A., Ahmad, M.S., Mustapha, A.: Sacrifice and Diligence in Normative Agent Environments. In: International Conference on Computer and Software Modeling (ICCSM 2010), Filipina (2010)

# Service Oriented Architecture
# for Semantic Data Access Layer

Chee Kiam Lee, Norfadzlia Mohd Yusof, Nor Ezam Bin Selan, and Dickson Lukose

MIMOS Berhad, Technology Park Malaysia, 57000 Kuala Lumpur, Malaysia
{ck.lee,norfadzlia.yusof,nor.ezam,dickson.lukose}@mimos.my

**Abstract.** This paper presents the Service Oriented Architecture (SOA)-based Semantic Data Access Layer (DAL) component approach which simplifies access to triple stores. Built with the benefits of web service and SOA, this component provides interoperability, loose coupling and programming-language independent access to different triple stores on various computer systems. Limitations of SPARQL query, SPARQL endpoint and the triples indexing mechanism as well as our solution are discussed. In this paper, particular attention is given to the performance comparison between our component and cache management, indexing facilities and native triple store vendors using the Lehigh University Benchmark (LUBM). The main contribution of our component is to provide an easy-one-stop common and reusable Application Programming Interface (API), to build high performance Semantic-based applications, without the need to develop yet another back-end system.

**Keywords:** Service Oriented Architecture, Data Access Layer, Semantic, Web Services, Extensible Markup Language.

# 1   Introduction

For many systems, persistent storage is implemented with different mechanisms, and there are marked differences in the Application Programming Interfaces (APIs) in different programming languages used to access these different persistent storage mechanisms [1]. Including the connectivity and data access code within business components introduces a tight coupling between the components and the data source implementation. Such code dependencies among components make it difficult and tedious to migrate the system from one type of data source to another or whenever APIs have been amended in newer versions.

Any system with persistent storage has client-server architecture if it handles data storage and retrieval in the database process and data manipulation and presentation somewhere else. Clients must use the provided connector that normally ties to certain programming languages like Java or C#. The communication between the client and server is normally through proprietary binary protocol on certain non-well known TCP port in the local network (i.e. intranet) environment. This client-server architecture cannot achieve interoperability, thus it cannot accomplish components

D. Lukose, A.R. Ahmad, and A. Suliman (Eds.): KTW 2011, CCIS 295, pp. 93–102, 2012.

using different types of computer systems, operating systems, and application software, interconnected by different types of local and wide area networks [2].

In this paper, we tackle the common problem faced by the database system. In Section 2, we discuss some related works that have been run by others. In Section 3, we propose solution models to the challenges described in Section 1, by elaborating on its system architecture in Section 4. This is followed by Section 5 to outline the set of SPARQL queries for the experiments to be conducted and its results. Finally we conclude with some future enhancement ideas in Section 6.

## 2    Related Works

SPARQL endpoint [3] is a conformant SPARQL protocol service as defined in the SPARQL Protocol for RDF specification. It normally resides outside firewall and can be accessed via clients implemented in any programming language from anywhere using standard HTTP protocol. A well-known example of SPARQL endpoint is http://dbpedia.org/sparql. SPARQL endpoint is able to provide interoperability as it allows clients that are implemented in any programming language to access any RDF data sources either in local or Internet networks. However, this solution requires developers to have knowledge on new domain areas like SPARQL. Furthermore, there are results that require compound operations and these require more round trips in network traffic. For instance, to retrieve RDF class hierarchy, it requires multiple calls of SPARQL to construct the hierarchy tree. This is an expensive operation and should be reduced to just one call. We have encountered the above-mentioned issues with our component presented in this paper. Semantic Reasoning is another effort to derive facts that are not expressed in ontology or in knowledge base explicitly. One of the example is rdfs:subClassOf that is used to state that all the instances of one class are instances of another. Some common approaches for providing reasoning include Frame-logic (F-logic) [4] or Description Logic (DL) [5]. However, enabling reasoning in SPARQL is a heavy operation and should be avoided in low-end machine or in enormous knowledge base. Our paper covers a method to discover sub-class relationship without performing real reasoning.

## 3    Solutions Models

In this section, we present KBInterface, a component that is built to comply with the Semantic Technology Platform (STP) [6], by leveraging best practices in enterprise architecture like layering design, caching, indexing, SOA and web services. At the highest and most abstract level, the logical architecture view of any system can be considered as a set of cooperating components grouped into layers [7]. The common three layers are Presentation Layer, Business Layer and Data Access Layer. This paper will only focus on Data Access Layer (DAL). DAL is a layer of a computer program which provides simplified access to data stored in persistent storage of some kind, such as a triple store [8]. By introducing an abstraction layer between Business Layer and triple store, it can achieve loose coupling and make each layer and component have little or no knowledge of each other. The DAL creates and manages all connections to all triple stores from different providers. With this, users of our platform can switch triple stores easily without impacting Business Layer.

Web services are becoming the technology of choice for realizing SOA. Web services simplify interoperability and, therefore, application integration. They provide a means for wrapping existing applications so developers can access them through standard languages and protocols [9]. Web Services Description Language (WSDL) is an XML format for describing network services [10]. Our system exposes services in the DAL layer as Web Services. Due to the fact that XML is a platform and programming-language independent way to represent data, the user can use it with virtually any programming language [11].

## 4    Architecture Overview

By implementing a service-oriented architecture, it can involve developing applications that use services, making applications available as services so that other applications can use those services, or both. As illustrated in Fig. 1, all requests from the external clients are made to the Delegator. Then the Delegator requests the location of the free instance of the KBInterface, from the Lookup registry, that is capable of processing the current request. After processing triples retrieved from triple store, KBInterface sends the notification to the Lookup registry and returns the response to the Delegator. There is no dependency on vendor-specific database connectivity protocols. KBInterface provides a connection to various triple-stores, cache mechanism and full-text search capability index file. These mentioned features will be discussed in more detail in the following section.

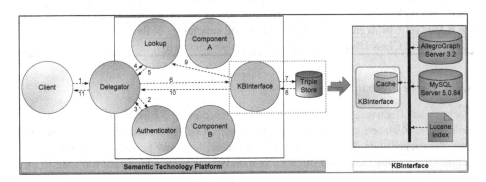

**Fig. 1.** KbInterface as Internal Component in STP

### 4.1    Connection to Various Triple-Stores

KBInterface can be connected to two different triple stores, AllegroGraph [12] or Jena/MySQL [13]. Due to triple stores specific APIs, KBInterface has two different implementation classes. User may select which implementation class to use by just modifying the configuration.properties file (Fig. 2). Switching of triple store is transparent to end users due to the fact that the exposed WSDL remains unchanged.

```
#Implementation file of the KBI for Allegrograph
Implementation=kbi.impl.allegro.KBInterfaceImpl
#Implementation file of the KBI for RDBMS
#Implementation= kbi.impl.rdbms.common.RdbmsImpl
```

**Fig. 2.** Configuration File for KBInterface

## 4.2    Predefined Queries

Without requiring users to have any SPARQL knowledge, KBInterface allows them to immediately discover the power of Semantic Web technology by invoking list of predefined queries in Table 1. These predefined queries have been optimized compare to direct SPARQL query call. For instance, to retrieve RDF class hierarchy, it requires multiple calls of SPARQL to construct the hierarchy tree. With "KB Do Get Class Hierarchy" predefined query, it can be reduced to just one call. Nevertheless, KBInterface allows advanced users to construct their own SPARQL queries that have been enhanced with caching functionality.

**Table 1.** List of Predefined Queries

| Query | Description |
|---|---|
| **KB Get All Triples** | To retrieve all the triples. |
| **KB Get Base Namespace** | To retrieve the base namespace. |
| **KB Do Search** | To perform a free-text searching. |
| **KB Do Get Parents** | To retrieve superclass of the given concept. |
| **KB Do Get Children** | To retrieve all subclasses of the given concept. |
| **KB Do Get Class Hierarchy** | To retrieve class hierarchy. |

## 4.3    Cache Mechanism

Cache is an effective approach for improving the round-trip time for request-response exchanges. KBInterface uses Apache Commons Collections LRUMap [14] as the base for cached data. It adds expiration of cached data and periodic clean-up of the expired data. Implementation of cache is mainly for triples which take too long to be retrieved when it doesn't even change that often.    Caching helps to improve performance by distributing query workload from triple store backend to cheap front-end systems. Statistical data about performance measurement among KBInterface with and without cache as well as a triple stores vendor is available at Section 5.1.

## 4.4    Lucene Indexing

Most of the triple store engines come with indexing capability. For example, AllegroGraph has the AllegroGraph.setIndexFlavors method when building the data store, which is able to support different types of indices [15], e.g. the spogi index first sorts on subject, then predicate, object, graph, and finally id. This index certainly helps to improve the speed of data retrieval operations in most cases. In this paper, we introduce extra indexing outside the triple store that has better performance for some of the daily usages. Apache Lucene [16] is a high-performance, full-featured text search engine library. KBInterface integrates with Lucene to accelerate the searching process. An external tool is built to generate the Lucene index file.

We found that Lucene index can provide the following advantages to KBInterface especially for non-realtime update or read-only ontologies. The Lucene index file is reusable and sharable among different triple-stores on various operating systems, i.e. built once run everywhere. The query performance is stable and predictable across different triple-stores. In some cases, retrieving result for simple query is more costly in SPARQL query than Lucene index. For example, a code snippet for KBInterface to return subjects that are graduate students is listed in Fig. 3 while its matching SPARQL query is as shown in Fig. 4.

```
List<String> matchers = new ArrayList<String>();
matchers.add("subject");

List<TriplesInfo>  triples  =  kbi.getTriples("Lubm-50",
"ub:GraduateStudent ", "predicate", matchers, 1, 50000);
```

**Fig. 3.** KBInterface getTriples Web Method

```
SELECT ?subject WHERE {
 ?subject rdf:type ub:GraduateStudent .}
```

**Fig. 4.** Query 15

```
SELECT ?subject WHERE {
 ?subject rdf:type ub:GraduateStudent .} LIMIT 100
```

**Fig. 5.** Query 16

```
SELECT ?subject WHERE {
 ?subject rdf:type ub:GraduateStudent .}
LIMIT 100 OFFSET 18
```

**Fig. 6.** Query 17

Query 16 in Fig. 5 is a typical query that is used in applications with Graphical User Interface (GUI) or page in Web browser, where it has limited screen to display all subjects. Instead of retrieving all results in one shot, which is an expensive operation, either in terms of CPU processing and network bandwidth, it is always advisable to just retrieve the exact amount of result to just fit onto the display screen. Query 17 in Fig. 6 is another typical query that is used to support pagination, where it divides results into discrete pages. These queries are supported in KBInterface. The performance comparison for returning subjects that are students is available in section 5.2.

Current SPARQL query specification does not allow regular expression on non-string literal subject or predicates. Fortunately, in KBInterface, we are able to apply regular expression on any place. For example, KBInterface is able to query all students who have graduated from the university, at either doctoral, masters or undergraduate level which have predicates given as

<http://www.lehigh.edu/%7Ezhp2/2004/0401/
univ-bench.owl#**undergraduate**DegreeFrom>,

<http://www.lehigh.edu/%7Ezhp2/2004/0401/univ-bench.owl#**masters**DegreeFrom>

or

<http://www.lehigh.edu/%7Ezhp2/2004/0401/univ-bench.owl#**doctoral**DegreeFrom>

The code snippet in Java when invoking KBInterface getTriples Web Services method is as shown below, where * indicates wildcard:

```
List<String> matchers = new ArrayList<String>();
matchers.add("subject");
matchers.add("predicate");
matchers.add("object");
List<TriplesInfo> triples =
kbinterface.getTriples("Lubm-50", "ub:*DegreeFrom",
"predicate", matchers, 1, 50000, true);
```

**Fig. 7.** KBInterface getTriples Web Method

The regular expression on predicate is not possible with SPARQL query. One of the workaround is to model the ontology to ensure there is a relationship amongst these properties. For LUBM data, the above properties are subclass of http://www.lehigh.edu/%7Ezhp2/2004/0401/univ-bench.owl#DegreeFrom,      which then can retrieve the desired result by enabling reasoning. However, this is an intensive operation.

```
SELECT ?subject ?predicate ?object WHERE {
?subject ?predicate ?object .
?subject ub:degreeFrom ?object . }
```

**Fig. 8.** Query 18

In case user has no privilege to model and create new relationship into the ontology, one of the workaround is to assume all predicates have labels that are well defined and create a query like below,

```
SELECT ?subject ?predicate ?object WHERE {
?subject ?predicate ?object .
?predicate rdfs:label ?label .
FILTER regex(?label, "degree from", "i") .}
```

**Fig. 9.** Query 19

Bear in mind, rdfs:label is not a compulsory field and its value can be badly defined such that itdoes not reflect onto the property URI. The performance results for comparison between Lucene index and Query 18 in Fig. 8 is shown in Experiment 2 in the next section.

Lucene indexing can reduce network round-trip time between application server and triple store, as it can sit reside immediately on the same host as the application server. Besides, since Lucene index is stored in a physical standalone file and does not require services provided by remote triple store server, it makes itself highly portable and can be transferred easily through the network. By locating index file on client side, this can achieve better performance and is able to save server resources by decentralizing server processing.

# 5    Experiments and Results

A series of tests were designed to examine the response time of KBInterface. We ran the experiments on LUBM(50) triple-store which has 6,657,853 number of triples. The platform for the tests was a 3.4GHz Pentium(R) D 32-bit CPU, with 2GB of memory running Windows XP Professional Service Pack 3. All the experiments were run with AllegroGraph version 3.2 and KBInterface version 3.1. The results were reported in seconds. We conducted two series of experiments, where one series compared caching performance and the other compared indexing performance.

## 5.1    Experiment 1: KBInterface Caching

We ran 14 LUBM queries [17] using KBInterface, KBInterface with cache enabled and directly through AllegroGraph API. The summarized results of the experiment are shown in Table 2 and Fig. 10.

**Table 2.** Summary of Experiment_1 Results

| Query | No. of triples | KBInterface | KBInterface Cache | AllegroGraph API |
|---|---|---|---|---|
| Query 1 | 4 | 19.1586 | 0.4379 | 1.1828 |
| Query 2 | 130 | 292.5735 | 0.3075 | 282.6720 |
| Query 3 | 6 | 0.4155 | 0.2869 | 0.0514 |
| Query 4 | 34 | 0.9685 | 0.3343 | 0.7075 |
| Query 5 | 719 | 4.9649 | 0.2727 | 4.7678 |
| Query 6 | 519842 | 9.7609 | 0.2686 | 9.2274 |
| Query 7 | 67 | 0.4090 | 0.3213 | 0.1415 |

**Table 2.** (*Continued*)

| Query 8 | 7790 | 46.7516 | 0.2699 | 46.3629 |
|---|---|---|---|---|
| Query 9 | 13639 | 2199.9223 | 0.2667 | 3805.5548 |
| Query 10 | 4 | 0.2659 | 0.3187 | 0.0344 |
| Query 11 | 224 | 0.3244 | 0.2567 | 0.1082 |
| Query 12 | 15 | 9.4510 | 0.2556 | 9.3415 |
| Query 13 | 228 | 0.4584 | 0.3137 | 0.2181 |
| Query 14 | 393730 | 5.3571 | 0.2540 | 4.8613 |

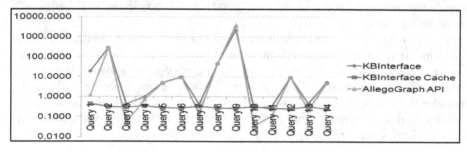

**Fig. 10.** Summary of Experiment_1 Results

Based on the above findings, the results for KBInterface (without cache) and AllegroGraph API were about the same. By enabling caching for KBInterface, the corresponding response time was able to be reduced drastically starting from the second iteration onwards as the result could be directly retrieved from the cache unlike before. It is a known fact that caching can certainly improve performance. The result here further conclude that triple stores require caching especially for complex queries like Query 9 with an almost 3000 s response time reduction.

## 5.2    Experiment 2: KBInterface Lucene Indexing

We ran another set of queries on AllegroGraph with SPOGI, POSGI, OSPGI, GSPOI, GPOSI, GOSPI indices[1] and KBInterface with Lucene indexing feature.

**Table 3.** Summary of Experiment_2 Results

| Query | No. of Triples | AllegroGraph w/ indices | KBInterface w/ Lucene index |
|---|---|---|---|
| Query 15 | 50000 | 30.8118 | 5.0603 |
| Query 16 | 100 | 8.7089 | 3.1110 |
| Query 17 | 100 | 6.2705 | 3.1879 |
| Query 18 | 50000 | 97.7918 | 5.2133 |
| Query 19 | 50000 | 3275.6107 | 5.2133 |

---

[1] S stands for the subject URI, P stands for the predicate URI, O stands for the object URI or literal, G stands for the graph URI and I stands for the triple identifier (its unique id number within the triple store).

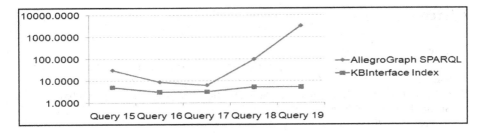

**Fig. 11.** Summary of Experiment_2 Results

By comparing results of Query 15, Query 16 and Query 17 which has the similar query (to return graduate students) but with different maximum number of returned triples, the KBInterface indexing result outperforms especially when the number of returned triples is huge. Take for instance when comparing across Query 16 and Query 15; when number of triples was increased from 100 for Query 16 to 50000 for Query 15, AllegroGraph experienced an increment of about 22 s compared to KBInterface which only incurred an additional 2 s to execute. For Query 18 that requires reasoning, KBInterface as shown in Fig. 11 again outperforms. Most significantly, the results show that for the same number of triples, AllegroGraph had a substantial increase in time (up till about 91 s) for Query 18 (with reasoning) even with its built-in indexing enabled as opposed to the KBInterface which experienced nearly zero increment in time. For workaround query that enables regular expression on predicates like Query 19, KBInterface indexing performance still stands ahead clearly.

## 6    Future Works and Conclusion

Object-Relational Mapping (ORM) in computer software is a programming technique for converting data between incompatible type systems in object-oriented programming languages [18]. With ORM, developers can avoid writing boilerplate programming code to persist data to triple store. JPA is a specification for managing Java objects, most commonly with a Relational Database Management System (RDBMS) and it is the industry standard for Java ORMs. We plan to investigate the possibility of integrating our system with Empire [19], an implementation of the Java Persistence API (JPA) for RDF and the Semantic Web. We are also interested in improving our component as a platform for linked data which exposes publically available data sets according to the following suggestions by Bizer, C et al. [20].

We believe the system proposed in this paper provides a technologically feasible and cost efficient solution to meet the performance and maintainability challenges in triple stores. By leveraging benefits of Web Services and integration with best practices in enterprise architecture like layering design, caching, indexing and etc, it allows our system to evolve and adapt to industry changes easily.

# References

1. Core J2EE Patterns - Data Access Object,
   http://java.sun.com/blueprints/corej2eepatterns/Patterns/Dat aAccessObject.html (last Visited: April 2011)
2. O'Brien, J., Marakas, G.: Introduction to Information System. McGraw-Hill (2005)
3. SPARQL endpoint, http://semanticweb.org/wiki/SPARQL_endpoint (last Visited: April 2011)
4. Query, A.: Inference Service for RDF, http://www.ilrt.bris.ac.uk/ discovery/rdf-dev/purls/papers/QL98-queryservice/ (last Visited: June 2011)
5. Baader, F., et al.: Theory, Implementation, and Applications. In: The Description Logic Handbook, p. 622. Cambridge University Press (2003)
6. Biring, N.S., Lukose, D.: SOA Framework. P00084 Malaysia (2007)
7. Microsoft: Microsoft Application Architecture Guide, 2nd edn. Microsoft Press (2009)
8. Triple Store, http://www.w3.org/2001/sw/Europe/events/ 20031113-storage/positions/rusher.html (last Visited: April 2011)
9. Motahari Nezhad, H.R., et al.: Web Services Interoperability Specifications. IEEE Computer 39(5), 24–32 (2006)
10. Christensen, E., et al.: Web Services Description Language (WSDL) 1.1, http://www.w3.org/TR/wsdl (last Visited: April 2011)
11. Bray, T.: Extensible Markup Language (XML) 1.1, http://www.w3.org/TR/xml11/ (last Visited: April 2011)
12. AllegroGraph RDFStore Web 3.0's Database, http://www.franz.com/agraph/allegrograph/
13. Jena Semantic Web Framework - Documentation Overview. Jena Semantic Web Framework, http://jena.sourceforge.net/documentation.html (last Visited: May 2011)
14. Apache Commons. Commons Collections, http://commons.apache.org/collections/ (last Visited: May 2011)
15. Overview of Indexing, http://www.franz.com/agraph/ support/learning/Overview-of-Indexing.1html (last Visited : June 2011)
16. Apache Lucene - Overview, http://lucene.apache.org/java/ docs/index.html (last Visited: May 2011)
17. 14 test queries, http://swat.cse.lehigh.edu/projects/lubm/query.htm (last Visited: May 2011)
18. Object-Relational Mapping, http://en.wikipedia.org/wiki/ Object-Relational_Mapping (last Visited: April 2011)
19. Empire, RDF for JPA, https://github.com/clarkparsia/Empire (last Visited: April 2011)
20. How to Publish Linked Data on the Web, http://www4.wiwiss.fu-berlin.de/ bizer/pub/LinkedDataTutorial/ (last visited: June 2011)

# A Novel Image Segmentation Technique
# for Lung Computed Tomography Images

Chin Wei Bong[1], Hong Yoong Lam[2], and Hamzah Kamarulzaman[2]

[1] School of Computer Science, Universiti Sains Malaysia, Penang, Malaysia
[2] Department of Cardiothoracic Surgery, Penang Hospital, Ministry of Health, Malaysia
cwbong@cs.usm.my, drlamhy@yahoo.ca, mkhamzah@gmail.com

**Abstract.** The aim of this paper is to propose and apply state-of-the-art fuzzy hybrid scatter search for segmentation of lung Computed Tomography (CT) image to identify the lung nodules detection. It utilized fuzzy clustering method with evolutionary optimization of a population size several times lower than the one typically defined with genetic algorithms. The generation of an initial population spread throughout the search space promotes diversification in the search space; the establishment of a systematic solution combination criterion favors the search space intensification; and the use of local search to achieve a faster convergence to promising solutions. With the appropriate preprocessing for lung region extraction, we then conduct the enhanced clustering process with hybrid scatter search evolutionary algorithm (HSSEA) followed with false positive reduction and nodules classification. The proposed approach has been validated with expert knowledge and it achieved up to 80% sensitivity.

**Keywords:** Evolutionary computing, clustering, soft computing, image segmentation.

## 1    Introduction

According to statistics from the American Cancer Society, lung cancer is the primary cause of cancer related deaths in the United States [4]. In Malaysia, lung cancer is the leading cause of cancer deaths accounting for 19.8% of all medically certified deaths due to cancers [15]. Pulmonary Nodule is a mass of tissue located in the lung [2]. It appears as round, white shadows on CT image [3]. Mass of 25mm or larger nodules is more likely to be lung cancer. The mortality rate for late lung cancer is higher than that of other kinds of cancers around the world. Thus, early detection of lung nodules is extremely important for the diagnosis and clinical management of lung cancer [12].

Computed tomography (CT) is the most commonly used diagnosis technique for detecting small pulmonary nodules because of its sensitivity and its ability to represent/visualize a complete three-dimensional structure of the human thorax [13]. Currently, nodules are mainly detected by one or multiple expert radiologists or clinicians inspecting CT images of lung. Each CT scan contains numerous sectional images that must be evaluated by a radiologist in a potentially fatiguing process. Recent research, however, shows that there is inter-reader and intra-reader variability in the

D. Lukose, A.R. Ahmad, and A. Suliman (Eds.): KTW 2011, CCIS 295, pp. 103–112, 2012.

detection of nodules by expert [1]. Therefore, segmentation of lung nodules plays an important role in automated detection of lung nodules in chest computed tomography.

The rest of this paper is organized as follows. Section 2 presented related works. Section 3 shows our proposal. Experimental results are presented in Section 4. Finally, we conclude the paper in Section 5.

## 2     Related Work

In lung CT image segmentation, usually region-based image features for the suspicious nodules are extracted according to the resulting contour by nodule segmentation. However, in the current nodule detection setting [13], 5% –17% of the lung nodules in their test data was missed due to the preprocessing segmentation, depending on whether or not the segmentation algorithm was adapted specifically to the nodule detection task. The latest approach available for image segmentation included threshold method and morphologic methods [12]. For instance, [10] have proposed a dot-enhancement filter for nodule candidate selection and a neural classifier for False Positive (FP) reduction. Besides, [11] proposed a CAD system for identifying lung nodules in 2-D chest radiographs that consists of using a weighted mean convergence index detector and an adaptive distance-based threshold algorithm to segment the detected nodule candidates. Many other groups have also presented systems and performance studies for detecting nodules [3, 7, 16, 17]. Although those CAD systems appear promising, further tuning/optimization of the different module parameters could improve the overall performance of the CAD system. The difficulties for detecting lung nodules in radiographs include [8, 14]: (i) Nodule sizes will vary widely: Commonly a nodule diameter can take any value between a few millimeters up to several centimeters, (ii) Nodules exhibit a large variation in density – and hence visibility on a radiograph – (some nodules are only slightly denser than the surrounding lung tissue, while the densest ones are calcified). (iii) As nodules can appear anywhere in the lung field, they can be obscured by ribs, the mediastinum and structures beneath the diaphragm, resulting in a large variation of contrast to the background. Thus, a valuable area of future work would be to improve the lung segmentation algorithm.

In this work, we try to exploit the benefits of applying scatter search (SS) to solve the lung segmentation problem for candidate nodule detection. SS fundamentals were originally proposed by Glover in 1977 and have been later developed in [17-18]. The main idea of this technique is based on a systematic combination between solutions (instead of a randomized one like that usually done in GAs) taken from a considerably reduced evolved pool of solutions named Reference set (between five and 10 times lower than usual GA population sizes). SS is also known as a population-based method that has recently been shown to yield promising outcomes for solving combinatorial and nonlinear optimization problems. Based on original formulation for combining decision rules and problem constraints such as the surrogate constraint method, SS uses strategies for combining solution vectors that have proved effective in a variety of problem settings.

Although SS has been applied in image registration [5, 6], it has not been used in medical image segmentation. Unlike GAs, SS components are designed considering a deterministic non-randomized scenario, encouraging a tradeoff between search intensification and diversification. To do so, we rely on: (i) the use of a fuzzy clustering method; (ii) the use of a population size several times lower than the one typically defined with GAs; (iii) the generation of an initial population spread throughout the search space, in order to encourage diversification; (iv) the establishment of a systematic solution combination criterion to favor the search space intensification; and (v) the use of local search to achieve a faster convergence to promising solutions (see Section 3).

# 3     A Fuzzy Hybrid Scatter Search Evolutionary Approach

This paper first proposes and applies a hybrid scatter search for lung CT image segmentation. Our proposal is based on fuzzy clustering approach with hybrid scatter search evolutionary algorithm (HSSEA), a stochastic optimization technique. Our fuzzy hybrid scatter search evolutionary approach is illustrated in **Fig. 1.** For the pre-processing, the lung regions are extracted using interactive binarization process. Then, the lung regions are segmented so that the candidate nodules are obtained. After that, false positive reduction is conducted before the candidate nodules are classified as nodule or non-nodule, cancerous or non-cancerous. The details of each process are discussed in the following subsections.

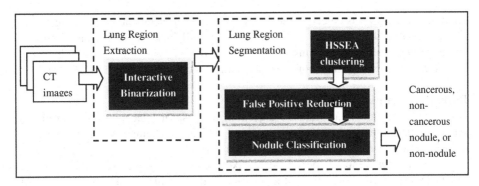

**Fig. 1.** The architecture of the proposed approach for the detection and segmentation of Lung CT scan for identifying the cancerous and non-cancerous nodules.

## 3.1     Lung Region Extraction

As the lung is essentially a bag of air in the body, it shows up as a dark region in CT scans. This contrast between lung and surrounding tissues forms the basis for the majority of the segmentation schemes. We use a rule-based binarization and to identify the combined lung and airway volume, separate the left and right lung and remove the trachea and mainstem bronchi. This is followed by morphological processing to obtain lung volumes without holes and with smooth borders. In this preprocessing stage, we use user-interactive segmentation tools as the segmentation of lungs affected by high density pathologies that are connected to the lung border and there is a lack of contrast between lung and surrounding tissues. With the user intervention, the interactivity can reduce the failure in segmenting these pathological parts of the lung. A binary image is constructed such that all pixels with a corresponding lung region gray level less than the selected threshold are turned on and all other pixels remain off. Construction of lung regions is performed by delineating the on regions in this binary image (as depicted in Fig. 2).

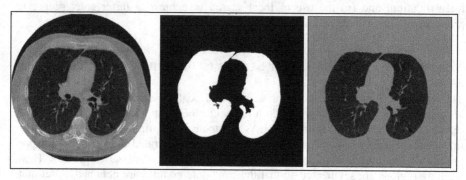

**Fig. 2.** Binarization of original CT image to normal lung region extraction

## 3.2     Lung Region Segmentation for Candidate Nodule Detection

After the lung region is detected, the following process is segmentation of lung region in order to find the candidate nodules. This step will identify the region of interest (ROIs) which helps in determining the cancer region. We use a fuzzy clustering algorithm with HSSEA that combines the characteristics of fuzzy c-means (FCM).

### Background of Fuzzy Clustering

The FCM algorithm is an iterative algorithm of clustering technique that produces optimal $c$ partitions, centers $V=\{v_1, v_2,..., v_c\}$ which are exemplars, and radii which defines these $c$ partitions [12]. Let unlabelled data set $X=\{x_1, x_2,..., x_n\}$ be the pixel intensity where $n$ is the number of image pixels to determine their memberships. The FCM algorithm tries to partition the data set $X$ into $c$ clusters. The standard FCM objective function is defined as follows:

$$J_m = \sum_{j=1}^{c} \sum_{i=1}^{n} u_{ij}^m \|x_i - v_j\|^2 \tag{1}$$

subject to $\sum_{j=1}^{c} u_{ij} = 1, 0 < \sum_{i=1}^{n} u_{ij} < n, and\ u_{ij} \in [0,1]; 1 \leq j \leq c; 1 \leq i \leq n.$. Here, $\{v_j\}_{j=1}^{c}$ are the centroids of the clusters   and the array      represent the fuzzy partition matrix and ‖.‖ denotes an inner-product norm (e.g. Euclidean distance) from the data point $x_i$ to the $j$th cluster center, and the parameter $m \in [1, \infty)$ is a weighting exponent on each fuzzy membership that determines the amount of fuzziness of the resulting classification. FCM algorithm starts with random initial $c$ cluster centers, and then at every iteration, it finds the fuzzy membership of each data point to every cluster using the following equation:

$$u_{ij} = \sum_{k=1}^{c} \left( \frac{\|x_i - v_j\|}{\|x_i - v_k\|} \right)^{-\frac{2}{m-1}} \tag{2}$$

Based on the membership values, the cluster centers are recomputed using the following equation

$$v_j = \frac{\sum_{i=1}^{n} u_{ij}^m x_i}{\sum_{i=1}^{n} u_{ij}^m} \tag{3}$$

The algorithm terminates when there is no further change in the cluster centers.

**Initialization of HSSEA**

First, the objective function in FCM has been incorporated in the HSSEA. The iteration of FCM has been adapted to HSSEA based on the scatter search template proposed in [4]. The image segmentation problem is often posed as clustering the pixels of the images in the intensity space. Clustering is a popular unsupervised pattern classification technique that partitions a set of $n$ objects into $K$ groups based on some similarity/dissimilarity metric where the value of $K$ may or may not be known a priori. Unlike hard clustering, a fuzzy clustering algorithm produces a $K \times n$ membership matrix $U(X) = [u_{kj}], k = 1, ..., K$ and $j=1,...,n$, where $u_{kj}$ denotes the membership degree of pattern $x_j$ to cluster $C_k$. For probabilistic non-degenerate clustering $0 < u_{kj} < 1$ and $\sum_{k=1}^{K} u_{kj} = 1, 1 \le j \le n$, unsupervised clustering method has high reproducibility because its results are mainly based on the information of image data itself, and it requires little or no assumption of the model, and the distribution of the image data.

**The Hybrid Scatter Search Evolutionary Clustering (HSSEA)**

The HSSEA approach is graphically shown in **Fig. 3**. The template defines five methods are described below:

*1) Diversification Generation Procedure:* The procedure is the same as that proposed in [4]. The goal is to generate an initial set $P$ of diverse solutions. This is a simple method based on dividing the range of each variable into a number of subranges of equal size; then, the value for each decision variable of every solution is generated in two steps. First, a subrange of the variable is randomly chosen. The probability of selecting a subrange is inversely proportional to its frequency count (the number of times the subrange has already been selected). Second, a value is randomly generated within the selected range. This is repeated for all the solution decision variables.

*2) Improvement procedure:* This procedure is to use a local search algorithm (a *simplex* method) to improve new solutions obtained from the diversification generation and solution combination methods (see Fig. 3). The improvement method takes an individual as a parameter, which is repeatedly mutated with the aim of obtaining a better individual.

*3) Reference Set Update procedure:* The reference set is a collection of both high-quality and diverse solutions that are used to generate new individuals. The set itself is composed of two subsets, *RefSet1* and *RefSet2*, of size $p$ and $q$, respectively. The first subset contains the best quality solutions in $P$, while the second subset should be filled with solutions promoting diversity.

*4) Subset Generation procedure:* This procedure generates subsets of individuals, which will be used to create new solutions with the solution combination method. The strategy used considers all pairwise combinations of solutions in the reference set [8]. Furthermore, this method should avoid producing repeated subsets of individuals, i.e., subsets previously generated.

*5) Solution Combination procedure:* This procedure is to find linear combinations of reference solutions. The use of a simulated binary crossover operator (SBX) makes HSSEA more robust.

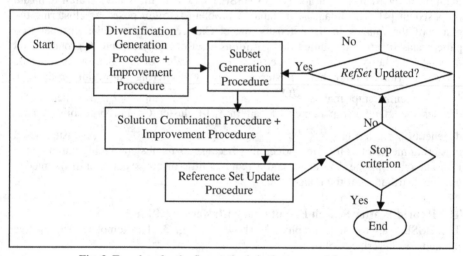

**Fig. 3.** Template for the five methods in the proposed framework

Initially, the diversification generation method is invoked to generate initial solutions, and each one is passed to the improvement method and the result is the initial set $P$. Then, a fix number of iterations are performed. At each iteration, the reference set is built, the subset generation method is invoked, and the main loop of the scatter search algorithm is executed until the subset generation method stops producing new subsets of solutions. Then, there is a restart phase, which consists of three steps. First, the individuals *RefSet1* in are inserted into $P$; second, the best individuals $n$ from the external archive, according to the crowding distance, are also moved to $P$; and, third, the diversification generation and improvement methods are used to produce new solutions for filling up the set $P$. The idea of moving $n$ individuals from the archive to the initial set is to promote the intensification capabilities of the search towards the global optimal found. The intensification degree can vary depending on the number of chosen individuals.

### 3.3    False Positive Reduction and Nodule Classification

After the necessary features are extracted, the following diagnosis rules can be applied to detect the occurrence of cancer nodule. A candidate nodule is characterized as "lung nodule" if the number of pixels within its ROI tagged as "nodule" is above some relative threshold. There are three rules [13] which are involved in threshold control are as follows:

**Rule 1:** Initially the threshold value T1 is set for area of region. If the area of candidate region exceeds the threshold value, then it is eliminated for further consideration. This rule will helps in reducing the steps and time necessary for the upcoming steps.

**Rule 2:** The threshold T2 is defined for value of maximum drawable circle. If the radius of the drawable circle for the candidate region is less than the threshold T2, then that is region is considered as non cancerous nodule and is eliminated for further consideration. Applying this rule has the effect of rejecting large number of vessels, which in general have a thin oblong, or line shape.

**Rule 3:** In this, the rage of value T3 and T4 are set as threshold for the mean intensity value of candidate region. Then the mean intensity values for the candidate regions are calculated. If the mean intensity value of candidate region goes below minimum threshold or goes beyond maximum threshold, then that region is assumed as non-cancerous region.

By implementing all the above rules, the maximum of regions which does not considered as cancerous nodules are eliminated. The remaining candidate regions are considered as cancerous regions. The sample result of lung region segmentation for candidate nodule detection and its classification is shown in Fig. 4.

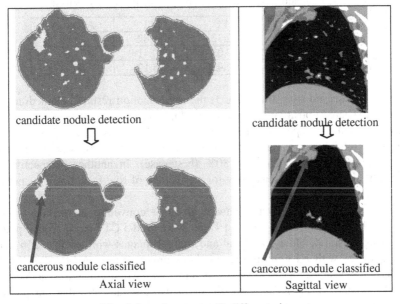

**Fig. 4.** Sample output with different views

## 4    Experiment and Result

Finally, for the SS, the initial diverse set P comprises *Psize* = 30 solutions and the *RefSet* is composed of the b = 12 best ones of them. All the experiments are performed on an Intel Core2Duo 2:66 GHz machine, with 2GB of RAM; while the codes are written using Matlab 2008a. The segmentation results are compared with segmentation with a clinician. In order to evaluate the spatial agreement between the manual delineation by clinical expert and the automatic segmentation, we chose to use the DICE coefficient or Similarity Index (SI):

$$SI = \frac{2(ED \cap Auto)}{ED + Auto}$$

where ED and Auto refer to the expert delineation and automatic segmentation respectively, while $(ED \cap Auto)$ refers to the overlap of $ED$ and $Auto$. The true positive fraction (TPF) is a measure of sensitivity and false positive faction (FPF) is measure of (1-specificity). Then, the number of false negatives (#*FN*) in the calculation of a third value: Accuracy.

$$TPF = \frac{TP}{TP + FN} \ , FPF = \frac{FP}{FP + FN} \ \text{and} \ Accuracy = \frac{TP+TN}{TP+FP+FN+TN}$$

Therefore, the sensitivity depends on the number of true positives (*TP*) and true negatives (*TN*). The specificity depends on the number of true negatives (*TN*) and false positives (*FP*).

Table 1. Validation results of lung nodule segmentation and detection

|  | SI | TPF | FPF | Accuracy |
|---|---|---|---|---|
| Patient 1 | 0.7462 | 72.32 | 31.82 | 73.02 |
| Patient 2 | 0.6272 | 52.03 | 40.41 | 71.03 |
| Patient 3 | 0.8071 | 79.92 | 25.12 | 80.11 |

As can be observed from the Table 1, the segmentation results obtained are very encouraging. There is positive correlation and high similarity between our automated segmentation and the expert segmentation provided which is reflected by SI values ranging from 62.7% (worst case) to 80.7% (best case). Accuracy is at an acceptable level between 71% (worst case) to 80% (best case). In another perspective, our method achieves up to 80% sensitivity for an average of five false positives per case for the datasets used (refer Fig. 5).

Lastly, we also compare our method with other conventional methods such as *K*-means and fuzzy *C*-means (FCM). In Table 2, we used 5 CT images from 3 patients from Penang Hospital, Malaysia. Axial and sagittal images with a 256 x 256 matrix are reconstructed with 1.1 mm thickness and 0.8 mm increment. All the datasets used in the CT images are in the Digital Imaging and Communication in Medicine (DICOM) standard format. The result for the 5 datasets indicates that our proposed method is outperformed as compared to other methods.

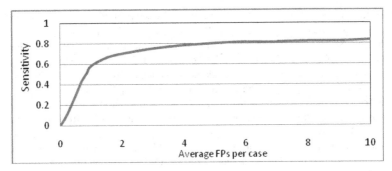

**Fig. 5.** Sensitivity versus average false positives per case for the datasets used.

**Table 2.** Comparison of Dice coefficient index for nodule segmentation for five CT images

| Methods | Coefficient index<br>Optimal no. of cluster | SI | | |
|---|---|---|---|---|
| | | $K$-Means | FCM | Proposed Method |
| DATASET 1 | 3 | 0.630672 | 0.630672 | 0.707766 |
| DATASET 2 | 4 | 0.559851 | 0.559851 | 0.70655 |
| DATASET 3 | 4 | 0.552605 | 0.552605 | 0.712029 |
| DATASET 4 | 5 | 0.64001 | 0.64001 | 0.705262 |
| DATASET 5 | 4 | 0.655314 | 0.655314 | 0.706172 |

# 5    Conclusion and Future Works

With the interesting properties and the recent successful outcomes achieved by the former SS strategy in other global optimization problems, the proposed use of SS components and the way they are assembled could solve the lacks presented by previous GA-based approaches. In this paper, we introduce a novel hybrid evolutionary framework, a modified version of scatter search technique for lung CT image segmentation problem with object detection. First, we conduct preprocessing step to extract the lung region. Then, the clustering process is performed with the proposed HSSEA. After that, the process continues with false positive reduction and nodules classification. Finally, the proposed approach has been endorsed by expert knowledge and its sensitivity reaches 80%. As future works, we suggest incorporating more advanced variants to the SS algorithm for its computation time with accurate result. It will be interesting to extend our proposal to other scenarios such as voxel-based multimodality, which would require minor changes over the proposed design, such as modifying the objective function.

# References

1. Armato, S.G., McLennan, G., McNitt-Gray, M.F., Meyer, C.R., Yankelevitz, D., Aberle, D.R.: Lung image database consortium developing a resource for the medical imaging research community. Radiology 232, 739–748 (2004)

2. Austin, J.H., Mueller, N.L., Friedman, P.J.: Glossary of terms for CT of the lungs: recommendations of the nomenclature. Radiology 200, 327–331 (1996)
3. Buhmann, S., Herzog, P., Liang, J., Wolf, M., Salganicoff, M., Kirchhoff, C., Reiser, M., Becker, C.: Clinical evaluation of a computer-aided diagnosis (CAD) prototype for the detection of pulmonary embolism. Academic Radiology 14(6), 651–658 (2007)
4. Cancer Facts and Figs 2009. The American Cancer Society (2009)
5. Cordón, O., Damas, S., Santamaría, J.: A fast and accurate approach for 3D image registration using the scatter search evolutionary algorithm. Pattern Recognition Letters 27, 1191–1200 (2006)
6. Cordón, O., Damas, S., Santamaría, J., Rafael Marti, R.: Scatter Search for the Point Matching Problem in 3D Image Registration. INFORMS Journal on Computing 20(1), 55–68 (2008)
7. Das, M., Muhlenbruch, G., Mahnken, A., Flohr, T., Gundel, L., Stanzel, S., Kraus, T., Gunther, R., Wildberger, J.L.: Small pulmonary nodules: effect of two computer-aided detection systems on radiologist performance. Radiology 241(2) (2006)
8. Gomathi, M., Thangaraj, P.: A Computer Aided Diagnosis System for Detection of Lung Cancer Nodules Using Extreme Learning Machine. International Journal Of Engineering Science And Technology 2(10), 5770–5779 (2010)
9. Gori, I., Fantacci, M., Preite Martinez, A., Retico, A.: An automated system for lung nodule detection in low-dose computed tomography. In: Giger, M.L., Karssemeiger, N. (eds.) Proceedings of the SPIE on Medical Imaging 2007: Computer-Aided Diagnosis, vol. 6514, p. 65143R (2007)
10. Hardie, R., Rogers, S., Wilson, T., Rogers, A.: Performance analysis of a new computer aided detection system for identifying lung nodules on chest radiographs. Medical Image Analysis 12(3), 240–258 (2008)
11. Messay, T.A., Russell, C.H., Steven, K.R.: A new computationally efficient CAD system for pulmonary nodule detection in CT imagery. Medical Image Analysis 14, 90–406 (2010)
12. Sluimer, I., Schilham, A., Prokop, M., van Ginneken, B.: Computer Analysis of Computed Tomography Scans of the Lung: A Survey. IEEE Transactions On Medical Imaging 25(4), 385–405 (2006)
13. Schneider, C., Amjadi, A., Richter, A., Fiebich, M.: Automated lung nodule detection and segmentation. In: Medical Imaging 2009: Computer-Aided Diagnosis, Proc. of SPIE (2009)
14. Vital statistics Malaysia: Department of Statistics, Malaysia (2005)
15. Wiemker, R., Rogalla, P., Opfer, R., Ekin, A., Romano, V., Bülow, T.: Comparative performance analysis for computer aided lung nodule detection and segmentation on ultra-low-dose vs. standard-dose CT. In: Proceedings of SPIE 6146, pp. 614-605 (2006)
16. Yuan, R., Vos, P., Cooperberg, P.: Computer-aided detection in screening CT for pulmonary nodules. American Journal of Roentgenology 186(5), 1280–1287 (2006)
17. Laguna, M., Martí, R.: Scatter Search: Methodology and Implementations in C. Kluwer Academic Publishers, Boston (2003)
18. Glover, F., Laguna, M., Martí, R.: Scatter search. In: Ghosh, A., Tsutsui, S. (eds.) Advances in Evolutionary Computation: Theory and Applications, pp. 519–537. Springer, Heidelberg (2003)

# Implementation of a Cue-Based Aggregation
# with a Swarm Robotic System

Farshad Arvin[1], Shyamala C. Doraisamy[2], Khairulmizam Samsudin[1],
Faisul Arif Ahmad[1], and Abdul Rahman Ramli[3]

[1] Computer Systems Research Group, Faculty of Engineering, University Putra Malaysia,
43400 UPM, Serdang, Selangor, Malaysia
farshadarvin@yahoo.com, {kmbs,faisul}@eng.upm.edu.my
[2] Department of Multimedia, Faculty of Computer Science & Information Technology,
University Putra Malaysia, 43400 UPM, Serdang, Selangor, Malaysia
shyamala@fsktm.upm.edu.my
[3] Intelligent Systems & Robotic Lab., Institute of Advanced Technology, University Putra
Malaysia, 43400 UPM, Serdang, Selangor, Malaysia
arr@eng.upm.edu.my

**Abstract.** This paper presents an aggregation behavior using a robot swarm. Swarm robotics takes inspiration from behaviors of social insects. BEECLUST is an aggregation control that is inspired from thermotactic behavior of young honeybees in producing clusters. In this study, aggregation method is implemented with a modification on original BEECLUST. Both aggregations are performed using real and simulated robots. We aim to demonstrate that, a simple change in control of individual robots results in significant changes in collective behavior of the swarm. In addition, the behavior of the swarm is modeled by a macroscopic modeling based on a probability control. The presented model in this study could depict the behavior of swarm throughout the performed scenarios with real and simulated robots.

**Keywords:** Swarm Robotics, Aggregation, Collective Behavior, Modeling.

## 1 Introduction

Coordination of the multiple robots for solving a joint problem can complete the task in a short period of time. Swarm robotics that is inspired from behavior of social insects is a new concept of multi-robots collaboration [1-2]. There are differences between swarm robotic and multi-robot researches [3]. These conceptual differences are explained by swarm robotics criteria [4]. Swarm robotics is a study on the collective behavior of large number of the simple homogeneous robots that use their limited perception to solve a joint task. Generally with decentralized control, small populations are not able to solve the problem hence swarm would improve the efficiency.

Since the swarm systems have strong roots in nature, swarm algorithms took inspiration from behavior of social insects and animals. Flocking [5], colonies [6], and aggregation [7-8] are examples of these behaviors which were implemented by swarm robots. Aggregation is a natural behavior of social insects and animals to find food or

D. Lukose, A.R. Ahmad, and A. Suliman (Eds.): KTW 2011, CCIS 295, pp. 113–122, 2012.

path [9]. In cue based aggregation, robots use environmental cues as the marker of aggregation spots [10]. The aggregation spot is called the optimal zone such as light and temperature for flies. Conversely, self-organized aggregation does not require cues [9] such as school of fish or aggregation of cockroaches [11]. Melhuish et al. [12] performed aggregation of robots around an infra-red (IR) transmitter. Each robot after reaching the marked zone with IR starts to play sound similar to frogs and birds. Size of group is estimated based on intensity of choruses. The aggregation behavior of honeybees is another inspired example of insects' aggregation which has been implemented by swarm micro robots [8]. Effects of several environment parameters were studied and it was observed that, the population size plays key role in aggregation performance. In our previous study, several effective parameters on swarm performance for a modified aggregation method that took inspiration from honeybee aggregation, were analyzed [13].

Modeling of a swarm behavior is the basic requirement in swarm researches to apply findings in different problems with differing scales. Commonly, swarm behaviors are explained with macroscopic modeling. Stochastic characteristic of swarm leads to use a probabilistic modeling to depict the collective behavior of swarm system [7, 14]. In cue-based aggregation, robots must provide a cluster around the cue. Robots change different algorithm states to accomplish the aggregation. Transitions between these states can be defined by probabilistic controls. Stock & flow is a macroscopic model of a swarm system that can be used to describe the behavior of swarm [15].

In this study, an audio based aggregation of swarm robots is studied. Groups of robots with different populations are deployed to provide a cluster near audio speaker following a simple control. There is not centralized control thus robots randomly start to execute a similar algorithm individually. Two aggregation methods which are original BEECLUST and *comparative* are implemented and performance of both methods is analyzed during experiments with real and simulated robots. In addition, a macroscopic model of the aggregation methods is presented which predicts the density of cluster during experiments.

## 2    Aggregation Method

Young honeybees prefer to aggregate in an optimal zone in their comb. The collective behavior of honeybees revealed that, the optimal zone is the part of comb with temperature in between 34 and 38°C [8]. They produce a cluster around the optimal zone. Hence, honeybees use a thermal cue to perform their aggregation. Based on aggregation behavior of honeybees, a clustering algorithm was proposed which is called BEECLUST [10]. In this algorithm, bees have random motion in their comb. When a bee meets another bee, it stops for a while and then starts to motion. The duration of this resting time depends on the temperature of the collision position. Collisions in a high temperature position results in long resting time.

For implementation of the aggregation method, we need several autonomous agents. As shown in Fig. 1, the aggregation algorithm has a simple control which relies on inter-robot collision. Robots in this algorithm move forward to meet other robots. They can detect other robots with their IR radiation. Collision in the high intensity position leads to longer waiting time hence chance to meet other robots will be increase. Therefore, most collisions will be occurred in high intensity area and robots produce a cluster.

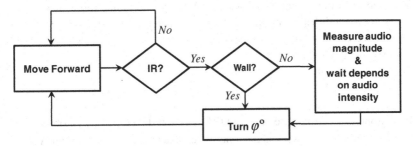

**Fig. 1.** Aggregation method (BEECLUST) based on collision between agents

Resting time has a key role in aggregation process time. In [10], a resting time model with respect to the sensors value was proposed. Robots in this study calculate the related time with respect to the intensity of the audio signals. Eq. 1 shows the resting time modeling of the swarm robots:

$$w = w_{max} \frac{S^2}{S^2 + \theta} \quad , \quad S = \frac{1}{4}\sum_{i=1}^{4} M_i \tag{1}$$

where $w_{max}$ is the maximum resting time for high intensity area that is occurred in source position and $S$ is the average of the captured values by robots microphones, $M_i$. $\theta$ is the parameter that indicates the steepness of the resting curve with respect to audio intensity. $w_{max}$ and $\theta$ must be estimated with empirical experiments for providing a fast aggregation. In this study, $w_{max} = 85$ sec and $\theta = 55000$.

After finishing the resting time, robots turn $\varphi$ degree and start to forwarding. The angle of rotation is different for aggregation methods. In original BEECLUST, robots turn randomly [10], however in the *comparative* method robots calculate a turning angle by the following formula [16]:

$$\varphi = \arctan\left(\frac{\sum_{i=1}^{4} M_i \sin(\beta_i)}{\sum_{i=1}^{4} M_i \cos(\beta_i)}\right) \quad , \tag{2}$$

where $\varphi$ is the estimated angular position of the source speaker, $\beta_i$ is the angular distance between $i$th microphone and the robot's head. $M_i$ is the captured audio signal's intensity from microphone $i$. This method increases the sensing ability of robots in estimating the direction of the audio source.

## 3    Robotics Platform

In this study, Autonomous Miniature Robot (AMiR) is utilized as the swarm robotic platform to perform aggregation scenario. AMiR is a modular robot that enables to connect several modules to realize different scenarios. In order to implement the audio based aggregation, an audio extension module has been developed to process the audio signals magnitude.

### 3.1    AMiR

AMiR (Autonomous Miniature Robot) [17] has been developed as an open-hardware swarm robotic platform. AMiR is designed with size of 70x70 mm and an AVR series micro-controller, ATMEGA-168, is deployed as the main processor to control all functions such as communication, trajectory, perception, power management, and user defined tasks. Processor has eight individual channels analog to digital converter (ADC) which are connected to the IR proximity sensors. Fig. 2 shows an AMiR robot that is equipped with an audio extension module.

**Fig. 2.** Autonomous Miniature Robot (AMiR) which is equipped with audio extension module

The robot is equipped with six IR proximity sensors that are used for obstacle detection with maximum range of 10 ±1 cm and inter-robot communication [18]. The actuator of the robot is two micro DC motors with internal gear with maximum speed of 10 cm/s [19]. AMiR is supplied with a 3.7 V lithium-polymer battery for autonomy of about 2 hours. A firmware library including basic functions of robot is developed. Open-source *gcc* compiler (C programming) is selected as the programming language of robot. In addition, AMiR is simulated with Player/Stage software to study of the large number of robots. Latest version of programming library, simulation tools, and compiler are available at: www.swarmrobotic.com.

## 3.2    Audio Extension Module

As the inputs of system, four condenser microphones in different directions are connected to the main processor. The output of microphones is amplified using an op-amp based amplifier. Output signals are connected to the individual ADC channels of the microcontroller. An ATMEGA-88 microcontroller with 8 MHz clock source was deployed as the main processor which captures sound samples of all microphones simultaneously. A basic task of this module is the estimation of given signals amplitude for each microphone.

## 3.3    Arena Configuration

In this study, arena is made out of white body plastic in size of 120x80 cm$^2$. Circular covered area by each robot with maximum perception is approximately 800 cm$^2$. Six AMiRs are used in this study, so the arena is setup two times larger than area covered by all robots (4800 cm$^2$). An audio speaker is placed in one side of the arena as the optimal zone marker. Fig. 3 shows the spatial distribution of the audio signals in the arena. In case of simulated robots, the arena is setup two times larger than real robot experiment which allows employing large populations.

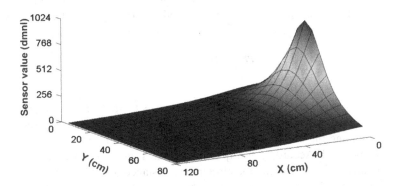

**Fig. 3.** Spatial distribution of the audio intensity in the robotic arena

# 4    Macroscopic Modeling of Aggregation

This section describes macroscopic behavior of aggregation for swarm system. During the aggregation, robots are in stop or free motion states. Based on the proposed Stock & Flow model in [15], we aim to utilize a modified version of this model that depicts the behavior of swarm system during an audio based aggregation.

The modeling software Vensim™ [20] is used to perform the model. Fig. 4 illustrates a simplified stock & flow model of the aggregation system. There are two transitions between the states which are: *join* and *leave*. Join is a probabilistic control that shows the tendency of the robots to join cluster. Conversely, leave shows the number of robots that are leaving the cluster at a certain time.

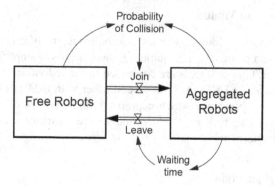

**Fig. 4.** Stock & flow model of aggregation scenario. Stocks are boxes (free motion and aggregated) and flows are double lines arrows that show transitions among stocks

$J(t)$ is the number of robots that are joining cluster. It is modeled by a probabilistic control with respect to the swarm density.

$$J(t) = F(t).P_c(t) \quad , \qquad (3)$$

where $F$ is the number of free robots and $P_c$ is the probability of the collision occur between robots. It is estimated based on occupied area by aggregated and free robots:

$$P_c(t) = \frac{F(t).(\pi R^2 + 2vR) + A(t).(\pi R^2)}{S_{sw}} \quad , \qquad (4)$$

where $A$ is the number of the aggregated robots, $R$ is the radius of sensing, and $S_{sw}$ is the size of arena. $P_c$ is dependent on the covered area by motion and stopped robots.

Robots joining the cluster wait until the resting time finishes. Hence, resting time should be modeled. According to Eq. (1), it depends on the intensity of the audio in collision point. With increasing the cluster size, distance for other robots to reach the source increases. The total covered area by the clustered robot is calculated with $k=A(t).\pi R^2$. Assuming the radial arrangement, the distance between a new joining robot and source is estimated by:

$$d(t) \approx \sqrt{\frac{2k}{\pi}} \quad . \qquad (5)$$

Therefore, the sensor value, $S$, in Eq. (1) and the number of leaving robots, $L$, are modeled by:

$$S(t) = \frac{I_s}{d(t)^2} \quad , \qquad L(t) = \frac{A(t)}{w(t)} \qquad (6)$$

where $I_s$ is the power of speaker and $L$ shows the leave transition.

# 5     Experiments

In order to investigate the effects of populations in aggregation time, groups of $N \in \{3,4,5,6\}$ for real robots and $N \in \{5,10,15,20,25\}$ for simulated robots in individual experiments were employed. Each experiment 20 times was repeated and the aggregation times were recorded. In this study, a fast aggregation of maximum number of participants is desired.

Fig. 5 shows the captured results for different population sizes in different aggregation methods. The observed results demonstrate that, with increasing the population size the aggregation time reduces hence performance of system increases. Two robots could not produce a stable aggregation that has been expected based on the swarm criteria [3]. In addition, comparative method shows fast aggregation in comparison with the original BEECLUST. It was because of the additional sensing ability of the comparative method that helps robots to move close to the audio source. Results of the large populations that were performed with simulated robots show significant improvement of the performance with comparative method.

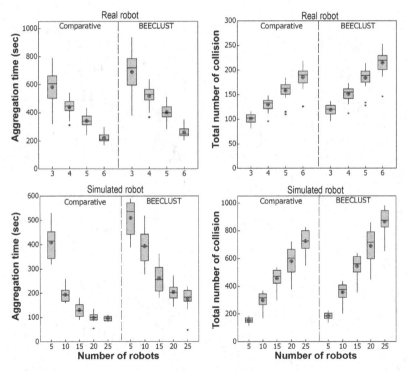

**Fig. 5.** Aggregation time and total number of inter-robot collisions for different populations with real and simulated robots

Since the aggregation methods are collision-based behaviors, the number of inter-robot collisions is an important parameter that could change the performance of the system. In our experiments, the numbers of collisions were counted individually with each robot and the total number of the collisions for each experiment was calculated. Total number of collisions for different experimental configurations is shown in Fig. 5 which increased with increase in population size. Results show that, aggregations with BEECLUST were accomplished with more inter-robot collisions.

In order to analyze the effects of population and aggregation method on performance of the swarm system, analysis of variance (ANOVA) test was used. Results of the ANOVA test showed that, an increase in population size significantly improve the performance. In addition, aggregation method used in this study had significant impact in performance analysis. However, the effect of populations was more than aggregation methods.

In case of modeling, size of the cluster during real and simulated robots experiments was tracked. Fig. 6 shows the aggregation density for different populations during experiments. Diagrams show that, the aggregations with comparative method are formed faster than BEECLUST. Observed experimental results for aggregation with 3 real robots showed a fluctuation indicating random behavior. Generally, the predicted values by macroscopic model were close to the observed results by real and simulated robots experiments.

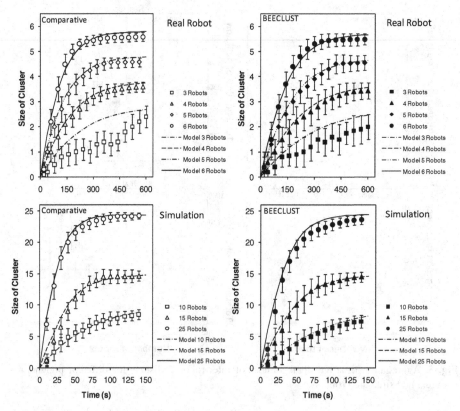

**Fig. 6.** Average size of clusters during experiments with real and simulated robots for comparative and BEECLUST methods

This study concludes that simple agents with a non-intelligent controller can solve a problem using a non-complex control. In addition, a simple modification on individual control of a swarm can significantly change the collective behavior of the system. Inter-robot collision played a key role in this aggregation hence any changes that result in increase probability of collision between agents, can improve the performance of swarm significantly. Despite the fact that swarm system requires a large number of agents to demonstrate its feasibility [2], it was shown to be possible implementing swarm experiments using small populations. However, the captured results in the small populations were more spread in comparison with the large populations. The implemented aggregation methods – BEECLUST and *comparative* – follow the swarm robotics definition as presented in [4].

# 6 Conclusion

In this paper, an aggregation method called BEECLUST was implemented with autonomous miniature robots (AMiR). The aggregation method was performed with a simple modification on sensing ability of the individuals. Effects of different populations and also the aggregation methods on performance of the swarm system were analyzed. Experiments were performed using real and simulated robots with different population sizes. Results showed that, an increase in population size results in increasing the number of inter-robots collisions of course improves the performance of the swarm system. In addition, a macroscopic model of the aggregation was presented with a simple stock & flow model. The model could depict the behavior of the aggregations and predicts the size of aggregate during the experiments. Finally, results of the performed experiments were statistically analyzed. The analysis of variance demonstrated that, population and aggregation method significantly improved the performance of the swarm system with different impacts.

# References

1. Şahin, E., Winfield, A.: Special issue on swarm robotics. Swarm Intelligence 2, 69–72 (2008)
2. Beni, G.: From Swarm Intelligence to Swarm Robotics. In: Şahin, E., Spears, W.M. (eds.) Swarm Robotics 2004. LNCS, vol. 3342, pp. 1–9. Springer, Heidelberg (2005)
3. Şahin, E.: Swarm Robotics: From Sources of Inspiration to Domains of Application. In: Şahin, E., Spears, W.M. (eds.) Swarm Robotics 2004. LNCS, vol. 3342, pp. 10–20. Springer, Heidelberg (2005)
4. Dorigo, M., Şahin, E.: Swarm robotics - special issue editorial. Autonomous Robots 17, 111–113 (2004)
5. Turgut, A.E., Çelikkanat, H., Gökçe, F., Şahin, E.: Self-organized flocking in mobile robot swarms. Swarm Intelligence 2, 97–120 (2008)
6. Melhuish, C., Sendova-Franks, A., Scholes, S., Horsfield, I., Welsby, F.: Ant-inspired sorting by robots: the importance of initial clustering. Journal of the Royal Society Interface 3, 235–242 (2006)

7. Soysal, O., Şahin, E.: A Macroscopic Model for Self-organized Aggregation in Swarm Robotic Systems. In: Şahin, E., Spears, W.M., Winfield, A.F.T. (eds.) Swarm Robotics Ws. LNCS, vol. 4433, pp. 27–42. Springer, Heidelberg (2007)

8. Kernbach, S., Thenius, R., Kernbach, O., Schmickl, T.: Re-embodiment of Honeybee Aggregation Behavior in an Artificial Micro-Robotic System. Adaptive Behavior 17, 237–259 (2009)

9. Soysal, O., Şahin, E.: Probabilistic aggregation strategies in swarm robotic systems. In: Proc. of the IEEE Swarm Intelligence Symposium, pp. 325–332 (2005)

10. Schmickl, T., Thenius, R., Moeslinger, C., Radspieler, G., Kernbach, S., Szymanski, M., Crailsheim, K.: Get in touch: cooperative decision making based on robot-to-robot collisions. Autonomous Agents and Multi-Agent Systems 18, 133–155 (2009)

11. Garnier, S., Jost, C., Gautrais, J., Asadpour, M., Caprari, G., Jeanson, R., Grimal, A., Theraulaz, G.: The embodiment of cockroach aggregation behavior in a group of micro-robots. Artificial Life 14, 387–408 (2008)

12. Melhuish, C., Holland, O., Hoddell, S.: Convoying: using chorusing to form travelling groups of minimal agents. Robotics and Autonomous Systems 28, 207–216 (1999)

13. Arvin, F., Samsudin, K., Ramli, A., Bekravi, M.: Imitation of Honeybee Aggregation with Collective Behavior of Swarm Robots. International Journal of Computational Intelligence Systems 4, 739–748 (2011)

14. Martinoli, A., Ijspeert, A.J., Mondada, F.: Understanding collective aggregation mechanisms: From probabilistic modelling to experiments with real robots. Robotics and Autonomous Systems 29, 51–63 (1999)

15. Schmickl, T., Hamann, H., Wörnb, H., Crailsheim, K.: Two different approaches to a macroscopic model of a bio-inspired robotic swarm. Robotics and Autonomous Systems 57, 913–921 (2009)

16. Gutiérrez, Á., Campo, A., Dorigo, M., Amor, D., Magdalena, L., Monasterio-Huelin, F.: An Open Localisation and Local Communication Embodied Sensor. Sensors 8, 7545–7563 (2008)

17. Arvin, F., Samsudin, K., Ramli, A.: Development of a Miniature Robot for Swarm Robotic Application. International Journal of Computer and Electrical Engineering 1, 436–442 (2009)

18. Arvin, F., Samsudin, K., Ramli, A.R.: Development of IR-Based Short-Range Communication Techniques for Swarm Robot Applications. Advances in Electrical and Computer Engineering 10, 61–68 (2010)

19. Arvin, F., Samsudin, K., Nasseri, M.A.: Design of a differential-drive wheeled robot controller with pulse-width modulation. In: Innovative Technologies in Intelligent Systems and Industrial Applications, pp. 143–147 (2009)

20. Vensim$^{TM}$, Ventana Systems, http://www.vensim.com/

# Parallel Web Crawler Architecture
# for Clickstream Analysis

Fatemeh Ahmadi-Abkenari and Ali Selamat

Software Engineering Department, Faculty of Computer Science & Information Systems,
University of Technology of Malaysia,
81310 UTM, Johor Baharu Campus, Johor, Malaysia
pkhoshnoud@yahoo.com, aselamat@utm.my

**Abstract.** The tremendous growth of the Web causes many challenges for single-process crawlers including the presence of some irrelevant answers among search results and the coverage and scaling issues. As a result, more robust algorithms needed to produce more precise and relevant search results in an appropriate timely manner. The existed Web crawlers mostly implement link dependent Web page importance metrics. One of the barriers of applying this metrics is that these metrics produce considerable communication overhead on the multi agent crawlers. Moreover, they suffer from the shortcoming of high dependency to their own index size that ends in their failure to rank Web pages with complete accuracy. Hence more enhanced metrics need to be addressed in this area. Proposing new Web page importance metric needs define a new architecture as a framework to implement the metric. The aim of this paper is to propose architecture for a focused parallel crawler. In this framework, the decision-making on Web page importance is based on a combined metric of clickstream analysis and context similarity analysis to the issued queries.

**Keywords:** Clickstream analysis, Parallel crawlers, Web data management, Web page importance metrics.

## 1 Introduction

A Web crawler either a single-process or a multi agent system, traverses the Web graph and fetches any URLs from the initial or seed URLs, keeps them in a queue and then in an iterated manner selects the first most important URLs for further processing based on a Web page importance metric. Search engines nowadays pay special attention into implementing a parallel crawler as a multi-processes system for achieving faster searching speed by the overall crawler and more coverage and download rate. The parallel agents may run at geographically distant location in order to reduce network load. Upon partitioning the Web into different segments, each parallel agent is responsible for one of the Web fractions [11]. An appropriate architecture for a parallel crawler is the one in which the overlapped download pages among parallel agents is low. Besides, the coverage rate of downloaded pages within each parallel agent's zone of responsibility is high. However, the quality of the overall

D. Lukose, A.R. Ahmad, and A. Suliman (Eds.): KTW 2011, CCIS 295, pp. 123–132, 2012.
© Springer-Verlag Berlin Heidelberg 2012

parallel crawler or its ability to fetch the most important pages should not be less than that of a centralized crawler. For achieving these goals, the link dependent parallel crawlers employ a measure of information exchange among parallel agents. Although this communication yields an inevitable overhead, achieving satisfactory trade-off among these objectives is the aim of link-based parallel crawlers for an optimized overall performance [10].

Moreover, there are two different classes of crawlers known as focused and unfocused. The purpose of unfocused crawlers is to search over the entire Web to construct the index. While a focused crawler limits its function upon a semantic Web zone by selectively seeking out the relevant pages to predefined topic taxonomy and avoiding irrelevant Web regions as an effort to eliminate the irrelevant items among the search results and maintaining a reasonable dimensions of the index [3], [12].

Since the bottleneck in the performance of any crawler is applying a robust Web page importance metric, this paper proposes architecture for a multi agent Web crawler that implements our previously proposed clickstream-based metric. This architecture performs most of page importance calculations either clickstream-based or context similarity-based and index construction processes offline in order to achieve an appropriate speed by the overall crawler. So in continue, first the existed Web page importance metric is discussed followed by a short review on our clickstream-based metric. Then the components of the proposed crawler architecture and its parallel agents are described.

# 2    Related Works

The related literature on dominant Web page importance metrics is discussed as follows;

## 2.1    Link-Dependent Web Page Importance Metrics

As stated earlier, a centralized crawler or each parallel agent of a parallel crawler retrieves URLs and keeps the links in a queue of URLs. Then due to the time and storage constraints, a crawler or a parallel agent must make a decision for selecting the most important $K$ URLs according to a Web page importance metric. There are diverse Web page importance metrics, each views the importance of a page from a different perspective [11]. The most well known category is the link-based metrics. *PageRank* as a modification to *Backlink count* that simply counts the links to a page, calculates the weighted incoming links in an iterated manner and considers a damping factor that presents the probability of visiting the next page randomly [4],[11].

*HITS* metric views the notion of Web page importance in the page's hub and authority scores. The authority score of a node in Web graph computed based on the hub scores of its parent nodes and the hub score of a node calculated by the authority scores of its child nodes. This metric has some drawbacks including the issue of topic drift, its failure in detecting mutually reinforcing relationship between hosts, and its shortcoming to differentiate between the automatically generated links from the citation-based links within the Web environment. Although the literature includes some modifications to *HITS* algorithm such as the research on detecting micro hubs, neglecting links with the same root, putting weights to links based on some text analysis approach or using a combination of anchor text with this metric, there is no evidence of a complete success of these attempts [5], [6], [7], [8], [9], [14].

## 2.2    Proposed Clickstream Based Importance Metric

The research on clickstream dataset has been mostly targeted toward achieving e-commerce objectives [13]. Upon employing the clickstream-based metric within the crawler, we will go beyond noticing the page importance in its connection pattern. Instead, the page credit computation performed according to a function of a simple textual log files in lieu of working with matrices of high dimensions. According to the link independent nature of clickstream-based metric, the need for link enumeration in the clickstream-based crawler is removes. As a result, the communication overhead is eliminated among parallel agents in order to inform each other of the existence of links in a parallel crawler. Moreover, upon employing the clickstream importance metric, the calculated importance of each page is precise and independent from the downloaded portion of the Web. Clickstream-based crawler is free from this defect too. Clickstream-based importance metric computed according to the total duration of all visits per each page multiplied by the number of distinct users visit that page during the observation span of time. By considering the number of distinct user sessions, the more different users visit the page at different time, the more important the pages will be. Therefore, the *LogRank* for a page is equal to the total duration of page-stays from different visits per that page multiplied by the number of users (equivalent to the number of user sessions) who have a visit to that page as illustrated in Equation (1). In Equation (1), $T$ is the time duration for log observation, $/d/$ is the number of days in $T$, $W=\{w_1, w_2, ..., w_i, ..., w_n\}$ is the set of all pages in a Web site, $U = \{u_1, u_2, ..., u_j, ..., u_n\}$ is the set of identified user sessions, $|U|$ is the number of distinct user sessions, $u_{wi}$ is the user session that contain the page $w_i$, $L$ is the length of each user session or the number of pages in the path of pages that user sessions include and $D_{w_i u_j}$ is the duration of page-stay of page $w_i$ in user session of $u_j$ [1], [2];

$$LR_{w_i} = \sum_{v=1}^{|d|} \left[ |u_{w_i}| \times \sum_{j=1}^{|U|} \sum_{l=1}^{L} D_{w_i u_j} \right] \qquad (1)$$

# 3    Operation Synopsis of the Clickstream-Based Multi Agent Crawler

In a crawler with parallel structure, the interior structure of each parallel agent and the modules of the crawler itself should be addresses. Since in our approach the employed Web page importance metric is based on clickstream analysis, so the architecture proposed in this paper not only guarantee the less job interference among parallel agents but also the crawler have the appropriate segments to address the employment of this Web page importance metric. We will present the result of our experiment on *UTM University* Web zone in section (4). Figure 1 shows the interior structure of each parallel agent and figure 2 represents the modules of the crawler itself. The proposed parallel crawler defined with four agents that could be easily developed to have more agents. Each module inside parallel agents described below followed by the description of each section of the overall multi agent Web crawler itself;

## 3.1    Architecture for Each Parallel Agent

- Crawler (C): Initializing by *Seed Partitionner* section of the crawler, the *Crawler* section of each parallel agent crawls the seed URLs according to Breath-First graph traversing approach and populates the *Primary Frontier* queue that is analyzed by *Duplication Detector* section later. The *Crawler* stops the process after a preset time constrain.
- Duplication Detector (D): This component checks the populated *Primary Frontier* to find the occurrence of any duplicated URL. When duplicated items found, this section removes them and copied the clean URL set into the Frontier queue. Page Scanner fetches addresses from *Frontier*, scans each Web page and saves the Web page content into centrally maintained *Index*.
- Distiller (L): This section is responsible to calculate each Web page importance based on the clickstream importance metric in offline manner. It first fetches Web pages address from *Frontier* queue. Then upon accessing server log files, it calculates the importance of all Web pages in each Web site. The result will form the *Ordered List* in each parallel agent. The *Ordered List* within each parallel agent represents the sorted list of the pure clickstream based ranked pages. These *Ordered Lists* form different parallel agents aggregated together to form the *Final Ordered List* of the overall parallel crawler.

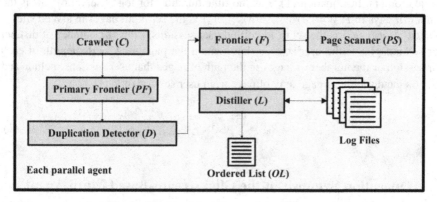

**Fig. 1.**   The interior structure of parallel agents

## 3.2    Architecture of the Overall Multi Agent Web Crawler

- Seed Partitionner (S): In order to start the crawling process, the crawler itself is responsible to provide the seed URLs to initialize the crawling process by each parallel agent. To do so, the *Seed Partitionner* module partitions the set of seed URLs among parallel agents. The seed URLs selected from highly ranked Web pages according to the statistics from previous crawl based on the clickstream ranking approach. The relationship between the *Seed Partitionner* and parallel agents is shown as connection labeled 1 in figure 2.

- Crawl History (CH): This component is one of the sections responsible to avoid page redundancy in the *Index* in order to prevent saving one page multiple times. Hence, when Web page content viewed and saved through the performance of *Page Scanner* module of each parallel agent, the correspondent URL added to the *Crawl History* queue if it is not exist there. The relationship between the *Crawl History* section and each parallel agent is shown as connection 2 in figure 2.

- Index: This repository includes all Web pages viewed so far by the *Page Scanner* section. The *Index* retained centrally because of avoiding Web page redundancy in different parallel agents. All Web pages saved as text files and the page length calculated based on word count for *BM* context relevancy checking measure by *Context Analyzer 1*. The relationship between each parallel agent and *Index* illustrated as connection 3 in figure 2.

- Dictionary: All distinct words that occurred in the indexed pages saved in *Dictionary*. Whenever a new page indexed, the page is scanned for new words and the dictionary is sorted to include them. These words used by the *Context Analyzer 1* in order to construct *BM Index*.

- Context Analyzer 1 ($H_{off}$): This component is responsible to calculate the context similarity of all words existed in *Dictionary* to all indexed documents based on *BM25* measure in offline manner for speeding up the whole searching process. The *BM25* Context similarity measure mostly known as *Okapi* is represented in Equation (2);

$$BM25(d_j, q) = \sum_{t_i \in d_j, q} A \times B \times C \qquad (2)$$

$$A = \ln\frac{N - df_i + 0.5}{df_i + 0.5} \quad B = \frac{(k_1 + 1)f_{ij}}{k_1(1 - b + b\dfrac{dl_j}{avdl}) + f_{ij}} \quad C = \frac{(k_2 + 1)f_{iq}}{k_2 + f_{iq}}$$

In Equation (2), $N$ is the total number of documents in the *Index* and $df_i$ is the number of documents that have the term $t_i$. $f_{ij}$ is the term frequency of the term $t_i$ in document $d_j$, $dl_j$ is the length of document $d_j$ (we calculated it based on word count), *avdl* is the average document length in the index and $f_{iq}$ is the term frequency of the term $t_i$ in the query $q$. $k_1$, $k_2$ and $b$ are parameters that set it to 1, 1 and 0.75. The value of $b$ differs among the *BM* variations. In *BM25*, it sets to 0.75 in order to prevent achieving high similarity score by the document of high length. In Equation (2), $A$, $B$, $C$ is calculated per each term and the final multiplication results in each term status in the whole index. We calculate $A$ and $B$ offline. $A*B$ represents each term similarity to every indexed documents regarding the fact that $C$ is equal to one in most cases.

- BM Index: The output from *Context Analyzer 1* ($H_{off}$) component results in the construction of *BM Index* in which we have a separated text file for each term. These files named the same as the term they include. These files consists the result of *BM25* of each term in *Dictionary* to all documents that we have in the *Index*. Table 1 depicts a portion of the result of *BM 25* approach based on our index of *UTM University* Web site for the term "Alliances".

- Context Analyzer 2 ($H_{on}$): The user issues a query through a Java applet interface. The query then passed to *Context Analyzer 2* ($H_{on}$) which works in online state. *Context Analyzer 2* searches the *BM25 Index* for each term $t_i$ in the query. Upon finding term(s), it will add the result of $A*B*C$ for the term(s) in the same intersect documents exist in the text files of *BM25 Index*. Then it passed the top ranked pages representing the highest context similarity score to the query to the *Result coordinator* section.

- Aggregator (G): Each parallel agent makes its own clickstream-based *Ordered List*. This *Ordered List* includes Web page name, its *Importance Value* and its *LogRank* from the most important one to the least. Since it is too time consuming for the crawler to search over all the *Ordered List* of the parallel agents in online manner, so we design the *Aggregator* section. The job function of the *Aggregator* section is to fetch all the *Ordered Lists* from the parallel agents offline and make a sorted *Final Ordered List*. Table 2 shows a portion of *Final Ordered List* that includes the final ranking of Web pages in our index of *UTM University*'s Web site. Connection 4 in figure 2 shows the relationship between each parallel agent and the *Aggregator* section.

- Result Coordinator (RC): This component receives the *Final Ordered List* produced by *Aggregator* section and the top context-based ranked pages produced by *Context Analyzer 2* in order to match them to announce the final answer set or *Search Answer Set* (*SAS*).

**Table 1.** A text file in *BM Index* constructed for the term "Alliances" ($df_i$= 12, A = 3.7897)

| File No. | $f_i$ | B | A*B | File Name |
|---|---|---|---|---|
| 1 | 1 | .8925 | 3.3823 | About UTM-about.txt |
| 23 | 1 | .6196 | 2.3481 | About UTM-Welcome.txt |
| 69 | 1 | .9889 | 3.7476 | Business-Partnerships & Collaborations.txt |
| 194 | 1 | 1.1335 | 4.2956 | Faculty of Computer Sceince-dean.txt |
| 414 | 1 | 1.1515 | 4.3638 | Global UTM-International Research Partnership.txt |
| 521 | 1 | .9857 | 3.7355 | Research-contact us.txt |
| 544 | 3 | 1.5236 | 5.774 | Research-Research Alliances.txt |
| 545 | 1 | 1.2035 | 4.5609 | Research-Research Centers & Institutes.txt |
| 546 | 1 | .9889 | 3.7476 | Research-Research Directory.txt |
| 547 | 1 | 1.0765 | 4.0796 | Research-Research Employment.txt |
| 549 | 2 | 1.4714 | 5.5762 | Research-Research Partnership.txt |
| 550 | 1 | .7673 | 2.9078 | Research-Research Scholarships.txt |

Table 2 also represent the first fourteen highly ranked pages of *UTM* Web site. Since *UTM* home page received the greatest number of visitors and the longest page-stay value of 16324151214 seconds * sessions, it has a considerable difference in importance to other Web pages in *UTM* Web site. Hence, table 2 represents top ranked pages from position two to position fifteen.

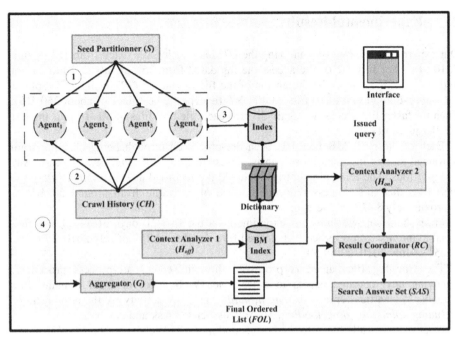

**Fig. 2.** Architecture of a clickstream based multi agent Web crawler

**Table 2.** -A portion of *Final Ordered List* represents the final ranking of Web pages in the *Index* of *UTM* Web pages

| Log Rank | Web Page ($d_j$) | Description | URL |
|---|---|---|---|
| 1 | PSZ | UTM library | http://www.utm.my/psz/ |
| 2 | Registrar | Registrar's office | http://www.utm.my/registrar/ |
| 3 | CICT | Centre for Information & Communication Technology | http://www.utm.my/cict/ |
| 4 | News | University news | http://www.utm.my/news/ |
| 5 | Staff | Needed links and information for all staffs | http://www.utm.my/staff/ |
| 6 | Faculties | Directory page to all UTM's faculties & schools | http://www.utm.my/faculties/ |
| 7 | Current Student | Needed links and information for current students | http://www.utm.my/currentstudent/ |
| 8 | About UTM | History and current state of the university | http://www.utm.my/aboututm/about-utm.html |
| 9 | Vice Chancellor | Vice chancellor' home page | http://www.utm.my/vicechancellor/ |
| 10 | FKA | Faculty of Civil Engineering | http://www.utm.my/civil/ |
| 11 | Events | University's Events | http://www.utm.my/events/ |
| 12 | RMC | Research Management Centre | http://www.utm.my/rmc/ |
| 13 | AIBIG | Artificial Intelligence and Bio Information research Group | http://www.utm.my/aibig/ |
| 14 | HEP | Office of Student Affairs and Alumni | http://www.utm.my/hep/ |

## 4     Experimental Results

Our experiment is based on analyzing the *UTM* server log file dated from 15th of July 2010 to 25th of July 2010. We access the log dated from 15th of July 2010 to 24th of September 2010. Due to the dimension of the file (about 7.5 *GB*), using a file splitter software we divided it to 19 files of 400 *MB* (except the last piece of about 250 *MB*). Then for faster file access, we again divided each file to 8 files of 50 *MB* plus the last smaller piece of each.

Each of these 50 MB files has been pre-processed through being cleaning from unwanted entries included those consist access to *.jpg, .gif, .png, .ico, .js, .css , swf* and *.pdf* files. The remaining entries included the recorded access to *.html* (*htm*) and *.php* files that this process reduced the size of each file from around 50 *MB* to approximately 8 *MB* on average.

From these set, we limit our experiment on the first 10 days of log observation equal to the first 24 pre-processed cleaned files dated from 15th of July 2010, 17:03:05 to 25th of July 2010, 23:59:59.

Then the log files again is processed through second pre-process procedure, removing any recorded access to outer space Web pages of *UTM* domain (like *Google*,...) , entries represents *search, mailto,* and any activity by the administrator including *uploading, downloading, My SQL* associated task and database.

After mapping all entries to the corresponding *UTM* Web page, the *Final Ordered List (T1)* formed which represent a list of all visited pages of *UTM* Web site during the mentioned period and ranked based on our clickstream-based approach. A portion of this file represented in table 2. The *Final Ordered List (T1)* constructed from four *Ordered Lists* form four parallel agents is our analyzed log-based resource upon which we fetch the Web page importance.

Table 3 shows our clickstream-based searching system's responses to three queries of *Automation Cybernetics* and *K-Economy*. The two queries are research alliances in *UTM University*. The column of *Search Result Set* includes the results of search represented by the file name of pages in answer set in order to have a description on page's content. The file name in the index is based on the *HTML title* tag and if not provided by the designer it is based on *HTML Meta title* tag. The last column is the page's URL. Our indexing and then search retrieving processes is according to *UTM University*'s Web site content in duration of January 2011 to April 2011.

## 5     Conclusion

In this paper, we propose an architecture for a multi agent Web crawler and its parallel agents that rely on a clickstream-based approach to rank Web pages. The proposed approach will perform the clickstream-based calculation and most of the context similarity computation offline. We divide the context similarity computation into two component functions as $H_{off}$ and $H_{on}$. The former will do most of the computation load offline while upon issuing a query, the latter can perform the rest of context similarity checking fast in online manner. Regarding the fact that there is no online communication among parallel agents and the *Aggregator* section construct the *Final Ordered List* to omit searching over the *Ordered List* of parallel agents, so the

*Result Coordinator* section is able to form the *Search Answer Result Set* (*SAS*) in an appropriated timely manner.

**Table 3.** The *Search Result Set* for two issued queries from our clickstream-based crawler

| | Query | Type | Search Result Set (SRS) | URL |
|---|---|---|---|---|
| 1 | *Automation Cybernetics* | Research Alliance | Automation Cybernetics (ACE)-Overview | http://www.utm.my/ace/ |
| | | | About UTM | http://www.utm.my/aboututm/about-utm.html |
| | | | Research UTM | http://www.utm.my/research/research.html |
| | | | Automation Cybernetics (ACE)-Awards | http://www.utm.my/ace/award.html |
| | | | Automation Cybernetics (ACE)-Introduction and Definition | http://www.utm.my/ace/introduction-a-definition.html |
| | | | Automation Cybernetics (ACE)-Ace Registration | http://www.utm.my/ace/ace-registration.html |
| | | | Research-Research Alliances | http://www.utm.my/research/index.php?option= com_content&task=view&id=61& Itemid=105 |
| | | | About UTM-Welcome | http://www.utm.my/aboututm/index.php?option= com_content&task=view&id=72& Itemid=116 |
| 2 | *K-Economy* | Research Alliance | K-Economy (RAKE) | http://www.utm.my/k-economy/ |
| | | | Research UTM | http://www.utm.my/research/research.html |
| | | | K-Economy-Commercialization | http://www.utm.my/k-economy/index.php?option=com_content&view=article&id=59:ctmg-research-group&catid=12:research-group&Itemid=129 |
| | | | Faculty of Biomedical Engineering and Health Science(FKBSK)-Home | http://www.biomedical.utm.my/ |
| | | | Research-K-Economy-Dean | http://www.utm.my/k-economy/index.php?option=com_content&view=article&id=67&Itemid=126 |
| | | | Faculty of Biomedical Engineering and Health Science-Research Alliance | http://www.biomedical.utm.my/V1/index.php?option=com_content&view=article&id=90&Itemid=77 |
| | | | Research-Research Alliances | http://www.utm.my/research/index.php?option=com_content&task=view&id=61&Itemid=105 |

**Acknowledgment.** The authors wish to thank Ministry of Higher Education Malaysia (MOHE) and University of Technology of Malaysia for funding the related research.

# References

1. Ahmadi-Abkenari, F., Selamat, A.: Application of Clickstream Analysis in a Tailored Focused Web Crawler. International Journal of Communications of SIWN, The Systemic and Informatics World Network (2010)
2. Ahmadi-Abkenari, F., Selamat, A.: A Clickstream-Based Focused Trend Parallel Web Crawler. International Journal of Computer. Applications (IJCA) 9(5), 1–8 (2010)
3. Bharat, K., Henzinger, M.R.: Improved Algorithms for Topic Distillation in a Hyperlinked Environment. In: Proceeding of the 21st ACM SIGIR Conference on Research and Development in Information Retrieval, pp. 104–111 (1998)
4. Brin, S., Page, L.: The Anatomy of a Large-Scale Hypertextual Web Search Engine. Computer Networks 30(1-7), 107–117 (1998)
5. Chakrabarti, S.: Mining the Web. Discovering Knowledge from Hypertext Data. Morgan Kaufmann (2003)
6. Chackrabarti, S.: Integrating Document Object Model with Hyperlinks for Enhanced Topic Distillation and Information Extraction. In: Proceeding of the 13th International World Wide Web Conference (WWW 2001), pp. 211–220 (2001)
7. Chackrabarti, S., Dom, B., Gibson, D., Kleinberg, J., Kumar, R., Raghavan, P., Rajagopalan, S., Tomkins, A.: Mining the Link Structure of the World Wide Web. IEEE Computer 32(8), 60–67 (1999)
8. Chackrabarti, S., Dom, B., Raghavan, P., Rajagopalan, S., Gibson, D., Kleinberg, J.: Automatic Resource Compilation by Analyzing Hyperlink Structure and Associated Text. In: Proceeding of the 7th International World Wide Web Conference, WWW 2007 (1998)
9. Chakrabarti, S., Van den Berg, M., Dom, B.: Focused Crawling: A New Approach to Topic Specific Web Resource Discovery. Computer Networks 31(11-16), 1623–1640 (1999)
10. Cho, J., Garcia-Molina, H.: Parallel Crawlers. In: Procceding of 11th International Conference on World Wide Web. ACM Press (2002)
11. Cho, J., Garcia-Molina, H., Page, L.: Efficient Crawling through URL Ordering. In: Procceeding of 7th International Conference on World Wide Web (1998)
12. Diligenti, M., Coetzee, F.M., Lawrence, S., Giles, C.L., Gori, M.: Focused Crawling using Context Graph. In: Procceeding of the 26th VLDB Conference, Cairo, Egypt, pp. 527–534 (2000)
13. Giudici, P.: Web Clickstream Analysis. In: Applied Data Mining, ch.8, pp. 229–253. Wiley Press (2003) ISBN: 0-470-84678-X
14. Kleinberg, J.: Authoritative Sources in a Hyperlinked Environment. Journal of the ACM 46(5), 604–632 (1999)
15. Liu, B.: Information Retrieval and Web Search. In: Web Data Mining, ch.6, pp. 183–215. Springer Press (2007) ISBN: 3-540-37881-2

# Extending Information Retrieval
# by Adjusting Text Feature Vectors

Hadi Aghassi and Zahra Sheykhlar

Iran University of Science and Technology, Tehran, Iran
hadi.aghassi@gmail.com
za_sheikhlar@comp.iust.ac.ir

**Abstract.** Automatic detection of text scope is now crucial for information retrieval tasks owing to semantic, linguistic, and unexpressive content problems, which has increased the demand for uncomplicated, language-independent, and scope-based strategies. In this paper, we extend the vector of documents with exerting impressive words to simplify expressiveness of each document from extracted essential words of related documents and then analyze the network of these words to detect words that share meaningful concepts related to exactly our document. In other words, we analyze each document in only one topic: the topic of that document. We changed measures of social network analysis according to weights of the document words. The impression of these new words to the document can be exerted as changing the document vector weights or inserting these words as metadata to the document. As an example, we classified documents and compared effectiveness of our Intelligent Information Retrieval (IIR) model.

**Keywords:** Text mining, Information retrieval, Semantic relationship, Social network analysis.

## 1 Introduction

With rapid evolution and large distribution of web data, extracting functional knowledge from many numerous and muddled sources has become a major aim of researchers. Text mining is an important technique for extracting information from textual concepts has been created to solve this problem. Information retrieval (IR) is the science of searching for documents, for information within documents and for metadata about documents, as well as that of searching relational databases and the World Wide Web [1]. Many tasks, for instance, indexing texts and search engines, topic detection and concept related tasks (such as text summarization, trend detection and so on), data mining tasks on text (such as clustering, classification and so on) use information retrieval to extract significant knowledge from text.

A standard text-mining task can be classified into two categories: (1) tasks that want to compare multiple documents, (2) tasks that want to compare multiple words or concepts. For each category, some specific similarity measures are useful. One main problem of both similarities is where some essential words are fallen out of use

D. Lukose, A.R. Ahmad, and A. Suliman (Eds.): KTW 2011, CCIS 295, pp. 133–142, 2012.

over the document. Many authors are not strongly familiar with the topic, especially in weblogs or other inexpert documents. For example, a text about "Obama" might not contain word "president". Another problem is using terms that are informal. Furthermore, some documents use acronyms or newfangled terms instead of being formal terms. This problem is very serious in chat mining where in many cases we cannot find words in dictionary.

Sometimes documents could not be represented as another general topic or as mixture of general topics. In other words, each topic consists of parts of other topics (not related to other topics). Another problem is related to words with multi form in a language. Linguistic processors should be used to do stemming and to distinguish the words, which are the same or related. Furthermore, documents might be written in multiple languages or might use some words of other languages among sentences. Stemming can be corrupting too. Linguistic computations and stemming often have high costs with high error rate and low accuracy.

Some other problems inhibit accurate computation of similarity. For example, a document might use an inordinate count of some words. This abnormal usage will cause similarity mistakes. In advertising or for raising rank of web pages we often encounter this problem. As another example, new documents or news usually use new terms, which never were seen before, or meanings of those terms are changed in time order. It will cause confusion in the computation of similarity. Furthermore, dependency of similarity on content of documents is not always good. It is clear that we can construct two sentences with same terms but without same meanings.

This paper addresses the above-mentioned problems. In our novel solution IIR, we extend term vector of documents in order to enhance precision of future text mining computations and simplify other text preprocessing tasks. We determine topic-scope of each document, adjust impression of its terms, and insert new terms, which strongly are related to its topic. We used social network analysis (with some changes) to find out terms that are similar to each document.

Our novel approach can often be used in text mining tasks, but certainly in some cases is not very beneficial. For evaluating our approach, we used weblog documents using two languages. We compared a simple classification task with and without our method. Furthermore, we evaluated our approach on Router dataset that is explained in section 5.

## 2    Related Works

Recently, many researchers have focused on Information Retrieval [1, 2]. Clearly, it is useful to apply social network analysis in IR. [3] has explored social annotations for IR. [4] uses semantics inference and social network analysis based on provided metadata and knowledge-base (i.e. ontologies) to establish a semantic enhanced information retrieval framework.

Recently, some papers have focused on the human-created hierarchies for the aim of designating words relations. Some significant instances are described as follows:

*WordNet*: is a semantic lexicon [5], which groups words into sets of synonyms, called synsets. WordNet records synsets in a hierarchy of conceptual organization.

*Roget's Thesaurus*: as well as WordNet, groups sets of words into sets of synonyms [6]. Although, it is geared more towards writing style than semantic information. *Wikipedia*: is composed of some *categories* and *portals* which are created by user.

Computing relations between words always depends on domain, and in specific domains often, we have variable relations between two similar words, particularly in cases, which we are involved in various types of domains.

In the [7], authors improved feature vector by using words' position and sequence. Therefore, the location of words in content is important as well. We inspired by this paper and determined N most relevant document to each document.

# 3     Finding Related Documents

In order to be able to find the most related terms (n-grams) to a specific document that were not used in the document, we firstly need to identify a group of related documents to that document. Practically, this problem is a query-based search problem in search engines. Therefore, the content of the document should be represented as a query in a search engine for detecting related documents.

Search engines' problem is a well-studied problem; it frequently arises in the analysis of convergence properties of Markov chains (e.g., see [8, 9, 10, 11]) or the problem of sampling a search engine's index [12, 13]. In [14] authors have an effective method for sampling information in social networks.

In view of the fact that our query is composed of contents of a document, length of the query is equal to length of the document, so we should use search engines that focus on large queries. Big society of search engines have paid attention to queries with a few terms, since important usage of them is the search service with a few terms. [15] has prepared an evaluation of usability in long query meta search engine.

One of the best solutions for this section is using online search engines (like Google[1] or Yahoo[2]). In this case, search engines use online, up to date and accurate methods to find related documents. Applying an online search engine needs all original documents to be available online and if you do not want to choose related words to a document from online web documents, target search engine had to index all documents which are going to be evaluated. Therefore, we did not use online search engines. One of the important advantages of online search engines is obtaining results quickly and without large computations, although it needs the results to be clarified (e.g. downloading, parsing or removing HTML markups).

In our approach, we rely on the top N (N = 30) documents retrieved for each query (each document) and analyze contents of these N documents for our goal in the next section. We use vector space model (VSM) for weighting terms and searching. A VSM [16] is a numerical document-by-term matrix V for a set of documents, where each element is an indicator of presence or absence of the word in the corresponding document either by a Boolean value or by a real-valued weight. For the real-valued model, TF-IDF (term-frequency inverse document frequency) is often used to take

---

[1] http://www.google.com
[2] http://www.yahoo.com

account of the appearance frequency of the word in the document set. The weight $w_{ij}$ of the $i$th term in the $j$th document is determined by

$$w_{ij} = TF_{ij} * IDF_i = TF_{ij} * \log(n/DF_i) \tag{1}$$

Where $TF_{ij}$ is the number of $i$th term's occurrences within the $j$th document, and $DF_i$ is the number of documents (out of n) in which the term appears. Using this model, the text similarity between two annotations $v_1$ and $v_2$, is simply calculated as cosine of the angle as following:

$$Sim(v_1, v_2) = Vv_1T \ Vv_2 \ / \ || \ Vv_1 \ ||2 \ || \ Vv_2 \ ||2 \tag{2}$$

The original VSM often suffers from inefficiency for massive real-world data. For this reason, the latent semantic analysis/indexing (LSA/LSI) reduces the VSM's dimensions using singular vector decomposition (SVD) [17].

## 4   Construction of Term Graph

After specifying N related documents, we construct terms graph by computing the similarity among words in these N documents. Constructing the graph helps us to distinguish more significant terms. This graph includes all terms that are used in N documents (except stop words, etc) and each term represents a vertex in the graph. Magnitude of each edge between two vertices is computed as a relation between words. We construct graphs and analyze them for each document.

Quantity of terms' relation is very important in our work, because it affects analysis of important terms. We use following formulation to estimate relatedness of two words $w_i$ and $w_j$:

$$Rel\left(w_i, w_j\right) = \frac{f'(w_i + w_j) + min\{\log f'(w_i), \ \log f'(w_j)\}}{f(w_i + w_j) + max\{\log f(w_i), \ \log f(w_j)\}} \tag{3}$$

Where $f'(w_i, w_j)$ is the frequency of all occurrences of $w_i$ and $w_j$ in the same document which is included in the N selected documents, $f'(w_i)$ or $f'(w_j)$ are frequencies of all occurrences of $w_i$ or $w_j$ in the N related documents, $f(w_i, w_j)$ is the frequency of all occurrences of $w_i$ and $w_j$ in the same document which is included of all documents, and $f(w_i)$ or $f(w_j)$ are frequencies of all occurrences of $w_i$ or $w_j$ in all documents.

Having found the relation of words, now we can strive to analyze the graph in the next step. Although, the computation of a relation between words is not so convenient yet, because of some presumptions that we considered. One of the important problems in the computation of $Rel(w_i, w_j)$ is when the computation of N related document is based on online search engines and does not have variables $f(w_i, w_j)$ and $f(w_i)$. In cases that presumptions of our solution are not regarded, we can often estimate the unclear variables. For instance, in online search engine problem we can estimate quantities of $f(w_i, w_j)$ with the number of found results for searching $w_i$ and $w_j$, and we use following relation for online search engines (such as Google):

$$Rel\left(w_i, w_j\right) = \frac{f'(w_i + w_j) + min\{\log f'(w_i), \log f'(w_j)\}}{g(w_i + w_j) + max\{\log g(w_i), \log g(w_j)\}} \tag{4}$$

Where $g(w_i, w_j)$ is the number of search results for $w_i$ and $w_j$ and in the similar manner, $g(w_i)$ is the number of search results for $w_i$.

It is important to note that in the computation of $Rel(w_i, w_j)$ a relation of words is not like other relations of words in the other researches. In most cases, a relation represents semantic or conceptual relations or relations that can be understood by human and understandability of a relation specifies the quantity of similarity between words, but here the percentage of variations of occurrences are more considerable.

## 5    Extracting Extended Words

After computing the relation between each two words, in this section we extract extended words for each document from constructed graph of that document. First, we introduce the similarity measure using relations of words.

### 5.1    Computing Words Similarity

In this section, we attain the similarity of two words. Note that similarity of two words is variable in respect of per document. This is because similarity depends on relations between words and the relations are computed from N similar documents for each document.

Every term that is related to other terms has similarities with respect to the quality and quantity of other relations of that term. For instance, the word "good" has many relations with other terms, but the word "stomach" has fewer relations. Therefore, we should consider the similarities of the "stomach" to be more valuable than those of the word "good". Because of this, we form graph model of words for better understanding and computing similarity.

To solve this problem, we have used lexical chains and computed comparative similarity of terms [18]. Here chains are composed of relations, which are computed in previous section. For this purpose, we computed chains of length $l$ for each pair.

$$C_l\left(t_i, t_j\right) = \left\{t_i \equiv t_1 \rightarrow t_2 \rightarrow \cdots \rightarrow t_l \equiv t_j\right\} \tag{5}$$

Here, $C_l$ is a chain of length $l$ from $t_i$ to $t_j$. By assigning $L$ as max value of $l$ we will have $1 \leq l \leq L$. With this presumption each pair of words have a set of chains of length $l$. Number of chains in each set is defined as $NC_l(t_i, t_j)$ which shows number of chains of length $l$ from $t_i$ to $t_j$. We computed $NC_l$ as following:

$$NC_l(t_i, t_j) = \sum_{Chain_k \in C_l} \left( \prod_{t_m \to t_n \in Chain_k} Rel(t_m, t_n) \right) \tag{6}$$

In other words, $NC_l(t_i, t_j)$ is the sum of all chains weights of length $l$ among two words where weight of a chain is obtained from multiplication of $Rel(t_m, t_n)$ for each edge in the chain.

Based on this relation, terms similarity is computed as following:

$$sim\left( t_i, t_j \right) = \sum_{l=1}^{L} \frac{NC_l(t_i, t_j)}{NC_l(t_j)} e^{-l} \tag{7}$$

Where $NC_l\left( t_j \right)$ is the total number of chains of length $l$ from every word to $t_j$. This means that, $NC_l\left( t_j \right)$ is the sum of all $NC_l(t_k, t_j)$ where $t_k$ represents all other words that have relation with $t_j$.

Computed similarity measure is a good comparative criterion for exhibiting relevance degree of word pairs. This measure tends to weaken relations of words with high usage frequency in literature. In addition, this measure computes similarity at L levels. Now we have a new word graph and it is time to indentify similar words.

## 5.2   Identifying Similar Words to Each Document

We computed a new term graph that its edges represent the similarity between terms and the nodes represent terms. In this section, we benefit from social network analysis to find the most similar terms that were not used in the specified document. To achieve our goal, we have changed some social network analysis as specified document terms make higher impact in determining significant terms.

In computing centrality measures for social networks, all nodes affect a node's centrality. In this article for computing centrality, only terms, which are in the document, have influence on other terms centrality. This means that we obtain terms related to the specific document by analyzing paths in the graph that begin from terms in the document.

We have selected two measures from all measures of social network analysis [19]. These measures are described as follows:

Degree: is the number of relations that are connected to each node. The simplest measure is degree centrality. We used degree as edges from words in the specific document to each word that do not exist in the document (d):

$$C_D(v) = \frac{\sum_{v_i \in d} Rel(v, v_i)}{n - 1} \tag{8}$$

Betweenness: vertices that occur on many shortest paths between other vertices have higher betweenness than those that do not. We used betweenness as the shortest paths that begin and end in terms in the specified document. The shortest path is a path between each pair of the graph such that the sum of the weights of its constituent edges is minimized. Betweenness is computed as following, where $\delta_{st}$ is the total shortest paths from $s$ to $t$, and $\delta_{st}(v)$ is the total shortest paths from $s$ to $t$ through $v$:

$$C_B(v) = \sum_{\substack{s \neq v \neq t \\ s \neq t \\ s, t \in d}} \frac{\delta_{st}(v)}{\delta_{st}} \tag{9}$$

It should be noted that computing centrality in this paper benefits from regarding just specific nodes in the graph and it differs from regular analysis of graph theory.

Using these measures, we computed the centrality for each term in the graph and obtained the importance of terms by the words in each document. Now we can show the $k$ words which are the most related to each document by arranging the most important words in the graph that do not exist in that document. In addition, it is possible to set real importance of each word in the document by considering the centrality of that word.

## 6    Experiments and Evaluation

We evaluated our new approach in predicting the most related terms weights using two types of data. First, we used weblog data in multi language. These data gathered from PersianBlog.ir weblog. Texts of this site were often in Persian. These data was valuable for us because: linguistic processes have many problems in this language (weak anthologies and thesauruses), informal form of writing has many differences with formal way, some of users use English letters to write Persian words (finglish), and in some domains texts are Multilanguage. Second data that is used in this article is Reuters 21578 texts categorized collection data set [20]. These texts are categorized into 20 groups. E-mail texts are good test data for our new approach.

We start describing our experiments over extended information retrieval by first data set. Multiple text mining methods were used to show effectiveness of extending features vector. Be noticed that we just have explained how better these methods are with our approach; furthermore, by our approach one can substitute or reduce other text mining methods.

First, we classified the data and compared the classification accuracy with and without extending feature vectors. Fig. 1 displays common classification methods that obtained higher accuracy with intelligent information retrieval (IIR). It is clear that functionality of classification became so better by using our approach and it was

useful for multilingual texts. Our second experience is related to clustering method
that is used for grouping these texts. We first clustered documents and have showed
results in Fig. 2, then we clustered terms and extracted topics. Although, it is hard to
evaluate extracted topics, we just applied Silhouette Coefficient [21] for evaluating
our approach. Besides, we experimentally perceived that extracted topics were very
significant than earlier for Persian blog texts. The average silhouette coefficient's
growth was 0.20 to 0.30.

The second data set that we evaluated was Reuters' data. We classified these data
with our new approach and showed in Fig. 3 how better classification methods were
with our new approach. This data set consists of 21578 messages taken from 20
newsgroups.

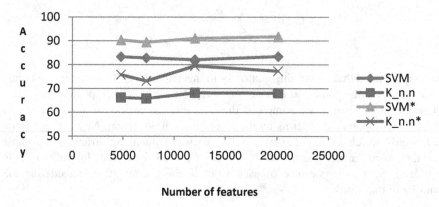

**Fig. 1.** Comparing classifiers with and without IIR (Persian Blog Data)

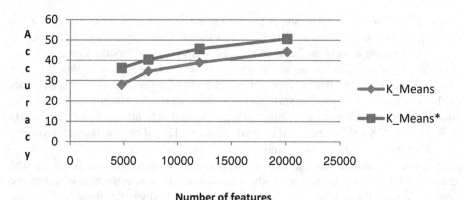

**Fig. 2.** Comparing clustering algorithm with and without IIR (Persian Blog Data)

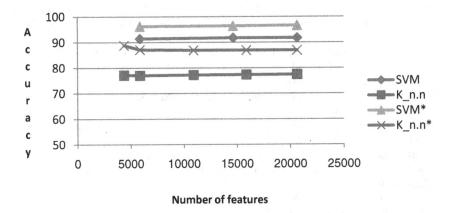

**Fig. 3.** Comparing classifiers with and without IIR (Reuters Data)

For all test vectors, we added 6 new features. In other words, for each document we had inserted six new terms that were more related to semantics of that document. Adding more features to every document can increase other related text mining computations, although, it will improve accuracy rate of text processes.

## 7  Conclusions

In this paper, we offered a new approach IIR for expanding features vector of documents and we showed how better it was for two sample data sets. By this approach, many texts related problems and machine misunderstandings are solved. As well, we can replace another text mining pre-processes and methods with high accuracy. In our approach, we extend term vector for each document by social network analysis and generating the term graph of related documents.

Although, it is clear that our approach is not suitable for every data. We estimate that some texts are not always appropriate to be extended. For example texts that are written in good linguistic instructions. In sum, objective of this paper is to improve effectiveness of IR by some additional computational costs.

Extending feature vector can be more usable with preparing additional tasks. In some cases, traditional text pre-processing tasks can be joined with this approach. As we mentioned earlier, it will be more effective if we adjust earlier features weights besides inserting new features. Finally, using online search engine for finding relevant N documents can be more accurate with less computation.

## References

1. Ricardo, A., Baeza, Y., Ribeiro, N.: Modern Information Retrieval. Addison Wesley Longman Publishing Co., Inc., Boston (1999)
2. Voorhees, E.M.: On test collections for adaptive information retrieval. IPM 44(6), 1879–1885 (2008)

3. Zhou, D., Bian, J., Zheng, S., Zha, H., Lee Giles, C.: Exploring social annotations for information retrieval. In: Proc. of the 17th Int. Conf. on World Wide Web, Beijing, China (April 2008)
4. Wang, W., Barnaghi, P.M., Andrzej, B.: Semantic-enhanced Information Search and Retrieval. In: Proc. of the 6th Int. Conf. on Advanced Language Processing and Web Information Technology (2007)
5. Pedersen, T., Patwardhan, S., Michelizzi, J.: WordNet Similarity - measuring the relatedness of concepts. In: AAAI, pp. 1024–1025 (2004)
6. Jarmasz, M., Szpakowicz, S.: Roget's Thesaurus and semantic similarity. In: Proc. of RANLP 2003, pp. 212–219 (2003)
7. Taheri Makhsoos, P., Kangavari, M.R., Shayegh, H.R.: Improving Feature Vector by Word's Position and. Sequence for Text Classification. In: Int. Conf. on IT, Thailand (March 2010)
8. Gkantsidis, C., Mihail, M., Saberi, A.: Random walks in peer-to-peer networks: algorithms and evaluation. Perform. Eval. 63(3), 241–263 (2006)
9. Aldous, D.: On the markov chain simulation method for uniform combinatorial distributions and simulated annealing. Probab. Engrg. Inform. Sci. 1(2), 33–46 (1987)
10. Teevan, J., Dumais, S., Horvitz, E.: Personalizing search via automated analysis of interests and activities. SIGIR (2005)
11. Mislove, A., Gummadi, K.P., Druschel, P.: Exploiting Social Networks for Internet Search. HotNets (2006)
12. Bar-Yossef, Z., Berg, A., Chien, S., Fakcharoenphol, J., Weitz, D.: Approximating aggregate queries about web pages via random walks. In: VLDB (2000)
13. Bar-Yossef, Z., Gurevich, M.: Random sampling from a search engine's index. In: WWW (2006)
14. Das, G., Koudas, N., Papagelis, M., Puttaswamy, S.: Efficient sampling of information in social networks. In: Proc. of the 2008 ACM Workshop on Search in Social Media, Napa Valley, California, USA (October 2008)
15. Isak, T., Spink, A.H.: Evaluating Usability of a Long Query Meta Search Engine. In: Sprague, S. (ed.) Proc. 40th Annual Hawaii Int. Conf. on System Sciences 2007 (HICSS 2007), Hawaii, USA (2007)
16. Salton, G., Wong, A., Yang, C.S.: A vector space model for automatic indexing. Communications of the ACM 18(11), 613–620 (1975)
17. Landauer, T.K., Foltz, P.W., Laham, D.: An Introduction to Latent Semantic Analysis. Discourse Processes 25(2&3), 259–284 (1998)
18. Hirst, G., Budanitsky, A.: Lexical Chains and Semantic Distance. In: EUROLAN, Romania (2001)
19. Wasserman, S., Faust, K.: Social Network Analysis: Methods and Applications. Cambridge University Press (November 1994)
20. Lang, K.: 20000 messages taken from 20 newsgroups, Dataset, http://www.cs.cmu.edu/afs/cs.cmu.edu/project/theo-20/www/data/news20.html
21. Kumar, R.: Cluster Analysis: Basic Concepts and Algorithms (2003), http://www.users.cs.umn.edu/~kumar/dmbook/ch8.pdf

# Tracking Multiple Variable-Sizes Moving Objects in LFR Videos Using a Novel genetic Algorithm Approach

Hamidreza Shayegh Boroujeni, Nasrollah Moghadam Charkari,
Mohammad Behrouzifar, and Poonia Taheri Makhsoos

Tarbiat modares University, 111-14115, Tehran, Iran
{shayegh,moghadam}@modares.ac.ir,
behroozi2010@gmail.com, pooniat@yahoo.com

**Abstract.** All available methods for tracking motion objects in videos have several challenges in many situations yet. For example, Particle filter methods cannot track objects that have variable sizes within frames duration efficiently. In this work, a novel multi-objective co-evolution genetic algorithm approach is developed that can efficiently track the variable size objects in low frame rate videos. We test our method on the famous PETS datasets in 10 categories with different frame rates and different number of motion objects in each scene. Our proposed method is a robust tracker against temporal resolution changes and it has better results in the tracking accuracy (about 10%) and lower false positive rate(about 7.5%) than classic particle filter and GA methods in the videos which contain variable size and small objects. Also it uses only 5 frames in each second instead of 15 or more frames.

**Keywords:** object tracking, far & variable-size objects, LFR videos, genetic algorithm, multi objective co-evolution GA.

## 1    Introduction

The use of video is becoming prevalent in many applications such as traffic monitoring, human tracking and detection, identification of anomaly behaviors in a train station or near an ATM, etc.[1]. While a single image provides a snapshot of a scene, the frame sequences of a video taken over time represent the dynamics in the scene; make it possible to understanding motion semantics in the video [2].

Tracking, the generating an inference about the motion of an object given a sequence of images, is one of the important and challenging problems in the video processing domain [3]. Good solutions to this problem can be applied to many applications. One of the tracking problems is object tracking in low frame rate videos. Tracking in low frame rate (LFR) video is an important requirement of many of real-time applications such as micro embedded systems and visual surveillance. Hardware costs, availability of only LFR data source, online processing requirements, etc. are the reasons of this importance [4]. This type of Tracking (from LFR videos) is an error prone problem especially in multi-target tracking with small and variable size objects. In these situations, the tracked objects may be very small; therefore tracking

D. Lukose, A.R. Ahmad, and A. Suliman (Eds.): KTW 2011, CCIS 295, pp. 143–153, 2012.

them with the appearance and dynamic features is not possible easily. In addition, the changes in the objects sizes in the sequences of frames, which are made by distance changes from camera, cause to tracking errors [5]. Some of the well known strong methods such as particle filter approaches, only are accurate is normal frame rate videos (15 to 30 fps) and they can't track objects with many changes in the size very well. In fact, there is abrupt motion and discontinuity in the LFR videos thus, the objects, especially those with variable features (size, appearance, etc.), cannot been tracked efficiently [1].

A few methods have been proposed for LFR video object tracking but none of them are efficient enough. In this paper, we present a novel genetic algorithm approach for tracking motion objects in the LFR videos that can track objects with variable and small sizes more efficient than other common approaches. The evolutionary approaches such as GA have been not used in tracking problem anymore. The main reason is the time complexity of such methods. We propose a novel GA method performing tracking process in parallel way for the whole motion objects. In this approach the dynamic of objects simulate in the dynamic of population and the appearance features of objects are put in the chromosomes. Thus we prepare dynamic and appearance features together but with different meanings. These two cases complement each other and solve the abrupt motion problem in LFR videos. In this method, the convergence of chromosomes in the population space during iterations is mentioned. The crossover and mutation operations change the population and chromosomes converge to the fitness values. The chromosomes of each detected object in each frame make the population for each object separately; and the evolutionary operations are done in parallel way for the whole detected motion objects; thus the time complexity is almost fixed always. In general, the strength of our method is that: 1) it can track objects in the LFR videos as well as HFRs. 2) It can track far and variable size objects better than other available methods since, it uses a GA approach that consider the dynamic of motions beside the appearance features.

In the rest of paper, in the section 2, the related works are presented. The proposed method discussed in the third section, the experiments and results are presented in the $4^{th}$ section and the conclusion will be discussed in the final section.

## 2     Related Works

Video object tracking techniques vary according to tracking features, motion-model assumption and temporal object tracking [6]. In general, the temporal object tracking methods can be classified into four categories: region-based [7], contour/mesh-based [8], model-based [9, 10], and feature-based methods [11],[12]. Two major components are distinguished in all of the tracking approaches; target representation and data association. The former is a bottom-up process dealing with the changes in the appearance of the object, while the latter is a top-down process dealing with the dynamics of the tracking [6].

Feature based methods such as KALMAN filter, particle filter and sift tracking have been used more than other methods. In these methods, there are two problems: feature selection and extraction is one of them. In the other hand which types of object features are supposed to use for tracking objects. Temporal features such as motion vectors, optical flows and so on, are more illustrative than spatial features such as colors, textures, etc. but the extraction of spatial features are easier. The second problem is the method of analyzing features for tracking objects. Particle filtering methods (PF), Kalman filtering and optical flow methods are more popular than others in this scope [13],[14].

Some of researches have been done in LFR tracking. Okuma et al. [16] use an offline-boosted detector to amend the proposal distribution of particle filters. Such a mixture proposal distribution can also be found in many other works such as [17], although not aimed at LFR video. Porikli and Tuzel [18] propose a mean shift method by optimizing around multiple kernels at motion areas detected by background modeling to track a pedestrian in a 1 fps camera-fixed video. These ideas can be concluded as using an independent detector to guide the search of an existing tracker when the target motion becomes fast and unpredictable. Another type of method that our proposed method is one of them too is to 'detect and connect' [1]. Such approaches are of potential to deal with LFR tracking, because they first detect the motion objects based on background subtraction or something like this and then construct trajectories by analysis of motion continuity, appearance similarity, etc.

As it has been noted previously, small object tracking is a difficult problem. Some researchers mention to this challenge on their works. Ohno et al [19] detect in football games the ball by assuming it is a small moving region characterized by a color different from players. Pin gali et al [20] use in tennis games color to distinguish between the ball and players. No more notable study doesn't available in this field.

Recently there has been a trend of introducing learning techniques into tracking problems, and tracking is viewed as a classification problem in the sense of distinguishing tracking target from the background. Representative publications include [22] – [24], have shown increased discriminative power of the tracker; however none of them have not been targeted at LFR tracking.

## 3    Multiple objects Tracking by Using a Co-evolution Multi-objective GA

We use a multi-objective co-evolution GA for tracking multi objects in the LFR videos. This approach match to such problems, because there are multiple goals for tracking (multi-objective) and the characteristics of the objects change in the time sequences thus they can't have a fixed value; as a result, it is a co-evolution problem. In this method, we don't want to find the best chromosome as the fittest value at the end of iterations. We want to see the similarities between chromosomes and the fitness values that each of them is presented one of the moving objects. In fact, we track the changes of chromosomes from one frame to the next and calculate the convergence of chromosomes in the population space.

## 3.1     Tracking Process

Figure 1 shows block diagram of the proposed method. Because of having multiple objects in each frame, the multi objective GA has been developed and because of dynamic changes in the frame sequences, we use a co-evolution method.

The appearance features of objects make chromosomes and populations. We assume that there are 'n' optimum (fitness) values that each of them present the average features of one of the detected moving objects. Through of the iterations, the chromosomes move toward the optimums and the new optimums are found for the next population in the next frame. These changes of optimums simulate the dynamic of objects. In each next frame, the feature points of each detected motion object make an independent population.

The chromosomes of each population converge to one the optimums through the iterations; thus, each motion object matches to one of the previous optimums based on convergence. This process continues to the end and the changes of chromosomes show the tracked objects thorough out the video frames.

The following steps are done for tracking multi objects:

a)  Remove the static background of frames by using a modified frame difference approach. Frames with only foreground pixels are given as output.
b)  Remove noisy sparse pixels from the first frame. Then label distinct regions as separate objects.
c)  Divide each object region to 20*20 rectangular regions. One chromosome for each of them is considered. (We use surrounded rectangles in this section).
d)  Construct each chromosome with the features of each region and then label it by the object ID. Then the populations are constructed.
e)  Calculate the fitness values by averaging the chromosomes of each object separately.

$$FV_i = \frac{\sum_{j=1}^{NumOf\ Regions} Chromosome_j^i}{Numof\ Regions} \tag{1}$$

f)  Perform the crossover and mutation for the population. Select the fittest chromosomes and remove the remains in the each iteration. In addition, remain the 2% of the worst non-fitted chromosomes because of the possibility of presence a new moving object. The fitness function is defined as the similarity distance between each chromosome and the best fitness value using Bhattacharya distance. For discrete probability distributions, the Bhattacharya distance is defined as:

$$D_B(p,q) = -\ln(BC(p,q)) \tag{2}$$

Where

$$BC(p,q) = \sum_{x \in X} \sqrt{p(x)q(x)} \tag{3}$$

is the Bhattacharyya coefficient.

g)  Calculate the new FVs (fitness values) based on convergence of chromosomes in the population space. At the end of iterations, the chromosomes converge to around of fitness values. Note that the number of FVs may change in this step based on the type of aggregations and the number of maximums in the population space.

h)  In the next frame, do the steps 'c' to 'f' for each moving object separately. At the end of 'f' step, the chromosomes belong to each object in the current frame aggregate around one of the maximums calculated in the last frame. Then the correspondence between the current observed objects and the last indexes of objects are done. There might be some chromosomes that don't converge around none of the last FVs. It means that the related object exits the scene. Also there might be some chromosomes that converge around the non-maximum point in the population space. It means that probably a new object comes to the current frame. Note that we save 2% of the worst chromosomes to find the new objects.

i)  Update FVs at the end of frame calculations.

j)  Do steps of 'h' to 'i' for the whole frames to the end.

In this regard, the objects are tracked in sequences of frames and the dynamic of objects and changes are found by FV updates (co-evolution). Also, the consideration of new and removed objects, are done by using the type of chromosome selection (choosing the number of worst).

## 3.2     The Parameters of GA

The GA methods have several parameters that should be selected carefully to obtain the best results. Usually these parameters are found practically. Thus we have implemented our method with several changes in all important parameters to reach to the best arrange of parameters.

Most of these parameters were found in the implementation and usually there are not any clear reasons for the superiority of some of them against others in theory. Our experiments with several combinations of different parameters, reach us to these parameters:

1-  The crossover rate: 65% and the mutation rate: 10% in each iteration.

Because of non-binary chromosomes, the crossover is done with 'one point random cut' from the beginning of one of the features in the chromosome.

2-  The mutations are done with the replacement of two random features of the selected chromosomes with some values in the same ranges of selected features.

3-  The type of chromosomes is considered as a set of numeric features. Here, each chromosome consists of some separate bytes that each of them shows a feature of motion object. A symbolic figure of a chromosome is shown in the figure 2.

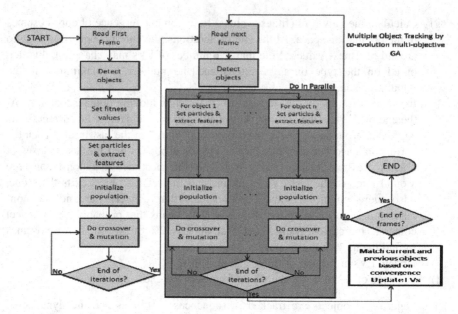

**Fig. 1.** The implementation diagram of proposed method. The colored rectangular section is done in the parallel. The tracking task that is a matching between current and previous objects is done in the white rectangle that acts based on convergence.

| Feature (Gene) name | Length of the Gene |
|---|---|
| Brightness values | 2 |
| CMYK average color | 1 |
| The softness of texture | 1 |
| The supermom distances in the color histogram | 1 |
| The number of supermom in the color histogram | 1 |
| The texture entropy | 1 |
| The color entropy | 1 |
| The light reflection value | 1 |
| The type of texture | 1 |
| RGB average color | 1 |
| The X-Y Direction | 1 |

**Fig. 2.** The genes of chromosomes used in the proposed approach. All of them except the last feature are the appearance features since the dynamic of objects consider in the fitness values changes in this method. The number in each row shows the feature length in bytes.

# 4    Experimental Results

We have tested our method on one of the well-known video datasets: PETS09 [28]. Then the results have been compared with a classic particle filter approach as the state-of-art method in this field and also with a GA method presented by Abe et al in 2008 [21]. We change two options for more accurate results: the frame numbers per second and the objects number per scene. The experimental results are shown in figures 3 and 4. The accuracy rate that is used is based on the Precision Criteria.

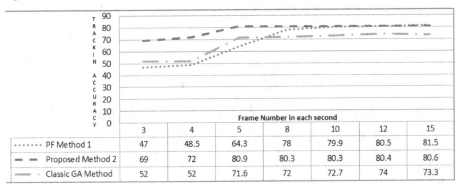

| | | 3 | 4 | 5 | 8 | 10 | 12 | 15 |
|---|---|---|---|---|---|---|---|---|
| ······· | PF Method 1 | 47 | 48.5 | 64.3 | 78 | 79.9 | 80.5 | 81.5 |
| ‒ ‒ | Proposed Method 2 | 69 | 72 | 80.9 | 80.3 | 80.3 | 80.4 | 80.6 |
| ‒‒‒‒ · | Classic GA Method | 52 | 52 | 71.6 | 72 | 72.7 | 74 | 73.3 |

**Fig. 3.** Proposed Method track objects with 80.9% of accuracy by only with 5 frames per second while in this fps, the particle filter accuracy is 64.3%. In general, the proposed method needs only 5 frames per second for the suitable accuracy and it is almost robust against temporal resolution changes (different fps) while the PF method, cannot track objects in LFR videos efficiently

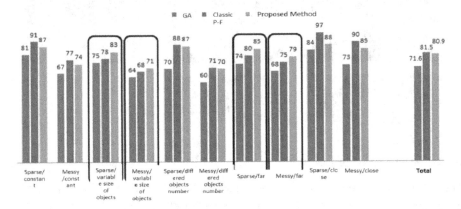

**Fig. 4.** tracking accuracy of objects in different situations with 3 different methods. In 4 of them that are shown by black rectangles, our proposed method has the better accuracy. In general, the accuracy of proposed method and the common PF method almost are the same. Our method is better than classic GA in this problem in all of situations. This figure shows that the accuracy of proposed method in the situations with far (small) and variable size objects is better than others. Note that this figure shows the best accuracy of each method that occurs in 15 fps. But in the lower frames per second like 5 fps, the proposed method has the better accuracy than others. It proves the efficiency of our method in LFR videos.

As it is shown in the experimental results, our proposed method has better accuracy in the two situations: when the objects are far from camera and their sizes are small and when the objects get close to camera and then go far from it. It means that the size of the same object is changed in the duration of video frames. In the other situations, the PF method has almost better efficiency in tracking.

Our method can work with only 5 frames in each second. Thus in the applications that exist only low frame rate video sequences, this method is better than PF algorithm. Additionally, if the objects move according to camera orthogonally, our method has a better accuracy. The other advantage of proposed method is that it can

**Fig. 5.** some results of tracking process: a-1 is frame number 245 and a-2 is frame number 469. a-3 shows the tracking results using proposed method that can detect F & G objects correctly (with some pixel detection errors). Note that F & G objects are present in the frame 245(a-1) but they are so far and they come closer in the frame 469. a-4 shows the tracking result using classic GA that lost the F & G object from frame 245 to 469 and get them a new incorrect index. a-5 is the result of PF method that made a mistake similar to classic GA. Our proposed method can detect object A from the frame number 260 (b-1) to frame number 452 (b-2). Note that object A is going far from camera and its size is variable. The PF method lost the object A (b-5) but the Classic GA track this objects although it is lost some others.

track different objects, in a parallel way. In the other hand, the GA for each object is done separately after the first frame. Thus we implement our method in a parallel way and thus the time complexity for a video with more than one motion object can reduce to the time complexity of a video with one motion object.

Some samples of videos that are tested in this research are shown in figure 5 with the output tracking results.

The classic GA that we used for comparison it with our method is an optimization algorithm that tries to optimize one chromosome of each moving object and if it succeeds, maps the object to the suitable index of fitness value. In the other hand, a moving object can be tracked if one of the chromosomes gains the optimized values after iterations with some thresholds [23, 25].

In general, the accuracy rate of the tracking by the PF method on the PETS 09 dataset is almost 81.5% and by the proposed GA method is almost 80%. However our proposed method has some advantages in some applications and some of situations. Note that we mention on the best result of each method that occurred in the 15 fps. If we draw a figure representing rates in the lower fps like 5 fps, proposed method was more accurate than two others in all of situations. This fact is shown in figure 3. This plot shows that our proposed method can be efficiently used in LFR videos while other methods don't have suitable criterions in the low frame rate videos.

## 5    Conclusions

In this paper, we propose a novel method for tracking objects in video sequences. Our method is based on evolutionary algorithms concept. However it has some differences with the normal concepts. This is a co evolution multi objective genetic algorithm method that doesn't try to optimize chromosomes. But try to check the changes of chromosomes in the duration of iterations and decide about the index of each object based on the position of convergence of their chromosomes in the population space. This method is more accurate and its tracking error is lower than PF in the situations that objects is far from camera and the distances of them change in different frames during video sequences. Additionally, our method can track objects in LFR videos as well as normal videos and only 5 frames of each second are required. The time complexity of this method is almost fix against number of objects in the scenes, since it can perform in a parallel way for all of objects in each frame. This method can be used in some other problems that have the same problem space such as action recognition or video indexing researches.

One of the important problems in this is the type of chromosomes and genes that can be better if we extract more useful features.

## References

1. Li, Y., Ai, H., Yamashita, T., Lao, S., Kawade, M.: Tracking in Low Frame Rate Video: A Cascade Particle Filter with Discriminative Observers of Different Life Spans. IEEE Transactions on Pattern Analysis and Machine Intelligence 30(10), 1728–1740 (2008)

2. Konrad, J.: Motion detection and estimation. In: Image and Video Processing Handbook, ch. 3.8. Academic Press (2000)
3. Rodríguez, P.: Monitoring of Induced Motions in a Shake Table Test via Optical Motion Tracking: Preliminary Results. In: Ibero American Conference on Trends in Engineering Education and Collaboration, October 27 - 28 (2009)
4. Zhang, T., Fei, S., Lu, H., Li, X.: Modified Particle Filter for Object Tracking in Low Frame Rate Video. In: Proc. 48th IEEE Conference on Decision and Control and 28th Chinese Control Conference, Shanghai, P.R. China, December 16-18 (2009)
5. Li, Z., Chen, J., Schraudolph, N.N.: An Improved Mean-Shift Tracker with Kernel Prediction and Scale Optimization Targeting for Low-Frame-Rate Video Tracking. In: Proceeding of the 19th International Conference on Pattern Recognition (ICPR 2008), Tampa, Florida, USA, December 8-11 (2008)
6. Rabiee, H.R., Asadi, M., Ghanbari, M.: Object tracking in crowded video scenes based on the undecimated wavelet features and texture analysis. EURASIP Journal on Advances in Signal Processing Archive 2008 (January 2008)
7. Kass, M., Witkin, A., Terzopoulos, D.: Snakes: Active contour models. Int. J. Computer Vision 1(4), 321–331 (1987)
8. Erdem, C.E., Tekalp, A.M., Sankur, B.: Video Object Tracking With Feedback of Performance Measures. IEEE Transaction for Circuit and Systems for Video Technology 13(4), 310–324 (2003)
9. Chen, Y., Rui, Y., Huang, T.S.: JPDAF based HMM for real-time contour tracking. In: Proc. IEEE Int. Conf. Computer Vision and Pattern Recognition, pp. 543–550 (2001)
10. Gu, C., Lee, M.: Semiautomatic Segmentation and Tracking of Semantic Video Objects. IEEE Transaction for Circuit and Systems for Video Technology (5), 572–584 (1998)
11. Wang, D.: Unsupervised video segmentation based watersheds and temporal tracking. IEEE Transaction for Circuit and Systems for Video Technology 8(5), 539–546 (1998)
12. Comaniciu, D., Ramesh, V., Meer, P.: Kernel Based Object tracking. IEEE Transactions Pattern Anal. Mach. Intell. 25(5), 546–577 (2003)
13. Karasulu, B.: Review and evaluation of well-known methods for moving object detection and tracking in videos. Journal of Aeronautics and Space Technologies 4(4), 11–22 (2010)
14. Carmona, E.J., Martínez-Cantos, J., Mira, J.: A new video segmentation method of moving objects based on blob-level knowledge. Pattern Recognition Letters 29(3), 272–285 (2008)
15. Thirde, D., Borg, M., Aguilera, J., Wildenauer, H., Ferryman, J., Kampel, M.: Robust Real-Time Tracking for Visual Surveillance. EURASIP Journal on Advances in Signal Processing 2007, Article ID 96568, 23 pages (2007)
16. Okuma, K., Taleghani, A., de Freitas, N., Little, J.J., Lowe, D.G.: A Boosted Particle Filter: Multitarget Detection and Tracking. In: Pajdla, T., Matas, J(G.) (eds.) ECCV 2004. LNCS, vol. 3021, pp. 28–39. Springer, Heidelberg (2004)
17. Liu, C., Shum, H.-Y., Zhang, C.: Hierarchical Shape Modeling for Automatic Face Localization. In: Heyden, A., Sparr, G., Nielsen, M., Johansen, P. (eds.) ECCV 2002. LNCS, vol. 2351, pp. 687–703. Springer, Heidelberg (2002)
18. Porikli, F., Tuzel, O.: Object Tracking in Low-Frame-Rate Video. In: SPIE Image and Video Comm. and Processing, vol. 5685, pp. 72–79 (2005)
19. Desai, U.B., Merchant, S.N., Zaveri, M., Ajishna, G., Purohit, M., Phanish, H.S.: Small object detection and tracking: Algorithm, analysis and application. In: Pal, S.K., Bandyopadhyay, S., Biswas, S. (eds.) PReMI 2005. LNCS, vol. 3776, pp. 108–117. Springer, Heidelberg (2005)

20. Lu, H., Wang, D., Zhang, R., Chen, Y.-W.: Video object pursuit by tri-tracker with on-line learning from positive and negative candidates. IET Image Processing 5(1), 101–111 (2011)
21. Abe, Y., Suzuki, H.: Multiple Targets Tracking Using Attention GA. In: Information Engineering and Computer Science, ICIECS 2009, International Conference on Computer Systems (2009)
22. Luque, R.M., Ortiz-de-Lazcano-Lobato, J.M., Lopez-Rubio, E., Palomo, E.J.: Object tracking in video sequences by unsupervised learning. In: Jiang, X., Petkov, N. (eds.) CAIP 2009. LNCS, vol. 5702, pp. 1070–1077. Springer, Heidelberg (2009)
23. Ross, D., Lim, J., Lin, R., Yang, M.H.: Incremental learning for robust visual tracking. Int. J. Comput. Vis. 77(1-3), 125–141 (2008)
24. Gerónimo, D., López, A.M., Sappa, A.D., Graf, T.: Survey of pedestrian detection for advanced driver assistance systems. IEEE Transactions on Pattern Analysis and Machine Intelligence 32(7), 1239–1258 (2010)
25. DeJong, K.A., Spears, W.M.: An Analysis of the Interacting Roles of Population Size and Crossover in Genetic Algorithms. In: Schwefel, H.-P., Männer, R. (eds.) PPSN 1990. LNCS, vol. 496, pp. 38–47. Springer, Heidelberg (1991)
26. Grefenstette, J.J.: Optimization of Control Parameters for Genetic Algorithms. IEEE Trans. Systems, Man, and Cybernetics SMC-16(1), 122–128 (1986)
27. Mahfoud, S.W.: Niching methods for genetic algorithms. 251 pp (a dissertation from the University of Illinois) (also IlliGAL Report No. 95001) (1995)
28. PETS 2009 Dataset, http://www.cvg.rdg.ac.uk/PETS2009/index.html

# From UML to OWL 2

Jesper Zedlitz[1], Jan Jörke[2], and Norbert Luttenberger[2]

[1] German National Library of Economics (ZBW)
[2] Christian-Albrechts-Universität zu Kiel

**Abstract.** In this paper we present a transformation between UML class diagrams and OWL 2 ontologies. We specify the transformation on the M2 level using the QVT transformation language and the meta-models of UML and OWL 2. For this purpose we analyze similarities and differences between UML and OWL 2 and identify incompatible language features.

## 1 Introduction

A class model following the Unified Modeling Language (UML) specification [11] often serves the purpose to express the conceptual model of information systems. UML's graphical syntax helps to understand the resulting models—also for non-computer scientists. Due to the rich software tool support UML class models lend themselves well to be used as a decisive artifact in the design process of information systems. The retrieval of data from information systems benefits from semantic knowledge. For this purpose it is desirable to provide an ontology for data. The commonly used language to describe ontologies is the Web Ontology Language (OWL).[15] Both modeling languages are accepted standards and have benefits in their areas of use. To leverage both languages' strengths it is usually necessary to repeat the modeling process for each language. This effort can be avoided by a transformation of a model from one language into the other.

This paper shows a transformation from a UML class model into an OWL 2 ontology by using OMG's Query/View/Transformation (QVT) [12] transformation language in conjunction with the meta-models for UML and OWL 2.

The paper is organized as follows: In section 2 we show some existing work on the transformation of UML and OWL. Section 3 explains our approach in general. Section 4 shows differences between UML and OWL 2. In section 5 we present our transformations en detail. Section 6 concludes and points out fields of future work.

## 2 Existing Work

A generic comparison of the seeming difference between models and ontologies is given in [1]. Differences between UML class models and OWL ontologies have

D. Lukose, A.R. Ahmad, and A. Suliman (Eds.): KTW 2011, CCIS 295, pp. 154–163, 2012.
© Springer-Verlag Berlin Heidelberg 2012

been studied in [7] and [9]. [5] contains an analysis of approaches for the transformation between UML and ontologies that have been published until 2003. Transformations from UML to OWL can be grouped into three categories:

**Extension of UML:** [2] presents an extension of UML to improve the description of (DARPA Agent Markup Language based) ontologies using UML. [14] presents a UML based graphical representation of OWL extended by OWL annotations.

**XSLT based approaches:** In [6] the transformation of a UML class diagram into an OWL ontology using Extensible Stylesheet Language Transformation (XSLT) is described. Additionally a UML profile is used to model specific aspects of ontologies.

**Meta-model based approaches:** [4] describes a meta-model for OWL based on Meta Object Facility (MOF) and a UML profile to model ontologies using UML. [3] is a preparation for the Ontology Definition Meta-model (ODM) specification. It presents a meta-model for OWL as well as a UML profile. [10] gives a transformation between a UML model and an OWL ontology using the Atlas Transformation Language (ATL). [8] uses MOFScript to perform a UML→OWL 2 transformation. However, their goal is the validation of meta-models. Therefore they insert several model elements into the ontology that are needed for this goal but complicate the usability of the ontology.

# 3  Our Approach

The goal of our work is not primary to create better OWL 2 ontologies from a UML class model but to provide a well-arranged transformation that can be used for further investigation on the transformations. Each of the mentioned approaches has one or more drawbacks—e.g. syntax-dependency, no declarative transformation. Therefore we combine several thoughts from the existing approaches and apply them to the up-to-date OWL 2 Web Ontology Language (OWL 2). OWL 2 offers several new model elements that facilitate the transformation. We will highlight these elements at the relevant points. These are the five fundamental ideas of our work:

**Restrict UML class models:** Because we are interested in the structural schema of an information system we do not have to take the behavioral schema into account.

**Meta-models on M2 level:** Instead of transforming elements of a M1-model[1] directly (like an XSLT-based transformation that works with the concrete syntax of M1-models) we describe the transformation using elements of the M2 meta-

---

[1] A four-layer meta-model hierarchy is used for UML and MOF: M0 = Run-time instances, M1 = user model, M2 = UML, M3 = MOF.[11, sec. 7.12].

models. Thus the transformation does not depend on the model instances. It only depends on the involved meta-models.

**Abstract syntax instead of concrete syntax:** By working with the abstract syntax our transformation becomes independent of any particular representation (like Functional-Style Syntax, Turtle Syntax, OWL/XML Syntax, Manchester Syntax, etc. )

**Declarative language:** Instead of processing the input model step by step we can describe how to map elements of one model on corresponding elements of the other language.

**Well-known transformation language:** We choose OMG's QVT Relations Language because it is declarative and works with MOF-based meta-models. The support by the OMG consortium and several independent implementations makes it future-proof.

# 4    UML and OWL 2

A complete transformation between a UML class model and an OWL 2 ontology is probably not possible. There are concepts in UML that do not exist in OWL. Also the semantic of some concepts differ which hinders a transformation. [1], [7] and [9] give detailed comparisons between UML and OWL. In the following we take a look at some of these differences and show possible solutions for a transformation.

**Global Names:** In UML it is specified that each model element has a unique name within its package. That assures that each model element is identified by its name and its position in the package hierarchy. A similar problem applies to names of class-dependent attributes. In UML the names of these attributes have to be unique only within the class they belong to. In contrast, in OWL all names are global. Therefore we have to take care to assign each model element with a name—in OWL called "Internationalized Resource Identifier" (IRI)—that is globally unique. These IRIs are generated during the transformation process.

**Unique Name Assumption:** UML uses a Unique Name Assumption (UNA) which states that two elements with different names are treated as different. OWL does not use a UNA. Therefore we have to explicitly mark elements as being different.

**Open-World vs. Closed-World Assumption:** In UML class models we work under a Closed-World Assumption: All statements that have not been mentioned explicitly are false. In contrast OWL uses an Open-World Assumption (OWA) where missing information is treated as undecided. These different semantics make it necessary to add various restrictions to the ontology during the transformation process from a UML model to an OWL ontology to preserve the original semantics of the model. We will highlight these problems later.

**Profiles:** UML has the concept of "profiles" which allow extensions of meta-model elements. There is no corresponding construct in OWL 2. In most cases UML profiles are used to define stereotypes to extend classes. The information of these stereotypes can be mapped to OWL by clever creation of some new classes and generalization assertions. However a large part of an UML profile is too specific and would require transformation rules adapted for the particular profile.

**Abstract Classes:** Abstract classes can not be transformed into OWL 2. If a class is defined as abstract in UML no instances of this class (objects) can be created. In contrast OWL has no language feature to specify that a class must not contain any individual. An approach to preserve most of the semantics of an abstract class would be the usage of a `DisjointUnion`. This would ensure that any individual belonging to a subclass would also belong to the abstract superclass. However, it does not prohibit to create direct members of the abstract superclass.

## 5    Transformations

### 5.1    Classes

The concepts of instances resp. individuals belonging to classes exists in both UML, and OWL 2. Because both concepts are similar a basic transformation can be done with little effort. As mentioned above the missing UNA in OWL 2 makes it necessary to insert a `DisjointClasses` axiom to the ontology for every pair of classes that are not in a generalization relation.

### 5.2    Specialization/Generalization

The concepts of specialization and generalization in UML and OWL 2 are similar. If $C'$ is a specialized sub-class of class $C$ and $i$ is an instance resp. individuum then in both cases we have $C'(i) \rightarrow C(i)$.

Therefore the transformation is straightforward. For each generalization relationship "$C$ is generalization of $C'$" (resp. "$C'$ is specialization of $C$") we add a `SubClassOf(` $C'$ $C$ `)` axiom to the ontology.

You can specify two kinds of constraints for generalizations in UML: completeness and disjointness. In a UML class diagram you can define these constraints

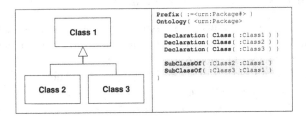

**Fig. 1.** A generalization relationship without constraints

by adding the keywords *disjoint* or *complete* near the arrowhead of the generalization arrow. A generalization without these keywords is not subjected to the two contains.

A generalization is disjoint if an instance of one sub-class must not be instance of another sub-class of the generalization. We transform this kind of constraint by adding a `DisjointClasses` axiom with all sub-classes to the ontology. This benefits from the fact that in OWL 2 a `DisjointClasses` axiom can make statements about more than two classes.

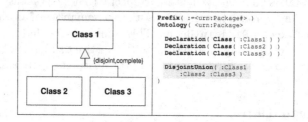

**Fig. 2.** A disjoint (but not necessary complete) generalization

If a generalization is disjoint and complete we can use a stronger axiom. The `DisjointUnion( C C'_1 ... C'_n)` axiom states that each individual of $C$ is an individual of exactly one $C_i$ and each individual of $C_i$ is also an individual of $C$. The `DisjointUnion` axiom is new in OWL 2. Although it is just syntactic sugar it makes the resulting ontology easier to read.

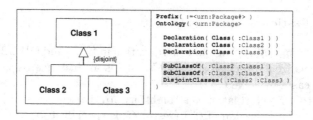

**Fig. 3.** A disjoint and complete generalization

## 5.3 Associations and Class-Dependent Attributes

In UML classes can be connected by associations and/or class-dependent attributes. In the UML meta-model both connection kinds are represented by the model element *Property*. Therefore it stands to reason that the transformation of both associations and attributes can be handled together.

Since the model element *Association* is a subclass of *Classifier* all associations in a UML class diagram are direct members of a package. Therefore an OWL 2 concept that is similar to an association is an object property that is also direct members of an ontology. Associations can be uni- or bi-directional. One directed association can be transformed into one object property. For a bi-directional association two object properties will be created—one for each direction. To preserve the information that both resulting object properties were part of one association an `InverseObjectProperties` axiom is added to the ontology.

The transformation of class-dependent attributes is more complex. There are no directly corresponding concepts in OWL 2 that allow a simple transformation. The main problem is that classes in OWL 2 do not contain other model elements which would be necessary for a direct transformation. The most similar concepts in OWL 2 for class-dependent attributes are object properties and data properties.

In both cases the decision whether a *Property* is transformed into an object property or a data property depends on the *type*-association of the *Property*: If it is associated with an instance of *Class*, an object property is needed. If it is associated with an instance of *DataType*, a data property is needed.

In OWL 2 properties do not need to have a specified domain or range. In that case domain and range are implicitly interpreted as `owl:Thing`. This corresponds to the OWA: if no further information is known about a relation it might exists between individuals of any two classes. However, to limit the properties similar to the closed-world assumption of a UML class diagram we have to add the appropriate domain and range axioms that lists the allowed classes and datatypes.

The range of a property can be determined easily: It is the name of the *Classifier* linked via the *type*-association of the *Property*. For the domain we have to distinguish between properties of class-dependent attributes and those of associations. For class-dependent attributes the *class*-association of a *Property* is set. That class is the needed domain. If the *Property* is part of an association (*association*-association of the *Property* is set) we have to choose the *type*-association of the other member end's property as domain.

The OWA allows to interpret two properties that have been transformed from distinct UML properties as one. To avoid that and to map UML's CWA best we

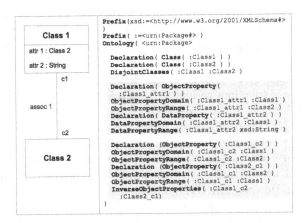

**Fig. 4.** Transformation of attributes and associations into object- and data properties

mark all properties that are not in a generalization relationship (i.e. a `SubProp-ertyOf` axiom exists for them) as disjoint. To do this we add OWL 2's `Disjoin-tObjectProperties` and `DisjointDataProperties` axioms to the ontology: For all pairs of UML *Property* elements we check if they were transformed into an OWL 2 property, are not identical, no generalization relationship exists between them, and they have not been marked disjoint before.

### 5.4   Association Inheritance

In UML class models it is not only possible to use generalization for classes but also for associations. One association can inherit from anther association. This can be transfered to the OWL 2 world using `SubPropertyOf` axioms. Since a bi-directional association is transformed into two `ObjectProperty` axioms an inheritance relation between two associations is also transformed into two `Sub-PropertyOf` axioms.

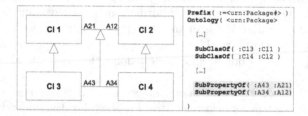

**Fig. 5.** Transformation of generalized associations

### 5.5   Cardinalities

In UML cardinalities can be added to associations and class-dependent attributes. A minimal and/or maximal occurrence can be a specified. OWL 2 also contains six different cardinality axioms for `ObjectProperty` and `DataProperty`: `ObjectMinCardinality`, `ObjectMaxCardinality`, and `ObjectExactCardinal-ity` as well as the respective axioms for a `DataProperty`.

Since an `ExactCardinality` axiom is just a abbreviation of a pair of `MinCar-dinality` and `MaxCardinality` axioms with the same number of occurrences a UML cardinality with an equal upper and lower limit can be transformed into an `ExactCardinality` axiom to make the ontology more readable. In the case that the upper limit is 1 the property can be marked functional by adding a `FunctionalObjectProperty` (resp. `FunctionalDataProperty`) axiom to the ontology.

It is not enough just to add cardinality axioms to the ontology. In UML a class is restricted by the use of cardinalities for its class-dependent attributes and associations. However, in OWL 2 properties are not contained in classes and therefore do not restrict them by their cardinality axioms. This problem can (partly) be solved by adding `SubClassOf` axioms to the ontology that create "virtual parent classes". The only purpose of these "virtual parent classes" is to inherit their cardinality restrictions to the actual classes involved in the associations and class-dependent attributes.

A reasoner can use these cardinality restrictions to check the consistency of the ontology. However, it is only possible to detect a violation of an upper limit. A violation of a lower limit it not possible to detect due to the OWA—there might be other individuals that are simply not listed in the ontology.

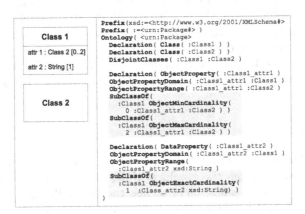

**Fig. 6.** Transformation of attributes and associations with cardinality constraints

## 5.6   Aggregation and Composition

*Aggregation* and *Composition* in UML are special kinds of associations between classes. There are a few restrictions for aggregations:

a) An object must not be in an aggregation association with itself.
b) An object must not be part of more than one composition.
c) An object of a class that is part of a composition must not exist without the class it belongs to.

The restriction (a) can be transformed to OWL 2 by adding an `IrreflexiveObjectProperty` axiom to the ontology. This axiom prohibits the individual to be connected to itself by the object property that represents the aggregation.

Restriction (b) can be enforced with a `FunctionalObjectProperty` or `InverseFunctionalObjectProperty` axiom. If the composition association is navigable bi-directionally the user is free to choose. Otherwise the following rules apply: If the association is navigable from 'part' to 'whole' a `FunctionalObjectProperty` is required. A connection from an individual of the 'part' class to more than one individual of the 'whole' class would make the ontology inconsistent. An `InverseFunctionalObjectProperty` is required if the association if navigable from 'whole' to 'part'.

The OWA makes a detection of (c) impossible—the individual might be part of a composition that is simply not listed in the ontology.

## 5.7  Datatypes

For the transformation of datatypes two cases have to be considered:

a) The datatype is one of the primitive UML datatypes like "Boolean", "Integer", or "String". OWL 2 uses datatype definitions provided by the XML Schema specification [16]. It is transformed into its corresponding datatype from XML Schema. The indicator that a datatype belongs to this category is its package name. For primitive UML datatypes the package name is "UMLPrimitiveTypes". In our implementation a function returns the corresponding name of the datatype from XML Schema.

b) The definition of the (user defined) datatype is given within the class model. We define a new datatype in the ontology by adding a `DatatypeDefintion` axiom.

## 5.8  Enumeration

In UML an *Enumeration* is used to create a datatype with a predefined list of allowed values. Although the graphical UML representation of *Enumeration* looks similar to the representation of a class they are different elements in the meta-model. Such an *Enumeration* can be transformed into a named datatype in OWL. First we have to declare the existence of the datatype. Then a definition of the datatype can be given using a `DatatypeDefinition` axiom. The allowed values of the *Enumeration* are listed in the `DataOneOf` statement.

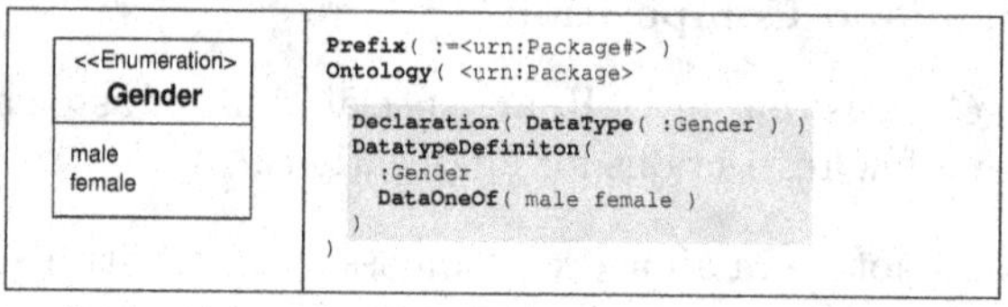

**Fig. 7.** Transformation of a classes with stereotype "Enumeration"

# 6  Summary and Outlook

We have developed a concept for a well-arranged transformation of a UML class model into an OWL 2 ontology that can be used for the analysis of the transformation rules itself as well as the semantics of UML and OWL 2. Our concept takes up several existing approaches and combines them. The use of QVT Relations Language enables us to describe the transformation declaratively and to use model elements of the meta-models. Therefore we do not depend on the concrete syntax.

We have analyzed similarities and differences on UML elements and OWL 2 elements in-depth. With this knowledge we have developed rules to transform UML class models into elements of an OWL 2 ontology. The developed implementation of an OWL 2 meta-model conforms to the specification of [15].

Currently we are working on the opposite direction, a transformation from an OWL 2 ontology into a UML class model. That will enable us to perform a better analysis of the transformations and the semantics by comparing the results of a bi-directional transformation.

# References

1. Atkinson, Guthei, Kiko: On the Relationship of Ontologies and Models. In: Proceedings of the 2nd Workshop on Meta-Modeling and Ontologies, pp. 47–60. Gesellschaft für Informatik, Bonn (2006)
2. Baclawski, K., Kokar, M.K., Kogut, P.A., Hart, L., Smith, J., Holmes III, W.S., Letkowski, J., Aronson, M.L.: Extending UML to Support Ontology Engineering for the Semantic Web. In: Gogolla, M., Kobryn, C. (eds.) UML 2001. LNCS, vol. 2185, pp. 342–360. Springer, Heidelberg (2001)
3. Brockmans, S., Colomb, R.M., Haase, P., Kendall, E.F., Wallace, E.K., Welty, C., Xie, G.T.: A Model Driven Approach for Building OWL DL and OWL Full Ontologies. In: Cruz, I., Decker, S., Allemang, D., Preist, C., Schwabe, D., Mika, P., Uschold, M., Aroyo, L.M. (eds.) ISWC 2006. LNCS, vol. 4273, pp. 187–200. Springer, Heidelberg (2006)
4. Brockmans, S., Volz, R., Eberhart, A., Löffler, P.: Visual Modeling of OWL DL Ontologies Using UML. In: McIlraith, S.A., Plexousakis, D., van Harmelen, F. (eds.) ISWC 2004. LNCS, vol. 3298, pp. 198–213. Springer, Heidelberg (2004)
5. Falkovych, Sabou, Stuckenschmidt: UML for the Semantic Web: Transformation-Based Approaches. Knowledge Transformation for the Semantic Web 95, 92–107 (2003)
6. Gasevic, Djuric, Devedsic, Damjanovic: Converting UML to OWL Ontologies. In: Proceedings of the 13th International World Wide Web Conference on Alternate Track Papers & Posters, pp. 488–489. ACM (May 2004)
7. Hart, Emery, Colomb, Raymond, Taraporewalla, Chang, Ye, Kendall, Dutra: OWL Full and UML 2.0 Compared (March 2004)
8. Höglund, Khan, Lui, Porres: Representing and Validating Metamodels using the Web Ontology Language OWL 2, TUCS Technical Report (May 2010)
9. Kiko, Atkinson: A Detailed Comparison of UML and OWL. Technischer Bericht 4, Dep. for Mathematics and C.S., University of Mannheim (2008)
10. Milanovic, Gasevic, Giurca, Wagner, Devedzic: On Interchanging Between OWL/SWRL and UML/OCL. In: Proceedings of 6th OCLApps Workshop at the 9th ACM/IEEE MoDELS, pp. 81–95 (October 2006)
11. Object Management Group: OMG Unified Modeling Language Infrastructure, version 2.3 (2010), http://www.omg.org/spec/UML/2.3/Infrastructure
12. Object Management Group: Meta Object Facility (MOF) 2.0 Query/View/Transformation (QVT) Specification 1.0 (2008), http://www.omg.org/spec/QVT/1.0/
13. Olivé: Conceptual Modeling of Information Systems (2007) ISBN 978-3-540-39389-4
14. Schreiber: A UML Presentation Syntax for OWL Lite (April 2002)
15. World Wide Web Consortium: OWL 2 Web Ontology Language Structural Specification and Functional-Style Syntax (October 2009)
16. World Wide Web Consortium: XML Schema Definition Language (XSD) 1.1 Part 2: Datatypes (December 2009)

# Completeness Knowledge Representation
# in Fuzzy Description Logics

Kamaluddeen Usman Danyaro[1], Jafreezal Jaafar[1], and Mohd Shahir Liew[2]

[1] Department of Computer & Information Sciences, Universiti Teknologi PETRONAS,
Seri Iskandar, 31750 Tronoh, Perak, Malaysia
[2] Department of Civil Engineering, Universiti Teknologi PETRONAS, Seri Iskandar,
31750 Tronoh, Perak, Malaysia
kudanyaro@dandali.com, {jafreez,shahir_liew}@petronas.com.my

**Abstract.** Semantic Web is increasingly becoming the best extension of World
Wide Web which enables machines to be more interpretable and present
information in less ambiguous process. Web Ontology Language (OWL) is
based on all the knowledge representation formalisms of Description Logics
(DLs) that has the W3C standard. DLs are the families of formal knowledge
representation languages that have high expressive power in reasoning
concepts. However, DLs are unable to express a number of vague or imprecise
knowledge and thereby cannot handle more uncertainties. To reduce this
problem, this paper focuses on the reasoning processes with knowledge-base
representation in fuzzy description logics. We consider Gödel method in
solving the completeness of our deductive system. We also discuss the desirable
concepts based on entailment to fuzzy DL knowledge-base satisfiability.
Indeed, fuzzy description logic is the suitable formalism to represent this
category of knowledge.

**Keywords:** Description Logics, Fuzzy Description Logics, Knowledge
Representation, Satisfiability.

## 1    Introduction

Semantic Web is increasingly becoming the best extension of World Wide Web
(WWW) where the Semantics from the available information are formally expressed
by logic-based engineering. Knowledge maintenance, creating knowledge, reuse of
components and integration of Semantic data are all governed by ontologies [1].
Ontology is an explicit representation of a particular domain of interest which
describes the representation of concepts and relationships [2]. This is in accordance
with the Gruber's [4] famous ontology definition that ontology is an explicit and
formal specification of a shared conceptualization.

Web Ontology Language (OWL) is used to explicitly represent and describe the
meaning of terms in vocabularies and relationships [5]. The machines interpretability
of information on Web has been currently standardized by World Wide Web
Consortium (W3C). The standard is perfect for use by applications that require or

D. Lukose, A.R. Ahmad, and A. Suliman (Eds.): KTW 2011, CCIS 295, pp. 164–173, 2012.

process the content of Web data rather than presenting information to humans. OWL has three main properties namely: OWL Lite, OWL DL and OWL Full. OWL properties have ability of expressive power in enabling the underlying knowledge representation formalism in Description Logics (DLs). DLs are the families of formal knowledge representation languages that have high expressive power in reasoning concepts. They are the best knowledge representation languages that create knowledge bases and specify ontologies [15]. Nevertheless, OWL Full contains the syntax of OWL and RDF (Resource Description Framework) which is considered as undecidable [1], simply because it would not impose restrictions when using transitive properties. Accordingly, OWL Lite, OWL DL and OWL Full correspond to $\mathcal{SHIF(D)}$, $\mathcal{SHOIN(D)}$ and $\mathcal{SROIQ(D)}$ respectively. The basic description of DL consists of atomic roles (properties or relations) and atomic concepts (classes). These often design and model the role description and complex concept description. Indeed, the standard propositionality of description logic is $\mathcal{ALC}$ where the letters means Attribute Language with Complement.

Vagueness or uncertainty in real-world knowledge has an intrinsic characteristic that is reflected across the Web [3]. Most of the concepts of knowledge modeling lack precise or well-defined criteria for distinguishing objects like hot, cold, tall, and young. These are also called world knowledge problem [9]. Although standard description logics offer substantial expressive power, but are restricted when dealing with crisp. The distinct roles and concepts cannot express the uncertain or vague knowledge. More clearly, classic ontologies are not suitable for tackling such problems of uncertainty, imprecision or vagueness formalisms [1]. To handle these problems, we employ fuzzy description logics using completeness reasoning procedure. Ordinarily, the requirement to handle and reason with uncertainty has been seen in numerous Semantic Web contexts [1, 3, 5, 17]. Therefore, an integral research direction with Semantic Web will be required to handle uncertainty.

On integrating uncertain knowledge into DLs, some important theories like: fuzzy sets, fuzzy logic, knowledge representation (KR) and inference algorithm are needed. Therefore, this paper contributes in bringing the detailed information about such theories. However, as a way to achieve algorithm, this paper provides one of the algorithms' properties. The property here means the completeness property of the fuzzy DL knowledge base satisfiability, which is the main objective the paper. Setting such procedure reduces the difficulties in fuzzy OWL ontology entailment to fuzzy DL knowledge-base satisfiability [16]. Nevertheless, the desirability of completeness property can be derived by the entailed sentence of inference algorithm [6].

The remaining of the paper is organized as follows. In section 2 we provide the preliminaries of the basic aspects in relation to fuzzy description logics. In section 3 we present the logic-based theories which consist of fuzzy set, fuzzy logic, and description logics. Subsequently, in section 4 we discuss the fuzzy description logics ($\mathcal{ALC}$) and then provide the syntax and Semantics of fuzzy $\mathcal{ALC}$. In section 5 we provide the decidability based on Gödel's completeness method. In doing that we attempt to provide the proof of the theorem that will help in reducing the problem of entailed fuzzy satisfiability knowledge-base. Finally, in section 6 we provide the conclusion and future work of the research.

## 2    Preliminaries

In this section we provide the backgrounds of some important aspects related to fuzzy DLs. We also present the main researchers' useful notations that have shown in Table 1. We consider $\mathcal{ALC}$ constructs based on their syntax and Semantics. We let A to be an atomic concept (atomic class), C, D be the concept descriptors (disjoint classes), R be an atomic role, and assume $x, y$ represent the individuals.

**Table 1.** Syntax and Semantics of Description Logics

| DL Constructor | DL Syntax | Semantic |
|---|---|---|
| Concept negation | $\neg C$ | $\Delta^I \backslash C^I$ |
| Concept conjunctions | $C \sqcap D$ | $C^I \cap D^I$ |
| Concept disjunctions | $C \sqcup D$ | $C^I \cup D^I$ |
| Top concept | $\top$ | $\Delta^I$ |
| Bottom concept | $\bot$ | $\emptyset$ |
| Atomic Concept | $A$ | $A^I \subseteq \Delta^I$ |
| Atomic negation | $\neg A$ | $\Delta^I \backslash A^I$ |
| Atomic role | $R_A$ | $R^I_A \subseteq \Delta^I \times \Delta^I$ |
| Inverse role | $R^-$ | $\{(y, x) \in \Delta^I \times \Delta^I : (x, y) \in R^I\}$ |
| Universal role | $U$ | $\Delta^I \times \Delta^I$ |
| Existential quantification | $\exists R.C$ | $\{x \mid \exists y, (x, y) \in R^I \text{ and } y \in C^I\}$ |
| Universal quantification | $\forall R.C$ | $\{x \mid \forall y, (x, y) \notin R^I \text{ or } y \in C^I\}$ |
| At least number restriction | $\geq n\ S.C$ | $\{x \mid \#\{y: (x, y) \in S^I \text{ and } y \in C^I\} \geq n\}$ |
| At most number restriction | $\leq n\ S.C$ | $\{x \mid \#\{y: (x, y) \in S^I \text{ and } y \in C^I\} \leq n\}$ |
| Nominals | $\{0_1, \ldots, 0_m\}$ | $\{0^I_1, \ldots, 0^I_m\}$ |

Consider the following rule such that

$$C, D ::= A \mid \top \mid \bot \mid \neg C \mid C \sqcup D \mid C \sqcap D \mid \rightarrow D \mid \forall R.C \mid \exists R.C$$

## 2.1    The $\mathcal{SHOIN(D)}$, $\mathcal{SRIOIQ}$, and $\mathcal{SHIF}$ Notations in $\mathcal{ALC}$ :

- $\mathcal{S}$ Stands for $\mathcal{ALC}$ plus role transitivity
- $\mathcal{H}$ Stands for role hierarchies (for role inclusion axioms)
- $O$ Stands for nominals that is for closed class with one element
- $\mathcal{I}$ Stands for inverse roles
- $\mathcal{N}$ stands cardinality restrictions
- $\mathcal{D}$ stands for data types
- $\mathcal{F}$ stands for role functionality
- $\mathcal{Q}$ stands for qualified cardinality restriction
- $\mathcal{R}$ stands for generalized role inclusion axioms

## 2.2    Definition

**Definition 1.** Let $x$ be an element of $\Delta^I$ (Interpretation domain) and $.^I$ be the fuzzy interpretation function then the fuzzy interpretation $I$ is a pair $I = (\Delta^I, .^I)$ such that

- for every individual $x$ mapped onto an element $x^I$ of $\Delta^I$,
- for every concept C mapped onto $C^I : \Delta^I \rightarrow [0, 1]$,
- for every role R mapped onto a function $R^I: \Delta^I \times \Delta^I \rightarrow [0, 1]$.

**Definition 2 (Bobillo et al. [1]).** A fuzzy knowledge-base satisfiability consists:

- Concept satisfiability C is $\delta$-satisfiable with respect to a fuzzy KB, $\mathcal{K} \cup \{(x : \geq \delta)\}$ is satisfiable, where $x$ is an individual not appearing in $\mathcal{K}$ newly.
- Entailment: Let $x : C \bowtie \delta$ be a fuzzy concept assertion entailed by KB $\mathcal{K}$ then $\mathcal{K} \models (x : C \bowtie \delta)$ iff $\mathcal{K} \cup \{(x : C \neg\bowtie \delta)\}$ is unsatisfiable.
- Supremum (sup). The greatest lower bound (glb) of a concept or role assertion $\tau$ is defined as $\sup\{\delta : \mathcal{K} \models (\tau \geq \delta)\}$.

# 3    Logic-Based Theories

## 3.1    Fuzzy Sets and Fuzzy Logic Theories

Fuzzy Set theory was initiated by Zadeh as an extension to classical concepts in order to capture the natural and intrinsic vagueness of the Set. As Zadeh [9] noted, fuzzy set theory preceded fuzzy logic in its broad sense. Moreover, fuzzy logic deals with precise logic of opproximate reasoning and imprecision. In a narrow sense, fuzzy logic is viewed as a kind of multivalued logics. The membership value of a set in a set theory is considered within the range of 0 and 1. While, in fuzzy logic the degree of truth of a statement between 0 and 1 is considered. However, in classical situation of fuzzy sets, 0 and 1 means no-membership and full-membership respectively [1].

**Definition 3.** Let $X$ be a fuzzy set defined by a membership function which assigns a real valued interval [0, 1] to all element $x \in X$ i.e. $\mu : X \rightarrow [0,1]$ then $\mu(x)$ gives a degree of an element $x \in X$ .

Consider the granule stage of fuzzy logic, which means the condition that has similarity, proximity, complexity, indistinguishability, and functionality of attribute-values [9]. This unique feature of granulation allows everything in fuzzy logic to be granulated. It shows that when moving from numerical to linguistics, fuzzy logic helps in generalizing the constraints of granular values. See Fig. 1 below:

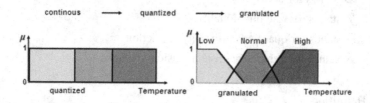

**Fig. 1.** Temperature granulation; Low, Normal and High are linguistic (granular) of temperature

**Example.** Suppose 41.5°C is given as human body temperature with respect to granulation then the granular value will be considered as "very high temperature". This is because the average human body temperature is between 36.1°C to 37.2°C. Also, in the case of blood pressure; if the blood pressure is about 160/80 then the granular value will be "high pressure". Therefore, humans deal with imprecision and complexity when they inspired graduated granulation.

The most visible and the widely used contribution of fuzzy logic is the concept of linguistic variable related to fuzzy logic [9]. Fuzzy logic is a better choice for approximate reasoning and imprecision.

## 3.2    The Operations in Fuzzy Logic

The operations of fuzzy logic are elucidated inform of mathematical functions over unit interval between 0 and 1 [3]. For instance, if [0, 1] is the interval of a t-norm t(x, y) which satisfies associatively, commutativity, and non-decreasing order with 1 as a unit element. Then, we consider the famous operations for t-norms and t-conorms (s-norms) with respect to Zadeh function, Lukasiewicz function, Product function, or Gödel function [1, 3, 10], see Table 2 for the families of fuzzy operations.

**Table 2.** Famous families of fuzzy operations

| Family | t-norm $x \oplus y$ | s-norm $x \otimes y$ | Negation $\ominus x$ | Implication $x \Rightarrow y$ |
|---|---|---|---|---|
| Zadeh | $\min\{x, y\}$ | $\max\{x, y\}$ | $1 - x$ | $\max\{1 - x, y\}$ |
| Lukasiewicz | $\max\{x + y - 1, 0\}$ | $\min\{x + y, 1\}$ | $1 - x$ | $\min\{1 - x + y, 1\}$ |
| Product | $x \cdot y$ | $x + y - x \cdot y$ | $\begin{cases} 1, & if\ x = 0 \\ 0, & otherwise \end{cases}$ | $\begin{cases} 1, & if\ x \leq y \\ 1/x, & otherwise \end{cases}$ |
| Gödel | $\min\{x, y\}$ | $\max\{x, y\}$ | $\begin{cases} 1, & if\ x = 0 \\ 0, & otherwise \end{cases}$ | $\begin{cases} 1, & if\ x \leq y \\ y, & otherwise \end{cases}$ |

## 3.3    Description Logics ($\mathcal{ALC}$)

As discussed above the historical and philosophical bases of mathematical logics gave birth to description logics (DL). DL is identified with its decidability nature, which is a fragment of first-order predicate logic. More importantly, it gives high impacts in knowledge representation paradigm. Nevertheless, DL gets more popular due to the advent of Semantic Web. In this sense, the established Tarski-style declarative Semantics enables it to capture the standard knowledge representation [10]. The knowledge representation formalism of description logics is governed by the W3C standard for OWL Semantic Web. However, DLs are the families of formal knowledge representation languages that have high expressive power in reasoning concepts.

The vocabularies or statements in $\mathcal{ALC}$ are divided into two groups. Namely: TBox statements and ABox statements. The TBox statements contain terminological (schema) knowledge and the ABox statements contain assertional (individual) knowledge [8, 13]. This particularization of ontologies brings the reasoning and knowledge bases together, which finally produces impact in language representation. The reasoning concepts of Semantic DLs that represent the OWL DL and OWL Lite are $\mathcal{SHOIN(D)}$ and $\mathcal{SHIF(D)}$ respectively [8, 12, 14]. They deal with the relationships of classes, membership and datatypes for the DLs knowledge-base concept expressions [11, 14]. In particular, $\mathcal{SHOIN(D)}$ works with the reasoning concepts of concrete datatypes and has a very expressive DL means.

Thus, DLs are unable to express a number of vague or imprecise knowledge and thereby cannot handle more uncertainties. As such, we attempt to bring fuzzy description logic KB satisfiability for reducing the complexity.

# 4    Fuzzy Description Logic

In this part we provide the syntax and Semantics of fuzzy $\mathcal{ALC}$. We also consider the Semantics given in [3, 8, 10].

## 4.1    Syntax of fuzzy $\mathcal{ALC}$

As described in section 3.1, the main focus of fuzzy set theory is to capture the notion of inherent vagueness of sets. Therefore, this process can be done by applying the concept membership function of the sets. Consider the definition 4 and the complex concepts given in section 2.

**Definition 4.** Assume the concepts are unary predicate and let roles to be binary predicates. Let A be a set of concepts and R be the set of abstract role names. Let abstract individual elements be $x$, y then the fuzzy concepts constructs formalize the knowledge-base structure of $\mathcal{ALC}$.

Consider the subsumption of fuzzy KB which consists of TBox and ABox say <T, A> then the concept inclusion axioms of fuzzy $\mathcal{ALC}$ is <C $\subseteq$ D, q> with q $\in$ C (0, 1]. Thus, the subsumption degree between C, D is at least q implies C = D iff <C $\subseteq$ D, 1> and <D $\subseteq$ C, 1>.

## 4.2    Semantics of Fuzzy $\mathcal{ALC}$

We consider the definition of interpretation given in section 2 above. We then extended the concept of classical fuzzy logic to fuzzy interpretation using membership function in [0, 1] interval.

**Definition 5.** Suppose $I = \{\Delta^I, \cdot^I\}$ is a fuzzy interpretation which contains a non-empty set $\Delta^I$ (domain), Let $R$, $C$ and $a$ be the abstract concept, abstract role and abstract individual element respectively. Also assume the interpretation function be $\cdot^I$ which assigns:

(a)  $C^I \colon \Delta^I \longrightarrow [0, 1]$,
(b)  $R^I \colon \Delta^I \times \Delta^I \longrightarrow [0, 1]$,
(c)  $a^I \in \Delta^I$.

Therefore, the fuzzy Semantic of $\mathcal{ALC}$ is based on the notion of interpretation. Classical interpretations are extended to the notion of fuzzy interpretations using membership functions that range over the interval [0, 1].

Now, using the $\cdot^I$ as the extension to non-fuzzy axioms given by Bobillo and Straccia [10],

$$(C \sqsubseteq D)^I = \inf_{x \in \Delta}{}^I C^I(x) \Rightarrow D^I(x)$$
$$(a : C)^I = C^I(a^I)$$
$$((a, b) : R)^I = R^I(a^I, b^I)$$

Thus, we have our fuzzy $\mathcal{ALC}$ Semantic. See Table 3 below:

**Table 3.** Semantics for fuzzy $\mathcal{ALC}$

$$
\begin{array}{l}
\top^I(x) = 1 \\
\bot^I(x) = 0 \\
(C \sqcup D)^I(x) = C^I(x) \oplus D^I(x) \\
(C \sqcap D)^I(x) = C^I(x) \otimes D^I(x) \\
(\neg C)^I(x) = \ominus C^I(x) \\
(\forall R.C)^I(x) = \inf_{y \in \Delta}{}^I R^I(x, y) \Rightarrow C^I(y) \\
(\exists R.C)^I(x) = \sup_{y \in \Delta}{}^I R^I(x, y) \otimes C^I(y)
\end{array}
$$

## 4.3    Fuzzy Axiom Satisfiability $\mathcal{ALC}$

Suppose the entailment of fuzzy interpretation $I$ be $I \models H$ and the satisfaction of fuzzy axiom $\tau$ (concept inclusion), then we define $I \models \langle \tau \geq q \rangle$, the assertion axiom iff $\tau^I \geq q$ [10].

Moreover, Consider an individual element say $x \in \Delta^I$ such that $C^\tau(x) > 0$ with interpretation $I$ iff a concept C is satisfiable. Note that, C $\sqcap$ ($\neg$C) is satisfiable under Zadeh family but is unsatisfiable under Lukasiewicz, Gödel or product families.

**Example.** Let $\mathcal{K}$ be a knowledge-base such that $\mathcal{K} = <\mathcal{T}, \mathcal{A}>$, $\mathcal{T}$ and $\mathcal{A}$ are terminological and assertional respectively. Suppose $\mathcal{T} = \emptyset$ and $\mathcal{A}$ {<(x, y) : R, 0.5>, <y : C, 0.7>}, sup $(\mathcal{K}, x : \exists R.C) = 0.7 \otimes 0.5$. Thus, the minimum value for $q$ $<\mathcal{T}, \mathcal{A} \cup \{x : \neg\exists R.C, 1 - q\}>$ is satisfiable. Obviously, applying the rules on two individual elements say x, y it implies that the value 0.35 is verified (glb = sup).

# 5    Completeness of the Reasoning Procedure for Fuzzy $\mathcal{ALC}$

In this part we show that logic is decidable under Gödel Semantic.

**Proposition 2.** A complete fuzzy $\mathcal{ALC}$ is satisfiable iff it is a model.

**Proof:** We consider only-if direction.

If a knowledge-base, KB is satisfiable then it is clear that fuzzy interpretation $I = \{\Delta^I, \cdot^I\}$. Let C be the constraint in the $\varphi : Var(C) \rightarrow [0, 1]$. Suppose we define $I$ in ABox, $\mathcal{A}$ as:

a)  $\Delta^I$ consist of all individual element in the domain $\mathcal{A}$
b)  $A^I(x) = \varphi(a_A(x))$ consist of all atomic classes in $\mathcal{A}$, with $\varphi(a_A(x))$ be the truth degree of variables in $a_{R(x, y)}$
c)  $R^I_o(x, y) = \varphi(a_{Ro(x, y)})$, where $R_o$ consist of all atomic roles, x, y individual elements in $\mathcal{A}$ and $\varphi(a_{Ro(x, y)})$ represent the degree of variable of $a_{Ro(x, y)}$).
d)  $x, y \in \Delta^I, R^I(x, y) = h_1 > h_2 = S^I(x, y)$

Now, using inductive technique, we show all concepts and roles that can be interpreted by $I$ such that $\{I, \varphi\}$ is a model.

Given that C as an atomic concept, then by definition we have $C^I$. C = $\neg$A implies that A contains $\{\neg A(x)\lambda\}$. But since $\mathcal{A}$ is complete and C has $x_{A(x)} = \neg\lambda$ and $x_{\neg A(x)} = \lambda$. So A(x) = $\neg\lambda$ contains the interpretation $A^I(x) = \neg\lambda$. Thus, applying concept negation, we have $(\neg A)^I(x) = \neg(\neg\lambda) = \lambda$ which implies the concept is obvious.

Next we show $I$ is satisfiable. Assume $I$ is not satisfiable, then without loss of generality $I \models R \sqsubseteq S$ iff $I' = I \square \{x : \exists R.(\exists R.B)\} \geq n, (x : \exists R.B) < n\}$ is unsatisfiable $\forall n \in (0, 1]$.

But, $\forall x, y \in \Delta^I R(x, y) \geq \sup_z t(R^I(x, z), R^I(z, y))$. Monotonically, for t-norms we apply $t(R^I(x, z), B^I(y)) \geq \sup_z t((R^I(x, z), t(R^I(z, y), B^I(y))) \forall x, y \in \Delta^I$. This implies the supremum of y. Assume the supremum or infimum does not exist then the value is undefined. Therefore,

$\sup_y t(R^I(x, y), B^I(y)) \geq \sup_z t((R^I(x, z), \sup_y t(R^I(z, y), B^I(y)))$

$(\exists R.B)^I(x) \geq \sup_z t(R^I(x, z) (\exists R.B)^I(z)) = (\exists R.(\exists R.B))^I(x)$, $I$ holds for model of the KB, hence is unsatisfiable.                                                           □

# 6    Conclusion and Future Work

We proved an essential step towards reducing the problem of imprecision. The method satisfies the operations in dealing with the fuzzy knowledge-base description logic paradigm. We provide the basis of Semantics and syntax that relates to fuzzy description logics. Also, we address the reasoning aspects on fuzzy $\mathcal{ALC}$ knowledge-bases. We further present the satisfiability in fuzzy knowledge-base reasoning for the completeness procedure. Significantly, this is the way of enhancing the strength of Semantic Web over the real world knowledge.

The approach taken in this paper allows interpretation of different types of imprecise knowledge that has relation to the real world applications. This differs from Józefowska et al. [17], where their proof is valid in query trie for skolem substitution only. Similarly, Zhao and Boley [3] proved the completeness but it is restricted to constraints sets. Thus, this shows the desirability of our completeness property – we can drive any entailed statement.

Moreover, the fuzzy $\mathcal{ALC}$ is expressed based on fuzzy subsumption of fuzzy concepts knowledge-bases that generalize modeling of uncertain knowledge.

Future direction of our research will be the designing of algorithm using type-2 fuzzy logic and to have practical implementation of the fuzzy reasoner. This will involve some technical designing decisions.

**Acknowledgements.** This work was partially supported by Universiti Teknologi PETRONAS and Carigali Sdn Bhd. We thank reviewers for their helpful comments.

# References

1. Bobillo, F., Delgado, M., Gómez-Romero, J., Straccia, U.: Fuzzy description logics under Gödel semantics. International Journal of Approximate Reasoning 50, 494–514 (2009)
2. Danyaro, K.U., Jaafar, J., Liew, M.S.: Tractability method for Ontology Development of Semantic Web. In: Proceedings of the International Conference on Semantic Technology and Information Retrieval, STAIR 2011, pp. 28–29 (2011)
3. Zhao, J., Boley, H.: Knowledge Representation and Reasoning in Norm-Parameterized Fuzzy Description Logics. In: Du, W., Ensan, F. (eds.) Canadian Semantic Web Technologies and Applications, pp. 27–53. Springer, Heidelberg (2010)
4. Gruber, T.R.: A Translation Approach to Portable Ontology Specifications. Knowledge Acquisitions 5, 199–220 (1993)
5. OWL Web Ontology Language Overview,
   http://www.w3.org/TR/owl-features/
6. Russell, S.J., Norvig, P.: Artificial Intelligence: A Modern Approach, 2nd edn. PearsonEducation, New Jersey (2010)
7. Royden, H.L., Fitzpatrick, P.M.: Real Analysis, 4th edn. Pearson, Prentice Hall (2010)
8. Hitzler, P., Krötzsch, M., Rudolph, S.: Foundations of Semantic Web Technologies. CRC Press, New York (2010)
9. Zadeh, L.A.: Is There a Need for Fuzzy Logic? Information Sciences 178, 2751–2779 (2008)

10. Bobilloa, F., Straccia, U.: Fuzzy Description Logics with General t-norms and Data Types. Fuzzy Sets and Systems 160, 3382–3402 (2009)
11. Lukasiewicz, T., Straccia, U.: Description Logic Programs Under Probabilistic Uncertainty and Fuzzy Vagueness. International Journal of Approximate Reasoning 50, 837–853 (2009)
12. Grimm, S., Hitzler, P., Abecker, A.: Knowledge Representation and Ontologies, Logic, Ontologies and Semantic Web Languages. In: Studer, R., Grimm, S., Abecker, A. (eds.) Semantic Web Services, pp. 51–105. Springer, New York (2007)
13. Yeung, C.A., Leung, H.: A Formal Model of Ontology for Handling Fuzzy Membership and Typicality of Instances. The Computer Journal 53(3), 316–341 (2010)
14. Lukasiewicz, T.: Probabilistic Description Logic Programs under Inheritance with Overriding for the Semantic Web. International Journal of Approximate Reasoning 49, 18–34 (2008)
15. Garcia-Cerdaňa, À., Armengol, E., Esteva, F.: Fuzzy Description Logics and t-norm Based Fuzzy Logics. International Journal of Approximate Reasoning 51, 632–655 (2010)
16. Stoilos, G., Stamou, G., Pan, J.Z.: Fuzzy extensions of OWL: Logical Properties and Reduction to Fuzzy Description Logics. International Journal of Approximate Reasoning 51, 656–679 (2010)
17. Józefowska, J., Lawrynowicz, A., Lukaszewski, T.: The Role of Semantics in Mining Frequent Patterns from Knowledge Bases in Description Logics with Rules. Theory and Practice of Logic Programming (TPLP) 2, 1–40 (2010)

# Random Forest for Gene Selection
# and Microarray Data Classification

Kohbalan Moorthy and Mohd Saberi Mohamad

Artificial Intelligence and Bioinformatics Research Group,
Faculty of Computer Science and Information Systems,
Universiti Teknologi Malaysia,
81310 Skudai, Johor, Malaysia ·
kohbalan@gmail.com, saberi@utm.my

**Abstract.** A random forest method has been selected to perform both gene selection and classification of the microarray data. The goal of this research is to develop and improve the random forest gene selection method. Hence, improved gene selection method using random forest has been proposed to obtain the smallest subset of genes as well as biggest subset of genes prior to classification. In this research, ten datasets that consists of different classes are used, which are Adenocarcinoma, Brain, Breast (Class 2 and 3), Colon, Leukemia, Lymphoma, NCI60, Prostate and Small Round Blue-Cell Tumor (SRBCT). Enhanced random forest gene selection has performed better in terms of selecting the smallest subset as well as biggest subset of informative genes through gene selection. Furthermore, the classification performed on the selected subset of genes using random forest has lead to lower prediction error rates compared to existing method and other similar available methods.

**Keywords:** Random forest, gene selection, classification, microarray data, cancer classification, gene expression data.

## 1 Introduction

Since there are many separate methods available for performing gene selection as well as classification [1]. The interest in finding similar approach for both has been an interest to many researchers. Gene selection focuses at identifying a small subset of informative genes from the initial data in order to obtain high predictive accuracy for classification. Gene selection can be considered as a combinatorial search problem and therefore can be suitably handled with optimization methods. Besides that, gene selection plays an important role preceding to tissue classification [2], as only important and related genes are selected for the classification. The main reason to perform gene selection is to identify a small subset of informative genes from the initial data before classification in order to obtain higher prediction accuracy. Classification is an important question in microarray experiments, for purposes of classifying biological samples and predicting clinical or other outcomes using gene expression data. Classification is carried out to correctly classify the testing samples according to the

D. Lukose, A.R. Ahmad, and A. Suliman (Eds.): KTW 2011, CCIS 295, pp. 174–183, 2012.
© Springer-Verlag Berlin Heidelberg 2012

class. Therefore, performing gene selection antecedent to classification would severely improve the prediction accuracy of the microarray data. Many uses single variable rankings of the gene relevance and random thresholds to select the number of genes, which can only be applied to two class problems. Random forest can be used for problems arising from more than two classes (multi class) as stated by [3].

Random forest is an ensemble classifier which uses recursive partitioning to generate many trees and then combine the result. Using a bagging technique first proposed by [4], each tree is independently constructed using a bootstrap sample of the data. Classification is known as discrimination in the statistical literature and as supervised learning in the machine learning literature, and it generates gene expression profiles which can discriminate between different known cell types or conditions as described by [1]. A classification problem is said to be binary in the event when there are only two class labels present [5] and a classification problem is said to be a multiclass classification problem if there are at least three class labels.

An enhanced version of gene selection using random forest is proposed to improve the gene selection as well as classification in order to achieve higher prediction accuracy. The proposed idea is to select the smallest subset of genes with the lowest out of bag (OOB) error rates for classification. Besides that, the selection of biggest subset of genes with the lowest OOB error rates is also available to further improve the classification accuracy. Both options are provided as the gene selection technique is designed to suit the clinical or research application and it is not restricted to any particular microarray dataset. Apart from that, the option for setting the minimum no of genes to be selected is added to further improve the functionality of the gene selection method. Therefore, the minimum number of genes required can be set for gene selection process.

## 2     Materials and Methods

There are ten datasets used in this research, which are Leukemia, Breast, NCI 60, Adenocarcinoma, Brain, Colon, Lymphoma, Prostate and SRBCT (Small Round Blue Cell Tumor). For breast, there are two types used which has 78 samples and 96 sample each with different class. Five of the datasets are two class and others are multi class.

The microarray datasets used are mostly multi class datasets. The description of the datasets such as the no of genes, no of patients, the dataset class and also the reference of the related paper for the dataset has been listed in the Table 1.

**Table 1.** Main characteristics of the microarray datasets used

| Dataset Name | Genes | Patients | Classes | Reference |
| --- | --- | --- | --- | --- |
| Adenocarcinoma | 9868 | 76 | 2 | [6] |
| Brain | 5597 | 42 | 5 | [7] |
| Breast2 | 4869 | 77 | 2 | [8] |
| Breast3 | 4869 | 95 | 3 | [8] |
| Colon | 2000 | 62 | 2 | [9] |
| Leukemia | 3051 | 38 | 2 | [10] |
| Lymphoma | 4026 | 62 | 3 | [11] |
| NCI60 | 5244 | 61 | 8 | [12] |
| Prostate | 6033 | 102 | 2 | [13] |
| SRBCT | 2308 | 63 | 4 | [14] |

## 2.1     Standard Random Forest Gene Selection Method

In the standard random forest gene selection, the selection of the genes is done by using both the backward gene elimination and the selection based on the importance spectrum. The backward elimination is done for the selection of small sets of non redundant variables, and the importance spectrum for the selection of large, potentially highly correlated variables. Random forest gene elimination is carried out using the OOB error as minimization criterion, by successfully eliminating the least important variables. It is done with the importance information returned from random forest.

Using the default parameters stated by [3], all forests that result from iterative elimination based on fraction.dropped value, a fraction of the least importance variables used in the previous iteration is examined. The default fraction.dropped value is 0.2 which allows for relatively fast operation is consistent with the idea of an aggressive gene selection approach, and increases the resolution as the number of variables considered becomes smaller. By default, the gene importance is not recalculated at each step as according to [15], since severe over fitting resulting from recalculating of gene importance. The OOB error rates from all the fitted random forests are examine after fitting all the forests. The solution with the smallest number of genes whose error rate is within standard errors of the minimum error rate of all forests is selected. The standard error is calculated using the expression for a binomial error count as stated below. The p resembles the true efficiency and n is the sample size.

$$Standard\ error\ =\ \sqrt{p(1-p) * \frac{1}{n}} \qquad (1)$$

The standard random forest gene selection method performs gene selection by selecting the smallest subset of genes with average out of bag (OOB) error rates between the smallest number of variables which is two and the subset with the number of variables that has the lowest OOB error rates. This strategy can lead to solutions with fewer genes but not the lowest OOB error rates.

## 2.2     Improvement Made to the Random Forest Gene Selection Method

Few improvements have been made to the existing random forest gene selection, which includes automated dataset input that simplifies the task of loading and processing of the dataset to an appropriate format so that it can be used in this software. Furthermore, the gene selection technique is improved by focusing on smallest subset of genes while taking into account lowest OOB error rates as well as biggest subset of genes with lowest OOB error rates that could increase the prediction accuracy. Besides that, additional functionalities are added to suite different research outcome and clinical application such as the range of the minimum required genes to be selected as a subset. Integration of the different approaches into a single function with parameters as an option allows greater usability while maintaining the computation time required.

**Automated Dataset Input Function.** In the current R package for random forest gene selection, the dataset format for input as well as processing is not mentioned and cause severe confusion to the users. Besides that, the method for inputting the dataset which is mostly in text file format required further processing to cater to the function parameters and format for usability of the gene selection process. Therefore, an automated dataset input and formatting functions has been created to ease the access of loading and using the dataset of the microarray gene expression based on text files input. The standard dataset format used for this package has two separate text files, which are data file and class file. These files need to be inputted into the R environment before further processing can be done. The method and steps for the automated dataset input is described in Figure 1. The steps have been created as an R function which is included inside the package and can be used directly for the loading of the dataset. The function takes two parameters, which are the data file name with extension and class file name with extension.

---

**Step 1:** Input data name and class name.
**Step 2:** Error checking for valid file name, extension and file existence.
**Step 3:** Read data file into R workspace.
**Step 4:** Data processing.
**Step 5:** Transpose data.
**Step 6:** Read length of class/sample.
**Step 7:** Read class file into R workspace.
**Step 8:** Create class factor.
**Step 9:** Load both data and class for function variable access.

---

**Fig. 1.** Steps required for the automated dataset input and formatting in R environment

**Selection of Smallest Subset of Genes with Lowest OOB Error Rates.** The existing method performs gene selection based on random forest to select smallest subset of genes while compromising on the out of bag (OOB) error rates. The subset of genes is usually small but the OOB error rates are not the lowest out of all the possible selection through backward elimination. Therefore, enhancement has been made to improve the prediction accuracy by selecting the smallest subset with the lowest OOB error rates. Hence, lower prediction error rates can be achieved for classification of the samples. This technique is implemented in the random forest gene selection method. During each subset selection based on backward elimination, the mean OOB error rate and standard deviation OOB error rate are tracked at every loop as the less informative genes are removed gradually. Once the loop terminates the subset with the smallest number of variables and lowest OOB error rates are selected for classification.

---

**While** backward elimination process = TRUE
    **If** current OOB error rates < = previous OOB error rates
        **Set** lowest error rate as current OOB error rates
        **Set** no of variables selected
    **End If**
**End While**

---

**Fig. 2.** Method used for tracking and storing the lowest OOB error rates

The subset of genes is located based on the last iteration with the smallest OOB error rates. During the backward elimination process, the number of selected variables decreases as the iteration increases.

**Selection of Biggest Subset of Genes with Lowest OOB Error Rates.** Another method for improving the prediction error rates is by selecting the biggest subset with the lowest OOB error rates. This is due to the fact that any two or more subsets with different number of selected variables with same lowest error rates indicates the informative genes level are the same but the relation of the genes that contribute to the overall prediction is not the same. So, having more informative genes can increase the classification accuracy of the sample.

The technique applied for the selection of biggest subset of genes with the lowest OOB error rates are similar to the smallest subset of genes with the lowest OOB error rates, except the selection is done by picking the first subset with the lowest OOB error rates from all the selected subset which has the lowest error rates. If there is more than one subset with lowest OOB error rates, the selection of the subset is done by selecting the one with highest number of variables for this method. The detailed process flow for this method can be seen in the Figure 3. This technique is implemented to assist researches that require filtration of genes for reducing the size of microarray dataset while making sure that the numbers of informative genes are high. This is achieved by eliminating unwanted genes as low as possible while achieving highest accuracy in prediction.

> **While** looping all the subset with lowest OOB error rate
>       **If** Current no of selected genes >= Previous no of selected genes
>             **Set** Biggest subset = Current number of selected genes
>       **End If**
> **End While**

**Fig. 3.** Method used for selecting the biggest subset of genes with lowest OOB error rates

**Setting the Minimum Number of Genes to be Selected.** Further enhancement is made to the existing random forest gene selection process by adding an extra functionality for specifying the minimum number of genes to be selected in the gene selection process that is included into the classification of the samples. This option allows flexibility of the program to suite the clinical research requirements as well as other application requirement based on the number of genes needed to be considered for classification.

The input for the minimum number of genes to be selected during the gene selection process is merged with the existing functions as an extra parameter input that has a default value of 2. The selected minimum values are used during the backward elimination process which takes place in determining the best subset of genes based on out of bag (OOB) error rates. At each time of a loop for selecting the best subset of genes, random forest backward elimination of genes is carried out by removing the unwanted genes gradually at each loop based on the *fraction.dropped* values selected. Therefore, as the no of loop increases, the no of genes in the subset decreases leaving the most informative genes inside the subset, as less informative genes are removed. The minimum no of genes specified is checked at each loop and if the total number of genes for a subset is less than the specified value, the loop is terminated leaving behind all the subsets.

---

**While** backward elimination process
        **If** length of variables <= to minimum required variables
                **Break**

---

**Fig. 4.** Method used for terminating the loop once the desired number of variables achieved

**Access Additional Functions through Parameters Input.** The automated dataset input have been created as a separate function to facilitate the loading and processing of the dataset to suite the data format required by the gene selection. The choice for selecting the smallest or biggest subset of genes with the lowest OOB error rates is integrated in a single function inside the random forest gene selection process and can be access by specifying the parameters. The minimum required variables can also be set by specifying the value at the function parameter. The default option is to select the smallest subset of genes and the minimum number of selected variables is two.

## 2.3    Performance Measurement

For gene selection using random forest, backward elimination using OOB error rates is used as the final set of genes is selected based on the lowest out of bag (OOB) error rates as random forest returns a measure of error rate based on the out-of-bag cases for each fitted tree. The classification performance of the microarray data using random forest is measured using .632 bootstrap method [16]. In this method, the prediction error rates obtained is used to compare the performance of the random forest in classification where lower error rates means higher prediction accuracy.

In the .632 bootstrap, accuracy is estimated as followed. Given a dataset of size n, a bootstrap sample is created by sampling n instances uniformly from the data (with replacement). Since the dataset is sampled with replacement, the probability of any given instance not being chosen after n samples is:

$$\left(1 - \frac{1}{n}\right)^n \approx e^{-1} \approx 0.368 \tag{2}$$

The expected number of distinct instances from the original dataset appearing in the test set is thus 0.632. The accuracy estimate is derived by using the bootstrap sample for training and the rest of the instances for testing. Given a number $b$, the number of bootstrap samples, let $c0_i$ be the accuracy estimate for bootstrap sample $i$. The .632 bootstrap estimates are defined as:

$$acc_{boot} = \frac{1}{b} \sum_{i=1}^{b} (0.632 \cdot c0_i + 0.368 \cdot acc_s) \tag{3}$$

Where $acc_s$ is the resubstitution error estimate on the full dataset (the error on the training set). The assessment method used has been able to populate and list the overall performance of the algorithm with other similar algorithms and techniques through prediction error rates calculation comparison.

## 3    Results and Discussion

In this section, the full result of all the options used is compared. For the bar chart, the result for each dataset is plotted against the accuracy, therefore the higher the values the lower is the error rates. Based on the Figure 5, the enhanced random forest gene selection performs better compared to standard method. Though, different options have different effects to the dataset being tested. Most of the datasets tested showed larger improvement in terms of accuracy achieved for classification when the subset of genes selected is larger.

However, some datasets with smaller subset of genes outperformed the larger subset of genes. This could be due to the effect of the informative genes, as more informative genes contribute to better classification accuracy. For example in leukemia dataset, selecting bigger subset or setting the minimum no of genes more than 10 has reduced the prediction accuracy as its possibility of low no of informative genes. The highest accuracy achieved for this dataset is by selecting smallest subset of genes which has only two genes selected as the subset. Hence, the gene selection options vary according to the dataset used.

Based on the three different options presented for the enhanced random forest gene selection, the first option which is selection of smallest subset of genes based on lowest OOB error rates is suitable for Breast 2 and Leukemia dataset as it provided the highest accuracy compared to other options. The second option using selection of biggest subset of genes based on lowest OOB error rates is suitable for Brain, Breast 3, Colon, Lymphoma, Prostate and SRBCT as it manage to achieve highest accuracy for these datasets using this option. Whereas, the third option which performs selection of smallest subset of genes based on lowest OOB error rates with minimum selected genes set to ten is suitable for Adenocarcinoma and NCI60 dataset as the accuracy achieved for these datasets is highest compared to other options.

The highest accuracy achieved for Adenocarcinoma dataset is 0.8371, for Brain dataset is 0.8197, Breast 2 dataset is 0.6718, Breast 3 dataset is 0.6682, Colon dataset is 0.8757, Leukemia dataset is 0.9418, Lymphoma dataset is 0.9620, NCI60 dataset is 0.7271, Prostate dataset is 0.9446 and SRBCT dataset is 0.9761. The huge improvement achieved in terms of the error rates differences between the standard random forest gene selection method and enhanced random forest gene selection method is from NCI60 dataset, where the differences of the error rates is 0.0801.

Further comparison with other available methods such as Diagonal Linear Discriminant Analysis (DLDA), K nearest neighbor (KNN) and Support Vector Machines (SVM) with Linear Kernel has been done as well. The comparison results have been included as supplementary page and can be downloaded at this link: http://www.mediafire.com/?y6m9edecsjd88xg.

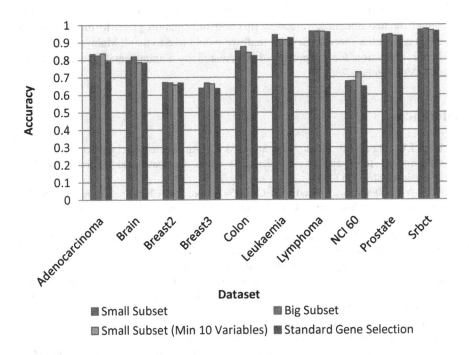

**Fig. 5.** Comparison between enhanced variables selection with three different options against standard gene selection method. A higher value indicates lower error rates

# 4    Conclusion

The proposed enhanced random forest gene selection has been tested with ten datasets and the outcome is as presented in the result and discussion section. There is improvement in terms of prediction accuracy for all datasets compared to the standard random forest gene selection. The gene selection plays an important role prior to classification and the way of selecting the subset of genes based on the type of dataset is also important in order to obtain lower error rates for classification. The option for selecting the smallest subset or bigger subset as well as setting the minimum required number of genes is the key factor in achieving higher accuracy in classification. For future works, additional options and functions can be integrated to suit other research works as well as clinical test by adding the features for selecting the range of genes or subset size to be selected for classification. This can be done by allowing the user to set the minimum number of genes as well as the maximum number of genes according to the requirement of the study. Hence, this enhanced random forest gene selection method provides the flexibility in determining the range of the genes in the subset as to how small or big the subset of genes required.

**Acknowledgements.** First of all, I would to thank my supervisor, Dr. Mohd Saberi B. Mohamad for his guidance, valuables tips and information given to me in order to make this research a success. Besides that, we would like to thank Malaysian Ministry of Higher Education for supporting this research by Fundamental Research Grant Scheme (Vot number: 78679).

# References

1. Lee, J.W., Lee, J.B., Park, M., Song, S.H.: An extensive comparison of recent classification tools applied to microarray data. Computational Statistics & Data Analysis 48, 869–885 (2004)
2. Chin, Y.L., Deris, S.: A Study On Gene Selection And Classification Algorithms For Classification Of Microarray Gene Expression Data. Jurnal Teknologi. 43(D), 111–124 (2005)
3. Díaz-Uriarte, R., Alvarez de Andrés, S.: Gene selection and classification of microarray data using random forest. BMC Bioinformatics 2006 7(3) (2006)
4. Breiman, L.: Bagging predictors. Machine Learning 26(2), 123–140 (1996)
5. Li, T., Zhang, C., Ogihara, M.: A comparative study of feature selection and multiclass classification methods for tissue classification based on gene expression. Bioinformatics 20, 2429–2437 (2004)
6. Ramaswamy, S., Ross, K.N., Lander, E.S., Golub, T.R.: A molecular signature of metastasis in primary solid tumors. Nature Genetics 33, 49–54 (2003)
7. Pomeroy, S.L., Tamayo, P., Gaasenbeek, M., Sturla, L.M., Angelo, M., McLaughlin, M.E., Kim, J.Y., Goumnerova, L.C., Black, P.M., Lau, C., Allen, J.C., Zagzag, D., Olson, J.M., Curran, T., Wetmore, C., Biegel, J.A., Poggio, T., Mukherjee, S., Rifkin, R., Califano, A., Stolovitzky, G., Louis, D.N., Mesirov, J.P., Lander, E.S., Golub, T.R.: Prediction of central nervous system embryonal tumour outcome based on gene expression. Nature 415, 436–442 (2002)
8. van't Veer, L.J., Dai, H., van de Vijver, M.J., He, Y.D., Hart, A.A.M., Mao, M., Peterse, H.L., van der Kooy, K., Marton, M.J., Witteveen, A.T., Schreiber, G.J., Kerkhoven, R.M., Roberts, C., Linsley, P.S., Bernards, R., Friend, S.H.: Gene expression profiling predicts clinical outcome of breast cancer. Nature 415, 530–536 (2002)
9. Alon, U., Barkai, N., Notterman, D.A., Gish, K., Ybarra, S., Mack, D., Levine, A.J.: Broad patterns of gene expression revealed by clustering analysis of tumor and normal colon tissues probed by oligonucleotide arrays. Proc. Natl. Acad. Sci. 96(12), 6745–6750 (1999)
10. Golub, T.R., Slonim, D.K., Tamayo, P., Huard, C., Gaasenbeek, M., Mesirov, J.P., Coller, H., Loh, M.L., Downing, J.R., Caligiuri, M.A., Bloomfield, C.D., Lander, E.S.: Molecular classification of cancer: class discovery and class prediction by gene expression monitoring. Science 286(5439), 531–537 (1999)
11. Alizadeh, A.A., Eisen, M.B., Davis, R.E., Ma, C., Losses, I.S., Rosenwald, A., Boldrick, J.C., Sabet, H., Tran, T., Yu, X., Powell, J.I., Yang, L., Marti, G.E., Moore, T., Hudson Jr., J., Lu, L., Lewis, D.B., Tibshirani, R., Sherlock, G., Chan, W.C., Greiner, T.C., Weisenburger, D.D., Armitage, J.O., Warnke, R., Levy, R., Wilson, W., Grever, M.R., Byrd, J.C., Botstein, D., Brown, P.O., Staudt, L.M.: Distinct types of diffuse large B-cell lymphoma identified by gene expression profiling. Nature 403(6769), 503–511 (2000)

12. Ross, D.T., Scherf, U., Eisen, M.B., Perou, C.M., Rees, C., Spellman, P., Iyer, V., Jeffrey, S.S., de Rijn, M.V., Waltham, M., Pergamenschikov, A., Lee, J.C., Lashkari, D., Shalon, D., Myers, T.G., Weinstein, J.N., Botstein, D., Brown, P.O.: Systematic variation in gene expression patterns in human cancer cell lines. Nature Genetics 24(3), 227–235 (2000)
13. Singh, D., Febbo, P.G., Ross, K., Jackson, D.G., Manola, J., Ladd, C., Tamayo, P., Renshaw, A.A., D'Amico, A.V., Richie, J.P., Lander, E.S., Loda, M., Kantoff, P.W., Golub, T.R., Sellers, W.R.: Gene expression correlates of clinical prostate cancer behavior. Cancer Cell 1, 203–209 (2002)
14. Khan, J., Wei, J.S., Ringner, M., Saal, L.H., Ladanyi, M., Westermann, F., Berthold, F., Schwab, M., Antonescu, C.R., Peterson, C., Meltzer, P.S.: Classification and diagnostic prediction of cancers using gene expression profiling and artificial neural networks. Nat. Med. 6, 673–679 (2001)
15. Svetnik, V., Liaw, A., Tong, C., Wang, T.: Application of Breiman's random forest to modeling structure-activity relationships of pharmaceutical molecules. Multiple Classier Systems 3077, 334–343 (2004)
16. Kohavi, R.: A study of cross-validation and bootstrap for accuracy estimation and model selection. In: Proceedings of the 14th International Joint Conference on Artificial Intelligence, Montreal, Quebec, Canada, August 20-25, pp. 1137–1143 (1995)

# A Novel Fuzzy HMM Approach for Human Action Recognition in Video

Kourosh Mozafari, Nasrollah Moghadam Charkari,
Hamidreza Shayegh Boroujeni, and Mohammad Behrouzifar

Faculty of Electrical and Computer Engineering,
Tarbiat Modares University, Tehran, Iran
kourosh.mozafari@ieee.org,
{moghadam,shayegh,m.behroozifar}@modares.ac.ir

**Abstract.** Hidden Markov Model (HMM) has been used in Human Action Recognition (HAR) since early 1980s. In this paper, an improvement of HMM using fuzzy concepts is proposed and used in HAR. HMM fuzzification is used widely in some research areas such as speech recognition and medicine, but it is used in HAR rarely. To increase the classification performance and decrease information losing in vector quantization, a fuzzy approach is used in HMM implementation. In feature extraction step, a human is represented by a skeletonization method which is effective and almost real-time. Experiments show that recognition rate increases about 6 percent using this approach. We call this approach Fuzzy Hidden Markov Model (Fuzzy HMM).

**Keywords:** Human Action Recognition, Hidden Markov Model, Fuzzy Hidden Markov Model, Skeletonization, Star Skeleton.

## 1 Introduction

Human Action Recognition (HAR) is an important research area in computer vision. The main goal of HAR is analysis the actions that occur in a video sequence. Vision-based HAR has found many applications such as surveillance systems, interactive systems, content based video retrieval, athletes' motion analysis, etc. This research area is scrimmaged with some problems such as self-occluding objects, well-defining actions and so on. Because of these problems and wide application area of HAR, this research area has been open yet.

We can divide Human simple Action Recognition into two main approaches: sequential approaches and space-time approaches. Input video is treated as 3-dimentional volume in space-time approaches. Space-time approaches are divided into three categories according to features used by them: volume, trajectories and local interested points. In sequential approaches, input video is viewed as a sequence of observations. Sequential approaches can be divided into two categories, according to methodology which they use: exemplar-based recognition methodologies and state model-based recognition methodologies. In state model-based approaches, human activity is represented as a model which composed of set of states. Then model is trained such that it corresponds to a sequence of feature vectors belonging to its activity class.

D. Lukose, A.R. Ahmad, and A. Suliman (Eds.): KTW 2011, CCIS 295, pp. 184–193, 2012.
© Springer-Verlag Berlin Heidelberg 2012

Hidden Markov Model (HMM) is a type of state model-based approaches which widely used in HAR. Many researchers have used HMM in HAR Since first use of HMM by Yamato et al [1] in 1992. Starner and Pentland (1995) [2] used standard HMMs to recognize American Sign Language (ASL). In addition to standard HMM, Variants of basic HMM also have proposed. Oliver et al. (2000) [3] proposed coupled HMM (CHMM) that modeled the human-human interaction. Natarajan and Nevatia (2007) [4] also proposed a new variety of HMM called coupled hidden semi-Markov Model (CHSMM). CHSMM is an extension of CHMM that considers the duration of an activity staying in each state.

An extension of HMM called Fuzzy HMM (FHMM) has been used widely In some research fields such as speech recognition and medicine. For example, a biomedical system based on fuzzy discrete hidden Markov model (FDHMM) is developed in [5]. In [6, 7] fuzzy HMM is used for speech recognition. [8] presented a hybrid approach for online Urdu script-based languages character recognition. An unconstrained Farsi handwritten word recognition system based on fuzzy vector quantization (FVQ) and HMM is proposed in [9] for reading city names in postal addresses. [10] used HMMs Incorporating Fuzzy Measures and Integrals for Protein Sequence Identification and Alignment.

Fuzzy HMM is used widely in above fields, but it used rarely in HAR. We can use fuzzy approach based on HMM in HAR in order to distinguish the similar actions. For example, some actions such as walk and run are similar; therefore recognition system maybe cannot recognize them well. In this paper we present a method for HAR using Fuzzy HMM. This Fuzzy HMM approach is used in biomedical system [5] and we apply this approach in HAR. We also use skeleton feature that is simple and almost real-time.

Rest of the paper is organized as follows: Section 2 describes an overview of proposed system. Section 3 describes the feature extraction method used in this paper. The HMM introduction and recognition and learning procedures are introduced in section 4. In section 5, Fuzzy HMM is described in detail and finally experiment results and conclusion are presented in section 6 and 7 respectively.

## 2    Method Overview

The process of this method has three steps: Feature Extraction, Learning by Fuzzy HMM and Test. Fuzzy HMM is used as learner in learning step. An overview of proposed HAR system is shown in Figure 1. In feature extraction step, each human in each frame is represented by a skeleton schema. Skeleton is structured by connecting centroid point to outstanding points of body. Each skeleton is described by a feature vector with five dimensions that each dimension's value is distance between centroid and an outstanding point. Learning step has two phases: generating output symbols (using clustering) and then training Fuzzy HMM using these symbols. Features are clustered into several clusters and each observation is assigned to all clusters using a fuzzy approach. Then, for each action, its observation sequences with their membership degrees are sent to a Fuzzy HMM as input. In test step, observation sequence of test video is given to the HMMs and HMM that has greatest log-likelihood determines the class of test action.

In next sections, each of these steps will be described by detail.

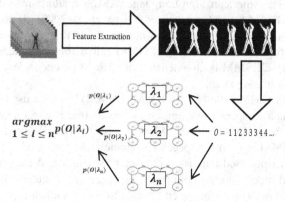

**Fig. 1.** Overview of proposed HAR system

# 3     Feature Extraction

Feature Extraction method which is used here is one that is proposed in [11, 12] with a little change. Feature extraction has two phases: background subtraction (finding human silhouette), and then using skeletonization technique. The flowchart of feature extraction phases is shown in Figure 2.

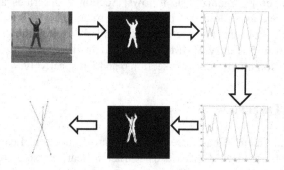

**Fig. 2.** Feature Extraction Process

## 3.1     Background Subtraction

The initial stage of the human motion analysis problem is the extraction of moving targets from a video stream. We use background subtraction method that is used in [13]. The notion is to use an adaptive background model to accommodate changes to the background while maintaining the ability to detect independently moving targets.

## 3.2    Skeletonization Technique

Finding the centroid of silhouette is the first step of extracting skeleton of human body. Suppose $(x_i, y_i), i = 1 \dots N$ are the coordinates of border pixels ($N$ is number of pixels in border of silhouette). $(x_c, y_c)$ is the coordinate of centroid of silhouette and is computed as below:

$$x_c = \frac{1}{N}\sum_{i=1}^{N} x_i, \quad y_c = \frac{1}{N}\sum_{i=1}^{N} y_i. \tag{1}$$

$$d_i = \sqrt{(x_i - x_c)^2 + (y_i - y_c)^2}. \tag{2}$$

Then, $d_i$, the distance between each border pixel $p_i(x_i, y_i)$ and centroid $(x_c, y_c)$, is computed and can be described as $d(i) = d_i$ function. To smoothing $d(i)$, we average height $(y_i)$ of $m$ pixels around each pixel $p_i$ that are located on border, and then obtained average value is set as new value for pixel height. Then local maxima points of $d(i)$ are determined. Five outstanding points of human body that are appeared in skeleton can be head, fingertips of two hands and legs. Therefore, if number of local maxima points of $d(i)$ is greater than five, parameter $m$ is modified and new local maxima points are computed until number of local maxima points is lower or equal than five. If number of local maxima points is lower than five, enough zero will be added as local maxima points in order to extend number of local maxima points to five. Then, skeleton is achieved by connecting local maxima points to centroid point (Figure 2). The 5-D vector which is composed of distances between these local maxima points and centroid are considered as feature vector for that frame.

In order to optimize the feature vector, we normalize each pair $(x_i, y_i)$ of border and centroid and then compute the feature vector. Normalization is done by dividing $x_i$ and $y_i$ of the pixels by width and height of human silhouette.

Furthermore, in addition to each observation sequence, its reverse sequence is stored too. For example, if a training video consists of person walks from left to right, the reverse sequence is helpful in recognition the person who walks from right to left in test phase, hence, reversed representative feature vectors are added to recognize the actions in counter direction.

## 4    Hidden Markov Model

A Hidden Markov Model is a statistical Markov model in which the system being modeled is assumed to be a Markov process with unobserved (hidden) state. In an HMM, the states are not observable, but when visit a state, an observation is recorded that is a probabilistic function of the state. In HMM, each activity is represented by a set of hidden states. In each time frame, human is located in one of the states and each state generates one observation (e.g. feature vector). In the next frame, system moves to another state considering transition probability between states. Suppose a discrete observation in each state from the set $\{V_1, V_2, \dots, V_M\}$:

$b_j(m) \equiv P(O_t = v_m | q_t = S_j)$

$b_j(m)$ is the observation, or emission probability that observed in $v_m, m = 1, \ldots, M$ in state $S_j$.

A HMM has the following elements:

1. N: Number of states in the model:

$S = \{S_1, S_2, \ldots, S_N\}$

2. M: Number of distinct observation symbols in the *alphabet*:

$V = \{v_1, v_2, \ldots, v_M\}$

3. State transition probabilities:

$A = [a_{ij}] where\ a_{ij} \equiv P(q_{t+1} = S_j | q_t = S_i)$

4. Observation probabilities:

$B = [b_j(m)] where\ b_j(m) \equiv P(O_t = v_m | q_t = S_j)$

5. Initial state probabilities:

$\pi = [\pi_i] where\ \pi_i \equiv P(q_1 = S_i)$

Given a number of sequences of observations, there are three basic problems:

1. (Evaluation problem) Given a model $\lambda$, we would like to evaluate the probability of any given observation sequence, $O = \{O_1\ O_2\ \ldots O_T\}$, namely, $P(O|\lambda)$.

2. (Decoding problem) Given a model $\lambda$ and an observation sequence $O$ we would like to find out the state sequence $Q = \{q_1 q_2 \ldots\ q_T\}$ which has the highest probability of generating $O$, namely, we want to find $Q^*$ that maximizes $P(Q|O, \lambda)$.

3. (Training problem) Given a training set of observation sequences, $\chi = \{O^k\}_k$, we would like to learn the model that maximizes the probability of generating $\chi$, namely, we want to find $\lambda^*$ that maximizes $P(\chi|\lambda)$.

Evaluation Problem: To overcome first problem, first we define forward variable $\alpha_t(i) \equiv P(O_1 \ldots O_t, q_t = S_i | \lambda)$ and backward variable $\beta_t(i) \equiv P(O_{t+1} \ldots O_T | q_t = S_i, \lambda)$. These variables can be calculated recursively as below:

- Initialization

$$\alpha_1(i) = \pi_i b_i(O_1). \tag{3}$$

$$\alpha_1(i) = \pi_i b_i(O_1). \tag{4}$$

- Recursion

$$\alpha_{t+1}(j) = \left[\sum_{i=1}^{N} \alpha_t(i) a_{ij}\right] b_j(O_{t+1}) \tag{5}$$

$$\beta_t(i) = \sum_{j=1}^{N} a_{ij} b_j(O_{t+1}) \beta_{t+1}(j). \tag{6}$$

The probability of the observation sequence is:

$$P(O|\lambda) = \sum_{i=1}^{N} \alpha_T(i) \,. \tag{7}$$

$$P(O|\lambda) = \sum_{i=1}^{N} \sum_{j=1}^{N} \alpha_t(i) a_{ij} b_j(O_{t+1}) \beta_{t+1}(j) \,. \tag{8}$$

Decoding Problem: Decoding Problem can be solved by Viterbi algorithm based on Dynamic Programming.

Training Problem: Training Problem can be solved using an Expectation Maximization (EM) algorithm such as Baum-Welch algorithm. To implement this algorithm, define $\gamma_t(i)$ as the probability of being in state $S_i$ at time t, given O and $\lambda$, which can be computed as:

$$\gamma_t(i) = \frac{\alpha_t(i) \beta_t(i)}{\sum_{j=1}^{N} \alpha_t(j) \beta_t(j)} \,. \tag{9}$$

Also, define $\xi_t(i,j)$ as the probability of being in $S_i$ at time t and in $S_j$ at time t+1, given the whole observation O and $\lambda$, which can be computed as:

$$\xi_t(i,j) = \frac{\alpha_t(i) a_{ij} b_j(O_{t+1}) \beta_{t+1}(j)}{\sum_k \sum_l \alpha_t(k) a_{kl} b_l(O_{t+1}) \beta_{t+1}(l)} \,. \tag{10}$$

Then, using an EM algorithm such as Baum-Welch, HMM parameters could be estimated, like below:

$$\hat{a}_{ij} = \frac{\sum_{k=1}^{K} \sum_{t=1}^{T_k-1} \xi_t^k(i,j)}{\sum_{k=1}^{K} \sum_{t=1}^{T_k-1} \gamma_t^k(i)} \,. \tag{11}$$

$$\hat{b}_j(m) = \frac{\sum_{k=1}^{K} \sum_{t=1}^{T_k-1} \gamma_t^k(j) 1(O_t^k = v_m)}{\sum_{k=1}^{K} \sum_{t=1}^{T_k-1} \gamma_t^k(j)} \,. \tag{12}$$

$$\hat{\pi}_i = \frac{\sum_{k=1}^{K} \gamma_1^k(i)}{K} \,. \tag{13}$$

## 5     Fuzzy HMM

In HMM, each observation vector is assigned to one cluster with degree of certainty 1. This means each observation vector is assigned only to one cluster. But, since assigning different observation vectors to the same cluster is possible and the observation probabilities of them become the same, consequently, classification performance maybe decreases. To overcome this problem, a form of fuzzification is used here. Distance between each observation vector to each cluster center is computed and inverse of this distance is considered as membership degree of observation vector to the cluster.

Suppose $cen(m)$ is center vector of cluster $m$, $O_t$ is observation at point $t$, $distance(a, b)$ is the distance between a and b, and $similarity(a, b) = (distance(a, b))^{-1}$, then membership degree of observation vector to cluster $m$, namely $u(m, t)$, can be defined as :

$$u(m, t) = similarity(cen(m), O_t)$$

According to above definitions, we can modify the Section 2 relations as below:

$$\alpha_1(i) = \pi_i \left[ \sum_{m=1}^{M} u(m, 1) b_i(O_1) \right].$$
(14)

$$\beta_T(i) = 1.$$
(15)

$$\alpha_{t+1}(j) = \left[ \sum_{i=1}^{N} \alpha_t(i) a_{ij} \right] \left[ \sum_{m=1}^{M} u(m, t) b_j(m) \right].$$
(16)

$$\beta_t(i) = \sum_{j=1}^{N} a_{ij} \beta_{t+1}(j) \left[ \sum_{m=1}^{M} u(m, t) b_j(m) \right].$$
(17)

$$P(O|\lambda) = \sum_{i=1}^{N} \alpha_T(i).$$
(18)

$$P(O|\lambda) = \sum_{i=1}^{N} \sum_{j=1}^{N} \alpha_t(i) a_{ij} \beta_{t+1}(j) \left[ \sum_{m=1}^{M} u(m, t) b_j(m) \right].$$
(19)

$$\xi_t(i, j) = \frac{\alpha_t(i) a_{ij} \beta_{t+1}(j) [\sum_{m=1}^{M} u(m, t) b_j(m)]}{\sum_k \sum_l \alpha_t(k) a_{kl} \beta_{t+1}(l) [\sum_{m=1}^{M} u(m, t) b_l(m)]}.$$
(20)

By comparing equations in this section with equations in section 4, we can see that usually $b_j(O_{t+1})$ is replaced by $\sum_{m=1}^{M} u(m, t) b_j(m)$. In other words membership degree of observation vector to clusters is involved in equations.

Then, the original equations of parameters $\hat{a}_{ij}$, $\hat{b}_j(m)$ and $\hat{\pi}_i$ are updated by using modified $\xi_t(i, j)$. At the last step of training phase, parameters are stored to be used in recognition phase. In recognition phase, the distance between each observation vector and all cluster centers are computed. Then, the membership degree of each observation to each cluster is obtained by using inverse of these distances. These membership degrees are used in modified forward algorithm. Then, test action's class will be determined using log-likelihood approach.

# 6    Experiments

## 6.1    Human Action Dataset

One of the datasets that has been used in many action recognition researches is Weizmann Blank[13]. This dataset that contains   90 low-resolution (180 x 144, deinterlaced 50 fps) video sequences showing nine different people, each performing 10 natural actions such as "run," "walk," "skip," "jumping-jack" (or shortly "jack"), "jump-forward-on-two-legs" (or "jump"), "jump-in-place-on-two-legs" (or "pjump"), "gallopsideways" (or "side"), "wave-two-hands" (or "wave2"), "waveone- hand" (or "wave1"), or "bend". In the training phase, we select 5 samples for each action randomly and other samples used for learning phase. This selection has been done 10 times.

## 6.2    Initialization

The initial state probabilities are initialized so that $\pi_1 = 1$, $\pi_i = 0$ $i = 2, 3, ..., N$. The number of states in HMM is assigned to 25. The state transition probabilities matrix A and observation probabilities matrix B are initialized as shown below.

$p_1 = 1 / (number\ of\ states)$

$$A = \begin{pmatrix} p_1 & \cdots & p_1 \\ \vdots & \ddots & \vdots \\ p_1 & \cdots & p_1 \end{pmatrix}$$

$p_2 = 1 / (number\ of\ observations)$

$$B = \begin{pmatrix} p_2 & \cdots & p_2 \\ \vdots & \ddots & \vdots \\ p_2 & \cdots & p_2 \end{pmatrix}$$

## 6.3    Experiments Results

The results have been shown in the forms of confusion matrix and column chart. Table 1 shows the confusion matrix. The rows and columns of confusion matrix represent the actions. The diagonal of confusion matrix shows the correctly recognized actions and other parts of matrix show the incorrectly recognized actions hence error classification. Figure 3 shows the recognition accuracy rate of HAR. The x-axis represents the actions and the y-axis represents the recognition accuracy. In x-axis, the recognition accuracy of our method and [12] are compared. Our experiments show that the total recognition accuracy rate is about 98.4% and it has improved about 6 percent as compared with other method.

According to Figure 3, It is obviously clear that [12] (using classic HMM) has not good recognition in similar actions like "run" and "walk" or "bend" and "Jump in place, well". Main reason of this problem is that feature vectors of these similar actions are very similar and these feature vectors have potential to locate in same cluster and have same symbol. But as it has shown in Figure 3, our fuzzy approach solves this problem significantly. Main reasons of this improvement are improvement in feature extraction method and especially fuzzification which locates the feature vectors in the different clusters with different membership degrees. Its effect is appeared in good recognition of similar actions like "run" and "walk" or "bend" and "Jump in place".

**Table 1.** Confusion matrix for HAR using Fuzzy HMM. C1-C10 are "walk", "run", "Jump", "Gallop sideways", " Bend", "One-hand wave", "Two-hands wave", "Jump in place", " Jumping Jack", "Skip"

|      | C1 | C2 | C3 | C4 | C5 | C6 | C7 | C8 | C9 | C 10 |
|------|----|----|----|----|----|----|----|----|----|------|
| C1   | 49 | 1  | 0  | 0  | 0  | 0  | 0  | 0  | 0  | 0    |
| C2   | 2  | 48 | 0  | 0  | 0  | 0  | 0  | 0  | 0  | 0    |
| C3   | 0  | 0  | 49 | 0  | 1  | 0  | 0  | 0  | 0  | 0    |
| C4   | 0  | 0  | 0  | 50 | 0  | 0  | 0  | 0  | 0  | 0    |
| C5   | 0  | 0  | 2  | 0  | 48 | 0  | 0  | 0  | 0  | 0    |
| C6   | 0  | 0  | 0  | 0  | 0  | 50 | 0  | 0  | 0  | 0    |
| C7   | 0  | 0  | 0  | 0  | 0  | 0  | 50 | 0  | 0  | 0    |
| C8   | 0  | 0  | 1  | 0  | 0  | 0  | 0  | 49 | 0  | 0    |
| C9   | 0  | 0  | 0  | 0  | 0  | 0  | 0  | 0  | 50 | 0    |
| C10  | 1  | 0  | 0  | 0  | 0  | 0  | 0  | 0  | 0  | 49   |

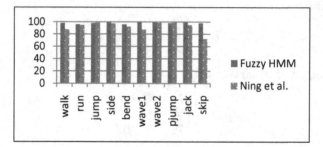

**Fig. 3.** Recognition accuracy rate of our HAR method. The x-axis represents the actions and the y-axis represents the recognition accuracy (%).

# 7    Conclusion

We propose a new human action recognition method by using a fuzzy approach in learning step. We also use Skeletonization technique which makes our method faster. A type of fuzzy clustering is used when we want assign an observation to a cluster. For this purpose, each observation is assigned to a cluster with a membership degree that represents degree of belonging observation to the cluster. Our method improves the recognition accuracy for the actions that are similar, such as "skip" and "run". In the future, we want to extend our method for a wider range of actions and also finding new approaches for fuzzification the HAR might produce some better results. As the first work, we test our approach on Weizmann dataset. It ensured us that our approach is promising. In the future works, we want improve our feature extraction method and fuzzification method such that our approach can be used in more realistic environments.

# References

1. Yamato, J., Ohya, J., Ishii, K.: Recognizing Human Action in Time Sequential Images Using Hidden Markov Models. In: Proc. Computer Vision and Pattern Recognition, pp. 379–385 (1992)
2. Starner, T., Pentland, A.: Real-time American Sign Language recognition from video using hidden Markov models. In: Proceedings., International Symposium on Computer Vision, November 21-23, pp. 265–270 (1995)
3. Oliver, N.M., Rosario, B., Pentland, A.P.: A Bayesian computer vision system for modeling human interactions. IEEE Transactions on Pattern Analysis and Machine Intelligence 22(8), 831–843 (2000)
4. Natarajan, P., Nevatia, R.: Coupled Hidden Semi Markov Models for Activity Recognition. In: IEEE Workshop on Motion and Video Computing, WMVC 2007, p. 10 (February 2007)
5. Harun, U., Ali, O., Ridvan, S., Ahmet, A.: A biomedical system based on fuzzy discrete hidden Markov model for the diagnosis of the brain diseases. Expert Systems with Applications 35(3), 1104–1114 (2008)
6. Cheok, A.D., Chevalier, S., Kaynak, M., Sengupta, K., Chung, K.C.: Use of a novel generalized fuzzy hidden Markov model for speech recognition. In: The 10th IEEE International Conference on Fuzzy Systems, vol. 3, pp. 1207–1210 (2001)
7. Tarihi, M.R., Teheri, A., Bababeyk, H.: A new method for fuzzy hidden Markov models in speech recognition. In: Proceedings of the IEEE Symposium on Emerging Technologies, September 17-18, pp. 36–40 (2005)
8. Razzak, M.I., Anwar, F., Husain, S.A., Belaid, A., Sher, M.: HMM and fuzzy logic: A hybrid approach for online Urdu script-based languages' character recognition. Knowledge-Based Systems 23(8), 914–923 (2010)
9. Dehghan, M., Faez, K., Ahmadi, M., Shridhar, M.: Unconstrained Farsi handwritten word recognition using fuzzy vector quantization and hidden Markov models. Pattern Recognition Letters 22(2), 209–214 (2001)
10. Niranjan, P.B., Chetty, M., Kamruzzaman, J.: Hidden Markov Models Incorporating Fuzzy Measures and Integrals for Protein Sequence Identification and Alignment. Genomics, Proteomics & Bioinformatics 6(2), 98–110 (2008)
11. Chen, H.S., Chen, H.T., Chen, Y.-W., Lee, S.Y.: Human action recognition using star skeleton. In: Proceedings of the 4th ACM International Workshop on Video Surveillance and Sensor Networks (VSSN 2006), pp. 171–178. ACM, New York (2006)
12. Li, N., Xu, D.: Action recognition using weighted three-state Hidden Markov Model. In: 9th International Conference on Signal Processing, ICSP 2008, pp. 1428–1431 (October 2008)
13. Fujiyoshi, H., Lipton, A.J., Kanade, T.: Real-Time Human Motion Analysis by Image Skeletonization. IEICE Trans. Inf. & Syst. E87-D(1) (January 2004)

# An Image Annotation Technique
# Based on a Hybrid Relevance Feedback Scheme

Mehran Javani[1] and Amir Masoud Eftekhari Moghadam[2]

[1] Department of computer Engineering, Behbahan Branch,
Islamic Azad University, Behbahan, Iran
[2] Department of Electrical Engineering and computer Science, Qazvin Branch,
Islamic Azad University, Qazvin, Iran
{M.javani,Eftekhari}@qiau.ac.ir

**Abstract.** Nowadays, with the advent of digital imagery, the volume of digital images has been growing rapidly in different fields; so there is an increasing requirement to effective image retrieval system. Hence, we need a more efficient and effective image searching technology. In this paper, we introduce a new scheme to image annotation in two stage. First semi-supervised k-means clustering with Mahalanobis similarity measure has been used. Second, a novel hybrid relevance feedback algorithm, *AHRFC* is proposed to narrow the gap between low-level image feature and high-level semantic and improve the accuracy of image annotation. The *AHRFC* algorithm is compound of three stages: (1) The images that the user knows irrelevant to cluster, are conducted to correct cluster by a *long-term RF*; (2) Regarding the images that the user knows relevant to cluster, we try to estimate feature weight of the clusters to provide a multiple similarity measure using a *re-weighting RF*; (3) To approach the exact place of the cluster centers, a *cluster center movement (CCM) RF* is used. Experimental results on the Corel database and satellite database taken from the Tehran city show the effectiveness of proposed methods in improving the retrieval performance.

**Keywords:** Content Based Image Retrieval (CBIR), Semantic Gap, Clustering, Relevance Feedback (RF).

## 1 Introduction

With rapid advances in the software, hardware, computer technology, cost reduction of data storage and capturing resources, the size of digital picture archives is increasing rapidly. Therefore, there is a great need for effective and efficient tools to image annotation and retrieval techniques. Most existing image retrieval systems, such as image search engines in Google and Yahoo, are textual based, in which images are searched by using the surrounding keyword, text, captions, etc [1]. In many applications such as satellite and medical image database there is no any text information to help retrieval stages thus we need the development of image retrieval techniques based on the image content. Many techniques have been developed for

D. Lukose, A.R. Ahmad, and A. Suliman (Eds.): KTW 2011, CCIS 295, pp. 194–205, 2012.
© Springer-Verlag Berlin Heidelberg 2012

content-based image retrieval (CBIR). CBIR is a process used to find images similar in visual features to a given query from an image database. It is usually implemented in two phases: image annotation and search images in database. In the annotation phase, each image according to its visual features is annotated with some machine learning algorithms. In the searching phase, when a user enters a query, a feature vector is extracted from query image content. Using a similarity measure, this vector is compared to the vectors in the features database. The images most similar to the query are returned to the user.

Machine learning algorithm used is image annotation can be divided into three categories of supervised, unsupervised and semi-supervised [2][5]. Supervised methods need annotation data to work. It should be note that annotation is a subjective, time consuming and cost ineffective process and prone to error [10-11]. Semantic concepts for the human being are very subjective so that they are not well defined. Due to training pattern and training time, supervised methods are used less frequently when the volume of data increases. Today, clustering methods are taken into account extreme consideration because training pattern is not necessary. Many clustering algorithms have been proposed in the literature for CBIR. In this paper we can divide clustering algorithm into seven groups such as hierarchical [6], graph theory [7-8], partitioning [9][11], Mixture models [7][10] [12], fuzzy clustering [7][9][12-13], neural network  and combinational methods. K-means is the most famous algorithm of partitioning methods. The K-means algorithm starts with K cluster centers (these centers are initially randomly selected). Each image in the data set is then assigned to the closest cluster centre. The centers are updated by means of the associated images. The process is repeated until criterion of mean square error is satisfied. The most important advantage of this method, that's made it popular, is time complexity O(n) [2] and It is very easy to implement. These algorithms have two disadvantages. First, it is sensitive to selection of initial cluster centers so that it might be trapped in a local optimum. Second, the performance of the algorithm depends on the selection of similarity measure. To overcome the first problem, semi-supervised clustering is used and the other one, a new multiple similarity measure is proposed to estimate an individual similarity measure for every cluster.

Clustering method, since unsupervised, it is very challengeable [16]. Semi-supervised method is proposed to removing some of the clustering challenge. In this method we need to feed small number of supervised data for clustering.

Images in database cannot adopt a fixed clustering for retrieval, since image retrieval is user dependent and time varying. The main reason of this problem is a gap between low-level image features and high-level semantic concept, which is the biggest problem in CBIR. This gap is known as semantic gap. The main contribution of this paper is reduce semantic gap and improved the efficiency of image annotation via semi-supervised K-means clustering and new hybrid relevance feedback algorithm-AHRFC. The system framework is shown in Figure 1.

Relevant feedback is an automatic process to identify the user intentions and learn the similarity concept from user's point of view. It used user's feedback to reduce the semantic gap between what queries says (low-level features) and what the user thinks. With user interaction, RF improved CBIR system performance considerably. RF algorithms are usually used to improve the accuracy of supervised CBIR system. On the other hand, RF algorithms are rarely paid attention in articles for unsupervised methods. In clustering based image annotation systems, some images may be assigned incorrect labels. In this paper, a new hybrid of three RF including *long-term RF*, *reweighting RF* and *cluster centers movement (CCM) RF* is introduced to improve the accuracy of image annotation and image database clustering.

The remainder of this paper is organized as follows: After reviewing of related works in the relevance feedback field is provided in Section 2. In section 3 introduced proposed system and experiments reported in Section 4. Finally, Section 5 conclusions of paper are presented.

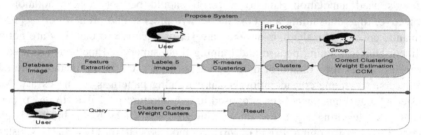

**Fig. 1.** Framework of the Proposed System

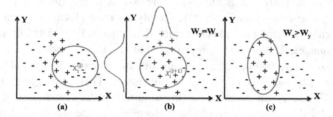

**Fig. 2.** a)initial Query b)reweighting is applied to stretch the neighborhood c) movement query with CCM

## 2     Related Works

The effort of researchers is developing RF techniques to achieve a higher performance. This is a most important stage in CBIR. For this purpose, try to best use of user feedback. In this section various methods of learning relevance feedback are reviewed and categorized.

### 2.1     Learning Methods for Relevance Feedback

As show in a Figure 2-a, CBIR system estimates the distance between database images and query image according to similarity measure and introduces $t$ images which has less distance with the query image to the user. The user recognized some of the images relevant and some of them irrelevant. Thus, a learning algorithm should be used to learn the subjectivity of the user. In general, RF learning method can be divided into supervised and semi-supervised or unsupervised methods.

### 2.1.1    Supervised Relevance Feedback

The supervised RF has a paradigm in various disciplines such as pattern recognition and machine learning. Since these systems are controlled by user feedback and called Human-Computer Interaction (HCI). HCI tries to simulate the human understanding system by using a non-linear separator analysis [19]. Three basic method to learn user opinion are *Query Point Movement (QPM)* [17][20][21] , *re-weighting* [17][20][22] and *Bayesian Inference(BI)* [17][20] . Some researcher usually used a hybrid of three method to achieve a great performance [1][20].

- *Query Point Movement:* It's supposed that there is an ideal point for every query in feature space. The purpose of retrieval process is to discover that point. QPM idea is movement query to positive example (PE) and far from negative example (NE) [17]. As shown in Figure 2-b query point, $X_i^{(j)}$ moved to $X_i^{(j+1)}$ by using Rocchio formula that would be included further relevant images.
- *Re-weighting:* the main purpose of re-weighting is giving more importance to some feature according to feedback samples. The evident feature should have less scattering on the class samples. In MARS [21] any feature is weight with regard to reverse variance $(1/\sigma)$. Comprehending user opinion can be challenging, of course. In Figure 2-c it is shown that giving much weight to X feature, more relevant images would be included.
- *Bayesian Inference:* another method for RF processing is probabilistic base model [17]. BI approach use Bayesian framework for deductive probabilistic estimation. Since probabilistic distribution on the whole database images after any feedback is updated, so the whole system can be improved.
- *Hybrid method:* researchers try to combine different available method to obtain an optimal method. For instance in [20] combine three method of QPM, re-weighting and BI, and using reinforced learning obtained an optimal feedback for query. In BALAS [1], QPM and BI are combined. Degrees of relevant/irrelevant images are determined by estimate online probability density functions.

### 2.2.2    Semi-supervised and Unsupervised Relevance Feedback

Since Semi-supervised and unsupervised are less dependent on user opinion, they are more popular. Especially in context like retrieval images on the web that asking user opinion are very time consuming in HCI. Hence Machine-Computer Interaction (MCI) system is an effective method. In [23] by training RBF neural network and use it in retrieval process. This system tries to simulate user opinion that is done RF in an unsupervised.

### 2.2    Short and Long Term Learning

Since different user may not have consistency in their idea about a single image at different times [24]. Therefore using multiple user opinion in multiple sessions is more useful. RF is divided into long term and short term when we face a couple of query sessions.

- *Short-term learning:* The traditional short-term relevance leaning scheme always starts a new query session without any assumption about the query formulations or feature weights images and repeated interaction with user until satisfied with the results.
- *Long-term learning:* This learning scheme keeps a global memory for each database image to store the last query formulation and feature weights images when the database image is previously used as a query. Thus, the relevance learning for a new query session can start from its previous state to speed up the learning process.

### 2.3    Relevance Feedback for Clustering

In recent years clustering has been a growing interest. In a great number of articles, relevance feedback is used for supervised learning. But RF learning methods in unsupervised method similar to proposed system are difficult to manage to rear instruction samples. RF is used in unsupervised system with different purposes to improve system. For instance, in CLUE[8] method is proposed that initially cluster image database based on graph theory and Ncut. Notwithstanding of the great applicability of this method, it cannot direct mapping between low level feature and high level semantic and computation of this method due to used graph-partitioning and Ncut for large dataset is very complicated. In SKKC [23] semi-supervised k-means clustering is used to annotate medical images, and then using relevance feedback for improved performance. But this method do not used any learning method in RF Stage and only try to correct clustering images. Thus this method needs many feedbacks.

## 3    The Proposed System and AHRFC

K-means algorithm is very suitable in applications that interactive with user because time complexity is O(n). But the main problem of k-means is sensitive to the selection of the initial cluster centers and may converge to local optima. To solve this problem, we use some training data to finding boundary of the clusters. But in many conditions finding boundary of cluster is not enough. Because diversity of data is large and cannot gain exact clustering only by use of low-level features and some data lie in adjacent clusters mistakenly. Therefore distance of image to adjacent cluster is less than correct cluster. To improve the accuracy of image annotation system and bridge the low-level and high-level semantic concepts AHRFC RF is introduced in following. Figure 3 shows the framework of our proposed system. First, user annotates some images. Then images are clustered with Mahalanobis similarity measure Equation (1).

$$S(F_i, F_j) = \sqrt{(Fi - Fj)^T \Sigma^{-1} (Fi - Fj)} \qquad (1)$$

$\Sigma$ is covariance matrix of whole images database that non-diagonal elements are set equal to zero. After clustering, the fuzzy membership values of each image to cluster centers are calculated according to distance image to each cluster. In last stage, performance system is improved with AHRFC scheme. In clustering stage some images maybe clustered mistakenly. Hence, we use relevance feedback to correct images clustering. Images in each cluster sorted according to distance to the cluster centers and images that have maximum distance are shown to user for relevance feedback process. In this regard, it is possible to have positive and negative examples and experiment shows that in this situation performance system is increase. The images set user relevant are relevant to cluster (Positive example), so system has recognized correctly. Images set user irrelevant are relevant to other clusters (Negative example) but system has recognized to this cluster. We used long-term RF for correct clustering. Thus, fuzzy membership value set zero for cluster that user know irrelevant and image allocate to cluster that have maximum membership value.

After each feedback, Cluster Center Movement (CCM) technique based on QPM is used to find ideal point for every cluster in feature space with Equation (2). Where $X_i^j$ and $X_i^{(j+1)}$ are previous and update cluster center respectively and $X_i^j$ meaning that a user submit the $i$th cluster in database image as the query and have $j$ RF iterations and $X_i^{(j+1)}$ is update cluster center in $(j+1)$ RF iterations. $\alpha$, $\beta$ and $\gamma$ are adjustment coefficient that we set $\alpha$ equal 1; $\beta$ and $\lambda$ are equal 0.33 empirically. Also, let the set of relevant images identified at the $j$th iteration be R, and the set of identified irrelevant images be N; $Z_k$ are images that belong to region R or N.

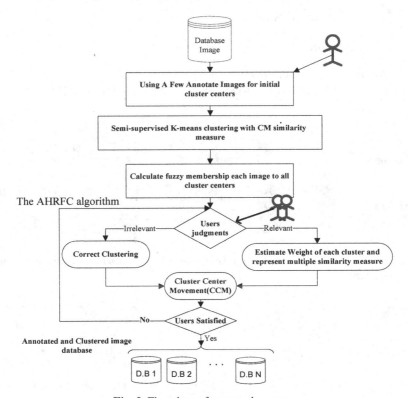

**Fig. 3.** Flowchart of proposed system

$$X_i^{(j+1)} = \alpha X_i^{j} + \beta \sum_{Y_k \in R} \frac{Z_k}{|R|} - \lambda \sum_{Y_k \in N} \frac{Z_k}{|N|} \tag{2}$$

Relevance feedbacks repeat until to the extent that satisfied user clustering result. This should be note that the user judgment for similarity between images is completely subjective and dependent on many factors [18]. The user preference is the most significant factor among all. For instance two objects with similar shapes but different colors seem similar by a user because the shape is more important feature while another user may consider them as different because color is more important feature to them [17]. Then we use images that user set relevant in several feedback to estimate feature weight of each cluster. Since important features in various clusters are different, Re-weighting RF is used to obtain suitable feature weight of each cluster according to scattering of features. We present multi-similarity measure in Equation (4) that we obtain weighing vector $W_j$ special for each cluster via relevance feedback. This formula, in response to the query image $I$, returned cluster $J$ that minimizes Equation (6). $Wj$ is the weight vector of cluster $j$ that has calculated in the relevance feedback phase.

$$Cluster(I) = \arg\min_{J \in (1..k)} \sqrt{(F_I - C_J)^T W_J (F_I - C_J)} \tag{3}$$

## 4    Experimental Result

### 4.1    Image Database

The performance of the image retrieval system is tested upon the following two databases: (a) SIMPLIcity[25] images and (b) Satellite images. The SIMPLIcity

|  |  |  |  |  |
|---|---|---|---|---|
| (1) Africa-100 | (2) Beach-100 | (3) Buildings-100 | (4) Buses-100 | (5) Dinosaurs-100 |
| (6) Elephants-100 | (7) Flowers-100 | (8) Horses-100 | (9) Mountains-100 | (10) Food-100 |

**Fig. 4.** Example Image of ten class in synthetic Dataset With cluster ID and number of image belong to clusters

(1) trees-520    (2) road and cars-712    (3) runway and desert-250    (4) Urban-870    (5) airplane-48

**Fig. 5.** Example Image of Five class in synthetic Dataset With cluster ID and number of image belong to clusters

database consists of 1000 images from 10 different categories. One sample image of each semantic group is shown in Figure 5. A satellite image of Tehran with size 3243 × 3243, shown in Figure 3 has been partitioned to 2400 sub-images with size 90×90, so that both adjacent sub-images have been overlapped by 50 percent. Therefore, Tehran satellite image database includes 2400 images with five semantic concepts. One sample image of each semantic group is shown in Figure 6. In our system we used 9 dimensions moment color, Inertia, energy, entropy, contrast, local homogeneity, correlation and difference moment extracted from GLCM matrix[22] [26] as texture and 9 Zernike moments shape features are much more common.

## 4.2    Experimental Results and Discussions

One of the major topics in machine vision and image retrieval is to determine the similarity of images based on their low level features. We have used Mahalanobis similarity measure with two reason 1) As shown in figure 6, Mahalanobis similarity measure has better accuracy rather than other similarity measure; 2) we can proposed multiple similarity measure by assigned different weighting vector with different semantic concepts.

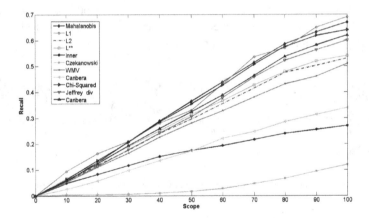

**Fig. 6.** Recall of clustering results for different similarity measures

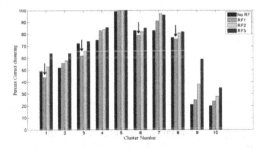

**Fig. 7.** Effective hybrid relevance feedback in clustering accyracy Corel Database

To examine the effect of hybrid relevance feedback method, accuracy of different clusters in three feedbacks is shown in Figure 7. As shown in Figure 7 for Corel database, clusters 1, 3, 6 and 8 after some feedback, accuracy of clusters decreases. According to the experiences of different people in several repeating, it is concluded that about 3 to 9 percent of user feedback images are wrong. If a user does mistake opinion about some image, he/she cannot recovery this opinion. But different users can recovery mistake of different users. In general we can mention three reasons for this problem.

- First, human being is sensitive, unpredictable and user judgment is subjective.
- Second, images in cluster boundary are very similar and user may mistakenly assign them correctly that caused decrease cluster accuracy. As shown in Figure 8, we have different image with different low-level feature for Africa concept in Corel dataset. A user may have different opinion in different time.
- Third, user may make mistake to put check mark or vice versa.

Figure 9 illustrates curve accuracy system in different relevance feedback for comparing with three RF Long term, Re-weighting, CCM and hybrid RF and SKKC system. These results are obtained from means of 100 random queries. The result of two databases show that increasing accuracy in hybrid method is approximately 19%, whereas for three methods alone, increasing accuracy is between 9 and 11 percent and for SKKC, increasing accuracy is approximately 9%.

**Fig. 8.** Differenf image of Africa Concept from Corel Database

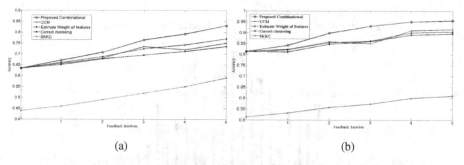

(a)                                          (b)

**Fig. 9.** Accuracy\Feedback iteration curve for proposed system and three component CCM, Re-weighting and Correct clustering with SKKC system a)Corel Database b)Satellite Database

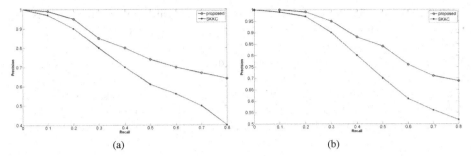

(a)                                                        (b)

**Fig. 10.** Precision\Recall curve for proposed system and SKKC system a) COREL Database b)Satellite Database

**Table 1.** Performance evaluation of proposed system, SNN and SKKC systems (a) Corel database; (b) Satellite database

| Method | Precision | Recall | F-measure |
|---|---|---|---|
| Proposed system | 0.825 | **0.821** | **0.822** |
| SNN | **0.839** | 0.658 | 0.739 |
| SKKC | 0.614 | 0.590 | 0.601 |

(a)

| Method | Precision | Recall | F-measure |
|---|---|---|---|
| Proposed system | 0.975 | **0.929** | **0.951** |
| SNN | 0.848 | 0.775 | 0.809 |
| SKKC | 0.652 | 0.685 | 0.668 |

(b)

(a)                                                        (b)

**Fig. 11.** Retrieval example from datasets for two queries evaluated. The top-left corner image is the query and top 8 retrieval results obtained by proposed scheme algorithm. (a) Africa, 6 matches out of 8; (b) Urban; 8 matches out of 8;

Curve precision-recall proposed system and SKKC system are shown in Figure 10. In both databases performance of proposed system is very better. The result of table 1 are show the proposed system is better than shared nearest neighbors (SNN) system [18] and SKKC system. The main reason of these improvements is AHRFC algorithm. To illustrate the performance of the proposed system, several examples are shown in Figure 11 with different semantics are picked as query images. For each query example, we examine the precision of the query results depending on the relevance of the image semantics.

## 5      Conclusion

In this paper, a new scheme for image annotation has been introduced. The proposed scheme consists of two steps: 1) A semi-supervised k-means clustering scheme; 2)

AHRC relevance feedback approach. The major contribution of this paper is to propose a new hybrid relevance feedback to improve the accuracy of image clustering approach. AHRFC relevance feedback scheme includes *long-term RF, re-weighting RF and CCM RF*. The proposed relevance feedback scheme increases the accuracy of image annotation about 19%. This hybrid relevance feedback is capable to improve every clustering method based on cluster center. Experimental results on Corel and Tehran satellite image database show the proposed system provide a significant improvement the accuracy of image database clustering and image annotation.

# References

1. Zhang, R., Zhang, Z.: BALAS: Empirical Bayesian Learning in the Relevance Feedback for Image Retrieval. Image and Vision Computing 24(3), 211–223 (2006)
2. Lee, J., Hwang, S., Nie, Z., Wen, J.: Query Result Clustering for Object-level Search. In: Proc. of KDD 2009, Paris, France, pp. 1205–1213 (2009)
3. Yip, K.Y., Cheung, D.W., Ng, M.K.: On Discovery of Extremely Low-dimensional Clusters Using Semi-supervised Projected Clustering. In: Proc. of the 21st International Conference on Data Engineering, ICDE (2005)
4. Dai, W., Xue, G., Yang, Q., Yu, Y.: Co-clustering Based Classification for Out-of-domain Documents. In: Proc. of the 13th ACM SIGKDD International Conference on Knowledge Discovery and Data Mining, KDD 2007, San Jose, California, USA, pp. 210–219 (2007)
5. Cao, F., Liang, J., Bai, L.: A New Initialization Method for Categorical Data Clustering. Expert Systems with Applications 36(7), 10223–10228 (2009)
6. Webb, A.R.: Statistical Pattern Recognition, 2nd edn. John Wiley & Sons, Ltd. (2002)
7. O-Duda, R., Hart, P.E., Stork, D.G.: Pattern Classification, 2nd edn. (2001)
8. Chen, Y., Wang, J.Z., Krovetz, R.: CLUE: Cluster-Based Retrieval of Images by Unsupervised Learning. IEEE Transaction on Image Processing 14(8), 1187–1201 (2005)
9. Das, S., Konar, A.: Automatic image pixel clustering with an improved differential evolution. Applied Soft Computing 9, 226–236 (2009)
10. Jain, A.K., Murty, M.N., Flynn, P.J.: Data clustering, a review. ACM Comput. Survay. 31(3), 264–323 (1999)
11. Chrng, C.H., Wel, L.Y.: An Evolutionary Computation Based onGA Optimal Clustering. In: Proce of the Sixth International Conference on Machine Learning and Cybernetics, Hong Kong, (2007)
12. Mitchell, T.M.: Machine Learning. McGraw-Hill Science Engineering Math. (1997)
13. Amato, A., Lecce, V.: A knowledge based approach for a fast image retrieval system. Image and Vision Computing 26, 1466–1480 (2008)
14. Sousa, F.M., Nascimento, S., Casimiro, H., Boutov, D.: Identification of upwelling areas on sea surface temperature images using fuzzy clustering. Remote Sensing of Environment 112, 2817–2823 (2008)
15. Xu, R., Wunsch, D.: Survey of Clustering Algorithms. IEEE Transactions on Neural Networks 16(3) (2005)
16. Sheng, W., Liu, X., Fairhurst, M.: A Niching Memetic Algorithm for Simultaneous Clustering and Feature Selection. IEEE Transactions on Knowledge and Data Engineering 20(7) (2008)

17. Kherfi, M.L., Ziou, D.: Relevance Feedback for CBIR: A New Approach Based on Probabilistic Feature Weighting With Positive and Negative Examples. IEEE Transaction on Image Processing 15(4), 1017–1030 (2006)
18. Wang, Y., Ding, M., Zhou, C., Zhang, T.: A Hybrid Method for Relevance Feedback in Image Retrieval Using Rough Sets and Neural Networks. Interbational Jurnal of Computational Cognition 3(1) (2005)
19. Muneesawang, P., Guan, L.: Multimedia Database Retrieval: a Human-Centered Approach. Signals and Communication Technology. Springer (2006)
20. Yin, P., Bhanu, B., Chang, K., Dong, A.: Integrating Relevance Feedback Techniques for Image Retrieval Using Reinforcement Learning. IEEE Transaction on Pattern Analysis and Machine Intelligence 27(10), 1536–1551 (2005)
21. Rui, Y., Huang, T.S., Mehrotra, S.: Content-based Image Retrieval with Relevance Feedback in MARS. In: Proc. in International Conference on Image Processing, Santa Barbara, CA, USA, vol. 2, pp. 815–818 (1997)
22. Liu, Y., Zhang, D., Lu, G., Ma, W.: A Survey of Content-based Image Retrieval with High-level Semantics. Pattern Recognition 40(4), 262–282 (2007)
23. Qiu, B., Xu, C.S., Tian, Q.: Efficient Relevance Feedback Using Semi-supervised Kernel - specified K-means Clustering. In: Proc of the 18th International Conference on Pattern Recognition 03, pp. 316–319 (2006)
24. Zhou, X.S., Huang, T.S.: Relevance Feedback in Image Retrieval: A Comprehensive Review. Multimedia System 8(6), 536–544 (2003)
25. Wang, J.Z., Wiederhold, J.: SIMPLIcity: Semantics-Sensitive Integrated Matching for Picture Libraries. IEEE Transactions on Pattern Analysis and Machine Inteligence 23(9) (2001)
26. Choras, R.S.: Image Feature Extraction Techniques and Their Application for CBIR and Biometrics Systems. International Journal of Biology Engineering 1, 6–16 (2007)
27. Broumandnia, A., Shanbehzadeh, J.: Fast Zernike Wavelet Moments for Farsi Character Recognition. Image and Vision Computing 25(5), 717–726 (2007)

# Using Semantic Constraints
# for Data Verification in an Open World

Michael Lodemann, Rita Marnau, and Norbert Luttenberger

Department of Computer Science, Christian-Albrechts-University in Kiel, Germany
{milo,rma,nl}@informatik.uni-kiel.de

**Abstract.** For the purpose of verifying given data sets with respect to semantic criteria, OWL-based ontologies from the problem domain can be combined with verification rules expressed in the *Semantic Web Rule Language* (SWRL). Whenever a verification fails the user needs support for identifying the SWRL rule atoms leading to the observed failure. We present an approach that derives so-called *Semantic Constraints* (SCs) from the verification rules in order to spot the errors within the given data set. This procedure has the advantage of continuously retaining the *Open World Assumption* (OWA) within the entire modeling process. SCs are derived from SWRL rules by a stepwise reduction of the rules' content. Though our approach is generic, we draw our examples from the domain of railway infrastructure planning.

**Keywords:** Semantic Verification, Semantic Constraints, Information Integration, Railway Modeling, OWL, SWRL, Protégé.

# 1  Introduction

By applying formal ontologies, various kinds of industrial design and implementation processes can benefit from their expressiveness, reasoning capabilities and tool support. We are focused on—but not restricted to—enhancing the planning process of railway infrastructures. Therefore the examples outlined in this paper are mostly railway related.

Contemporary German railway infrastructure planning comprises many tasks that are done manually. The machine readable representation of railway infrastructure planning data is a first step towards an improved process that in its full dimension should include a formal model which can be verified automatically. The benefit of a well-designed model not only resides in the ability to use tools in order to support the planning process on the whole, but also in the possibility to spot syntactic and semantic errors early within the planning process [1].

In computational applications ontologies can be used as an abstraction of heterogeneous data such as railway infrastructure planning data. Ontologies modeled in OWL [2] and SWRL [3] are usually subdivided into two parts: TBoxes contain all terminological axioms such as concept and property definitions as well as SWRL rules. The other part is formed by ABoxes which contain all assertional knowledge

D. Lukose, A.R. Ahmad, and A. Suliman (Eds.): KTW 2011, CCIS 295, pp. 206–215, 2012.

like individuals and property assertions. With SWRL rules terminological statements can be phrased that are not expressible in pure OWL. The rules are formed as implications consisting of an antecedent and a consequent part, thus following the description logic's principle of entailment. By definition SWRL is undecidable [4], but in practice decidable fragments of SWRL are often applied. Description Logic safe (DL-safe) rules for example restrict variable assignments within the rule atoms to named individuals [5] of the ontological model.

Within the antecedent of an SWRL rule certain combinations of OWL expressions outline a scenario which has to be met in order to let the corresponding consequent be processed. The consequent may consists of OWL expressions as well. With SWRL the ABox can be extended by new class or property assertions. However, with SWRL it is not possible to remove or to change prior assertions from OWL ontologies. The underlying *Open World Assumption* (OWA) assumes an ontological system with incomplete knowledge where every statement has to be defined true or false. There does not exist *Negation as Failure* (NaF). Missing assertions are handled as unknown and not as false like in ordinary *Closed World Assumption* (CWA) systems. This leads to difficulties when applying OWL and SWRL to verify data integrity: First, every positive and negative scenario has to be modeled, which leads to high complexity. And second, if a SWRL rule containing a verification statement fails, it cannot be stated why.

In order to introduce an example we consider a domain ontology which defines railway concepts like Track, Signal and so forth. Our methodology implies that we formalize planning directives as SWRL rules. This can be explained best by an example. The following statement of a railway infrastructure verification directive is considered: *A signal is correctly placed iff its position lies within the boundaries of the track it is related to.* The listing below shows this statement expressed in SWRL.

```
Signal(?s), Track(?t), isOn(?s,?t), position(?s,?p),
start(?t,?st), end(?t,?e), swrlb:greaterOrEqual(?p,?st),
swrlb:lessOrEqual(?p,?e) -> CorrectlyPlacedSignal(?s)
```

Notice that if this rule fires, individuals of the class *Signal* are classified to be *CorrectlyPlacedSignals* which is a subclass of *Signal*. There may exist other properties and subclasses of *Signal* which are not mentioned in the example above. As a result of the verification process, every individual of the class *Signal* has to be classified correspondingly.

It can be assumed that within the planning process this aim is not reached straightaway because there may be planning errors. This leads to certain rules like above not firing at all. In this example it cannot be determined why a *Signal* is not classified as *CorrectlyPlacedSignal*, i.e. it is unknown which specific atoms of the rule are not met.

The membership of individuals to certain classes is determined by the individuals' property assertions. In order to verify the data, these properties have to be examined. The planning errors which may occur can be classified as follows:

- Value range violations of data property assertions
- Nonexistent data property assertions
- Incorrect object property assertions
- Nonexistent object property assertions

It is desirable to support the planner to overcome planning errors. This can be accomplished by identifying fragments of the rule with which certain individuals do not comply. Intentionally this can be seen as a contradiction to the OWA to which OWL / SWRL ontologies underlie, because there does not exist NaF so that missing knowledge cannot be spotted. If it has not been stated explicitly that there is no assertion, in the OWA it is assumed there might be one. Therefore no consistent conclusion can be drawn. Within this paper we present an approach where such planning errors can be located while retaining OWA. This is achieved by a stepwise reduction of the information contents of a concept introduced as *Semantic Constraints* (SC). In the next section we present the idea of our approach. Design, implementation and the algorithmic idea are illustrated in section 2.1 followed by an overview of related work in section 3. Finally an outlook is given and a conclusion is drawn in section 4.

## 2     ABox Verification with Semantic Constraints

With common SWRL, conclusions about the existence of certain individuals with specific property assertions can be drawn. Term 1 states that there is at least one individual $I_1$ of class $C_1$ for which an individual $I_2$ of class $C_2$ and a property $p$ relating $I_1$ and $I_2$ exists.

$$\exists I_1 \in C_1 \, \exists I_2 \in C_2 \, (p(I_1, I_2)) \tag{1}$$

However, without the use of closure axioms in open world systems it cannot be stated whether all individuals comply with this assertion as there might be unknown individuals of class $C_1$. Anyhow, for verification purposes it is desirable to restrict the ABox to known class and property assertions in order to approve data integrity.

Our approach enables the user to check the static ABox against the corresponding TBox. The user is able to define SCs which are targeted to verify all individuals of a certain class of the TBox. These SCs act as invariants to which the individuals of the ABox have to hold in order to be positively verified. We enable the user to remain in the open world context while being able to verify the ABox which is restricted to the currently known assertions. An expression about all known individuals and their known assertions as outlined in term 2 can be formed.

$$\forall I_1 \in C_1 \, \exists I_2 \in C_2 \, (p(I_1, I_2)) \tag{2}$$

As mentioned in chapter 1 it is even more important to strictly detect the lack of object property assertions as formulated in term 3.

$$\forall I_1 \in C_1 \, \neg \exists I_2 \in C_2 \, (p(I_1, I_2)) \tag{3}$$

This is accomplished by reducing the SCs stepwise. Suppose an individual $I_1$ which does not hold to the unreduced SC containing the rule atom $(p(I_1, I_2))$. In the next step the rule atom $(p(I_1, I_2))$ is removed. When the individual $I_1$ now holds to this reduced SC it can be stated that the assertion $(p(I_1, I_2))$ is not present in the current ABox and therefore caused the error.

### 2.1    Design and Implementation

**Mutual Ontology Compositions.** Our approach of SC is implemented as a plugin for the Protégé Ontology Editor [6]. The plugin consists of several ontologies which are described within this chapter as well as the implementation of the verification algorithm outlined in the following chapters.

In figure 1 the ontological architecture of our approach is shown. The *Domain Ontology* contains the TBox and the corresponding ABox that is to be verified. Within the *Constraint Class Ontology* the necessary classes and properties of our approach are defined. It contains a class *Constraint* of which every SC is an individual. Within the ontology the following property assertions of the domain *Constraint* are defined.

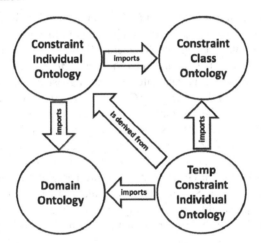

**Fig. 1.** Ontology architecture for the SC conception

- **isEnabled** - This flag marks whether this constraint will be processed during the verification process.
- **specifiedForClass** - A SC is always related to a class of the Domain Ontology. All individuals of that class are to be verified with respect to the specific SC.
- **ruleID** - This functional data property contains the *rdf:ID* of the SCs underlying SWRL rule. This SWRL rule contains the SC verification expression.
- **essential** - This nonfunctional property contains the expression of an atom of the SWRL rule. This atom is marked as essential for the proposition of the SC.

- **precondition** - This is a reference to other individuals of the Constraint class. Prior to being verified by this SC an individual has to be verified by the precondition constraints first.
- **preconditionGrade** - The functional *preconditionGrade* property is set by the verification algorithm in order to provide an acyclic constraint concatenation.

Apart from the class *Constraint* within the *Constraint Class Ontology* the data property *isVerifiedFor* is defined. It has as range *xsd:string* and is important for the verification process outlined in the next chapter.

As shown in figure 1 the *Constraint Class Ontology* is imported by the *Constraint Individual Ontology* which likewise imports the *Domain Ontology*. With the accessibility of the knowledge within the *Domain Ontology* SCs can be defined and saved in the *Constraint Individual Ontology*. During the verification phase the SCs are processed and reduced stepwise. Therefore the *Temp Constraint Individual Ontology* is created. Within the current verification step all relevant constraints and sub-constraints are stored in this ontology. As it imports the *Constraint Class Ontology* and the *Domain Ontology* the verification process can be applied to this ontology.

**Hierarchization of Semantic Constraints.** SCs can have other SCs as preconditions. This allows re-use of certain expressions and reduction of complexity regarding the individual SC. The introduction of SC hierarchization is accompanied with the demand for acyclicity. This is guaranteed by the *preconditionGrade* property of every SC. For every SC this grade is calculated by the highest *preconditionGrade* of its preconditions incremented by one. A SC has *Grade 0* if it does not have any preconditions. Regarding the example in figure 2, suppose the preconditions of *Constraint7* should be added. At first, without any preconditions, *Constraint7* is of *Grade 0*. Now *Constraint1* (of *Grade 0* as well) is added as a precondition which increments *Constraint7* to *Grade 1*. Then *Constraint6* (of *Grade 2*) is added which results in *Constraint7* being of *Grade 3*. Adding lower-graded SCs as preconditions is straightforward, but when a SC of a higher grade is to be added it is examined whether it or its preconditions currently contain the lower-graded SC as a precondition. If this is true the higher-graded SC cannot be added as a precondition itself in order to preserve acyclicity.

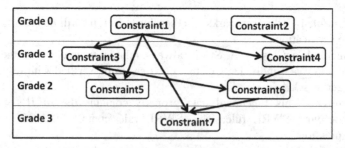

**Fig. 2.** Hierarchy of Semantic Constraints

**Verification Process.** During the verification every SC is processed. Every SC has the property *specifiedForClass*. Its value is the name of a *Domain Ontology* class whose complete individuals have to be verified by this specific SC. Furthermore the SC has the property *ruleID* which refers to a SWRL rule in which the actual verification expression is stored. Consider the following example:

```
Signal(?self), Track(?t), isOn(?self,?t),
position(?self,?p), start(?t,?st), end(?t,?e),
swrlb:greaterOrEqual(?p,?st),swrlb:lessOrEqual(?p,?e)
-> isVerifiedFor(?self, "SignalPlacementConstraint")
```

The algorithm adds the atom *Signal(?self)* according to the *specifiedForClass* property as well as the *consequent isVerifiedFor(?self, "SignalPlacementConstraint")*. This is the verification result. When processing this SWRL rule for every *Signal* which holds to the antecedent of the SC's SWRL rule the axiom *isVerifiedFor([SignalName], [ConstraintName])* as the consequent of the SC's SWRL rule is created. This marks an individual as positively verified.

**Listing 1.** Pseudo code describing the verification Process

```
foreach(SemanticConstraint c) {
   to = new TempConstraintIndividualOntology();
   to.import(domainOntology);
   cExpression = c.getSWRL_Expression();
   while(!allIndividualsVerified) {
      to.addRule(cExpression);
      axioms = SWRL-Processor.processRules(to);
      if(axioms.containAllIndividuals)
        allIndividualsVerified = true;
      else
        to.clearRules();
        cExpression =
          getNextCombination(c.getSWRL_Expression());
   }
 }
```

The verification algorithm is described as pseudo code in listing 1. While processing every SC, an empty *Temp Constraint Individual Ontology "to"* is created. It imports the *Domain Ontology*, thus having access to all its knowledge. As a next step, the SC's unreduced SWRL expression is assigned to a string variable. Now this expression is added to the *Temp Constraint Individual Ontology* in order to be processed by the *SWRL-Processor* in the next step. The axioms which are created during the SWRL processing are returned by the *SWRL-Processor*. As described above, for this example they consist of *isVerifiedFor* assertions for each positively verified *Signal*. Afterwards the algorithm checks whether all related individuals have been verified. If so, the algorithm sets the flag *allIndividualsVerified* and continues with the next SC. If not, the unreduced SWRL expression from the current SC is removed from the *Temp Constraint Individual Ontology*, is reduced and recombined (*getNextCombination*) and added again to this ontology in order to be processed with reduced information content. This is done until every individual of the focused class is verified. The reduction mechanism is described in the next chapter.

**Reduction of Semantic Constraints.** Whenever an individual cannot be positively verified by the original constraint expression it has to be determined why the verification for this individual fails. In order to spot the error the original SWRL expression is reduced and recombined atom by atom.

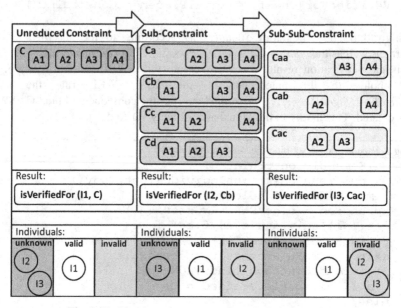

**Fig. 3.** Verification process - creation of sub-constraints

Figure 3 outlines a reduction process. First, all three individuals *I1*, *I2* and *I3* are sought to be verified by processing the original SC. As a result the axiom *isVerifiedFor(I1, C)* is created which means that the individual *I1* has been positively verified. *I2* and *I3* have not yet been positively verified, so there have been modeling errors. To spot the errors the original SC is reduced as expressed in the *Sub-Constraint* column of figure 3 and the verification process is repeated for the remaining individuals. As a result the atom *isVerifiedFor(I2, Cb)* is created. This means *I2* holds to the reduced SC *Cb*. Now it can be derived that the modeling error resides in the atom *A2*, which has been removed from the original SC in order to create *Cb*. As *I3* has not yet been verified, the reduction proceeds to the next iteration. Finally the axiom *isVerifiedFor(I3, Cac)* is created. So it can be concluded that the modeling errors for *I3* lie in the atoms *A1* and *A4*. The verification process is terminated when all individuals have been verified.

Our algorithm examines the SC before removing atoms within the reduction process. The atoms which are important for the meaning of the SC are marked as *essential* by the SC creator (see chapter *Ontological System Architecture*). These atoms are removed at the very end of the reduction process. Furthermore some atoms may rely on other atoms. Then both of them can be reduced in one step because solely they would be meaningless.

Due to the maximum amount of created sub-constraints the algorithm has a an efficiency of $O(2^n)$ where $n$ is the number of atoms within the SWRL rule. In practice this high amount may never be reached, because as described before in certain situations more than one atom is removed in one step. While modeling railway infrastructures, lots of data properties are used for positions, ranges, distances, etc.. Constraints dealing with these kinds of properties usually perform comparisons. So when removing the data property atom, the corresponding comparison atom is meaningless and is removed as well. This leads to a significantly smaller amount of sub-constraints.

As outlined in chapter *Hierarchization of Semantic Constraints* SCs can have preconditions which are SCs as well. Therefore an individual has to pass the precondition, before it is processed within the superordinate constraint. So there can be various reasons why an individual does not pass the verification. They are prioritized as follows:

1. Errors at essential atoms
2. Errors at preconditions
3. Errors at ordinary atoms

Our approach is implemented in a way that whenever an erroneous verification result is indicated, only the highest priority errors are shown. It cannot be ruled out that errors of lower priority have occurred as well.

## 3    Related Work

Regarding the examination of railway infrastructure data there are approaches in which the application of XML-schemas is proposed [7]. Nevertheless they only serve as a syntactic data validation technique. Semantic correlations between elements and attributes cannot be examined or validated with XML-schemas.

Motik, Horrocks and Sattler evolved an idea how OWL can be extended with integrity constraints in order to gain a better compatibility of ontologies and relational databases. In [8] and [9] they describe how constraints can be expressed as TBox atoms. Within TBox reasoning they behave like ordinary TBox atoms. Regarding ABox reasoning they can be used as model checking components in the sense of assertions of relational databases. This can be seen as closed world reasoning in a way that *Integrity Constraint Satisfaction* provides a possibility of proving that all mandatory assertions are fulfilled by the given ABox and TBox of the ontological knowledge base. Anyhow, *Integrity Constraint Satisfaction* postulates an extension of OWL thus differs from our approach.

Another approach resides in a SPARQL-based [10] rule and constraint language (SPIN) [11]. With SPIN, constraints which are related to specific concepts of the underlying knowledge base can be expressed. Each individual of the related concept is checked against the corresponding SPIN constraint. Negation is accomplished by the introduction of filters within the constraints. The constraint processing follows the principles of the CWA, so that wrong property values can be spotted without having to apply closure atoms. However, it is not possible to detect missing knowledge during SPIN constraint processing.

Tao, Sirin and McGuinnes present an approach of *Integrity Constraints* [12] which has capabilities to perform closed world reasoning with open world OWL ontologies. The underlying idea is to spot the OWL restrictions and integrity constraints in order to translate them into SPARQL queries. Typically the restrictions of related OWL concepts are used to classify certain individuals to be members of specific concepts. Following the open world assumption it is not possible to use these restrictions as validation constraints. By translating these restrictions into SPARQL queries, a validation mechanism can be obtained. These SPARQL queries are processed with respect to the Closed World Assumption (CWA). In their publication they address OWA related problems which we are facing as well and which will be focused during our future work. Although the approach of *Integrity Constraints* seems to be promising, it needs the extension of OWL and the application of a special reasoning software.

Standard SWRL enabled reasoners such as Pellet [13] have the ability to determine which atoms of a DL-safe rule written in SWRL lead to an inconsistent ontology. This may be seen similar to our idea of SC, but it leads to a completely different approach of phrasing the rule base. Our approach always assumes a consistent knowledge base. Apart from that the user is encouraged to phrase positive SWRL expressions only. No negative expressions or error cases have to be modeled as in ordinary open world systems. In our approach the verification failures are detected automatically without defining them manually.

## 4     Future Work and Conclusion

Apart from resolving OWA related challenges it is desirable to minimize the amount of reduced SCs. We are investigating whether a heuristic mechanism of prioritizing the reduction of certain atoms can lead to faster verification results and less sub-constraint creation. SWRL-built-ins for comparisons such as *greaterOrEqualTo* or mathematical operations seem to be appropriate reduction candidates.

Another optimization lies in the enhancement of expressiveness. Concerning our current approach of SCs only positive verification expressions can be formed. It is not possible to express error cases due to the assumption that an unreduced SC describes a condition where individuals pass the verification.

Another improvement may be the integration of grouping mechanism. The ability to define groups of SCs where an individual is positively verified whenever it passes only one SC of the whole group would lead to logical disjunctions and thus enhance expressiveness as well.

We have shown that our realization of SCs is a promising approach to combining the modeling liberties and advantages of ontologies in conjunction with inference mechanisms of open world reasoners with the pragmatism and the model checking capabilities of closed world systems. This can be achieved without an extension of OWL or non-standard reasoning software. Our approach is able to spot non-existent but essential assertions. Atoms leading to the failure of individual verifications can be identified, pointing the user towards modeling errors.

# References

1. Lodemann, M., Luttenberger, N.: Ontology-based Railway Infrastructure Verification – Planning Benefits. In: International Conference on Knowledge Management and Information Sharing (KMIS 2010) Proceedings, pp. 176–181. Scitepress (2010)
2. W3C OWL Working Group: OWL 2 Web Ontology Language Document Overview. W3C Recommendation (2009), http://www.w3.org/TR/owl2-overview/
3. Horrocks, I., Patel-Schneider, P.F., Boley, H., Tabet, S., Grosof, B., Dean, M.: SWRL: A Semantic Web Rule Language combining OWL and RuleML. W3C Member Submission (2004), http://www.w3.org/Submission/SWRL/
4. Horrocks, I., Patel-Schneider, P.F., Bechhofer, S., Tsarkov, D.: OWL Rules: A Proposal and Prototype Implementation. Journal of Web Semantics 1(3), 23–40 (2005) ISSN 1570-8268
5. Patel-Schneider, P.F.: Safe Rules for OWL 1.1 (Statement of Interest). In: SoI: OWL: Experiences and Directions 2008, OWLED 2008, Washington DC, USA (2008)
6. Stanford University: Protégé - An Open Source Ontology Editor. Stanford University School of Medicine (2011), http://protege.stanford.edu
7. Nash, A., Huerlimann, D., Schuette, J., Krauss, V.: railML - A Standard Data Interface for Railroad Applications. In: Proc. of the 9th International Conference on Computer in Railways, Comprail IX, vol. 15(5), pp. 233–240 (2004)
8. Motik, B., Horrocks, I., Sattler, U.: Bridging the Gap between OWL and Relational Databases. Journal of Web Semantics 7, 74–89 (2009), doi:10.1016/j.websem.2009.02.001
9. Motik, B., Horrocks, I., Sattler, U.: Adding Integrity Constraints to OWL. In: OWL: Experiences and Directions, OWLED 2007, Innsbruck, Austria (2007)
10. Prud'hommeaux, E., Seaborne, A.: SPARQL Query Language for RDF. W3C Recommendation (2008), http://www.w3.org/TR/rdf-sparql-query/
11. Knublauch, H., Hendler, J.A., Idehen, K.: SPIN - Overview and Motivation. W3C Member Submission (2011), http://www.w3.org/Submission/spin-overview/
12. Tao, J., Sirin, E., McGuinness, D.L.: Integrity Constraints in OWL. In: Proc.: Conference on Artificial Intelligence (AAAI 2010), Atlanta, USA (2010)
13. Sirin, E., Parsia, B., Cuenca Grau, B., Kalyanpur, A., Katz, Y.: Pellet: A practical OWL-DL reasoner. Journal of Web Semantics 5(2), 51–53 (2007), doi:10.1016/j.websem.2007.03.004

# Ontology Model for National Science and Technology

Michelle Lim Sien Niu[1], Nor Ezam Bin Selan[1],
Dickson Lukose[1], and Anita Binti Bahari[2]

[1] MIMOS Berhad,
Technology Park Malaysia, Kuala Lumpur, Malaysia 57000
{michelle.lim,nor.ezam,dickson.lukose}@mimos.my
[2] Malaysian Science and Technology Information Centre (MASTIC),
Level 4 Block C5, Complex C, Federal Government Administrative Centre,
Putrajaya, Malaysia 62662
anita@mosti.gov.my

**Abstract.** Heterogeneous data from various sources related to the National Science and Technology (NST) domain can be represented using ontological modelling. Data about researchers, research organizations, research projects, publications and research facilities are represented in Resource Description Framework (RDF), enabling all the data to relate to each other and can be easily made accessible to the user. Reusing standard existing vocabularies such as Friend-of-A-Friend (FOAF), Dublin Core Metadata Initiative (DCMI) and Semantically-Interlinked Online Communities (SIOC) also enable knowledge reusability and interoperability. This paper will talk about the techniques used to model the ontology for Malaysia Science, Technology and Innovation Knowledge Base system.

**Keywords:** National Science and Technology, Semantic Technology, Ontology Modelling, Resource Description Framework, Knowledge Base.

## 1    Introduction

Science and technology, together with engineering and innovation, play a fundamental role in the creation of wealth, economic growth and the improvement of the quality of life of the people [1] in the most competitive economies of the world today. Information about science and technology are widely available in the internet. However, they exist in heterogeneous data sources, such as different information systems, web pages of researchers, universities and organizations. To find the relevant information, users would need to go to various resources such as their own library's catalogue, reference databases, digital repositories and web pages [2].

There is a need to build a national repository for science and technology for relevant information to be collected, classified and aggregated. With the information consolidated, knowledge about research can be searched, visualised, presented and analysed rapidly. Research communities would have access to the vast and interlinked knowledge repository, and decision and policy makers would be able to make informed decisions.

D. Lukose, A.R. Ahmad, and A. Suliman (Eds.): KTW 2011, CCIS 295, pp. 216–225, 2012.
© Springer-Verlag Berlin Heidelberg 2012

In this paper, we described the Ontology Model of Malaysia's National Science and Technology that was adopted in the KRSTE[1] project. Section 2 starts with the knowledge repository for KRSTE to support the NST, follows by Section 3 that outlines the techniques used to address data heterogeneity, various schema of databases of the information sources, the reusability and performance. Section 4 makes a comparative analysis to other knowledge repository systems which are based on ontology models and Section 5 will talk about future works and additional enhancements that can be done to the model.

## 2    NST Knowledge Repository

KRSTE system provides a web portal that serves as Malaysia's resource centre for data, information and knowledge artifacts on Research and Development (R&D) in Science and Technology. KRSTE uses AllegroGraph [3] as the triple store and the contents in the KRSTE Knowledge Base are classified according to the Malaysian Research and Development Classification System (MRDCS) [4].

Fig. 1 depicts the high level model of the KRSTE Knowledge Base (KB), which comprised of the KRSTE Ontology, Bridge Ontology and the Artifacts Ontology.

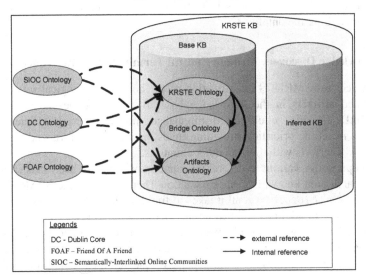

**Fig. 1.** High Level Ontology Architecture

KRSTE Ontology consists of MRDCS Ontology that leverages on three other popular vocabularies - SIOC [5], Dublin Core [6] and FOAF [7]. FOAF is used to describe information about person, while metadata about artifacts are described using

---

[1] KRSTE stands for Knowledge Resources for Science and Technology Excellence, a project funded by Ministry of Science, Technology and Innovation, Malaysia under the 9th Malaysia Plan. available at `https://krste.my`

the Dublin Core vocabulary. SIOC vocabulary is used to describe information about online communities such as posts, forums, sites, etc. The Artifacts Ontology consists of a collection of metadata obtained from the various sources, such as database repositories, word documents, power point presentations, websites, images, emails, etc. The Bridge Ontology serves as the global schema to link related concepts in the KRSTE Ontology and Artifacts Ontology.

As of May 2011, KRSTE knowledge base consists of 3,629 concepts and close to 14 million triples. New data, information and artifacts are continuously added and updated into the knowledge base through periodical scheduled processes. These processes includes automation of database conversion to RDF using conversion codes written in Lisp; extraction and conversion of PDFs and HTMLs data formats using a clipping tool, and finally resolving and merging URIs based on defined algorithms at the final stage of updating the knowledge base.

# 3    NST Ontology Model

The NST Ontology was modelled to handle the discovery of the knowledge through a central terminological reference, heterogeneity of data types and formats, reusability and performance. The following sub sections provided the techniques that were adopted in modelling the ontology.

## 3.1    Knowledge Discovery through Central Terminological Reference

KRSTE project uses MRDCS classification as the core terminological reference for all the related artifacts in the knowledge base. By utilizing these classifications, relevant R&D activities could be distinguished, hence ensuring efficiency and effectiveness in setting priorities, providing funds, maximizing efforts and comparing Malaysian R&D efforts with those in other countries.

Fig. 2 depicts the various types of information which has been classified to a specific research area known as "Approximation Theory". A simple SPARQL query (*?subject rdf:type :F1010101* ) is all it takes to find out all the related information for this research area. In addition, the taxonomic hierarchy inference can contribute to a richer knowledge discovery, in this example, with inference turned on, information for the research area "Mathematics" can also be discovered even if there is no direct relationship between that data and "Mathematics" research area. This is the benefit of using the transitive property[8] of rdfs:subClassOf.

Inference on the Semantic Web can be characterized by discovering new relationships [9]. In KRSTE, we used inference to discover non-explicit knowledge, for example, finding all the related research areas that are relevant to a particular organization.

Fig. 3a depicts a part of the MRDCS Class Hierarchy and its relationship with an organization (*Org_MIMOS_Berhad*). For example, when a query is executed to find the organizations which are related to *F1050600*, no results will be returned as there is

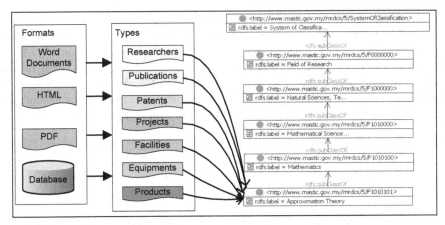

**Fig. 2.** Organizing Artifacts based on MRDCS Classification

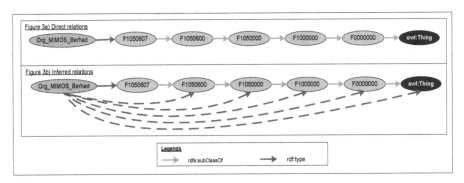

**Fig. 3.** MRDCS Class Hierarchy and Inferencing

no direct relation between *Org_MIMOS_Berhad* and *F1050600*. However, executing the same query with inference turned on will indicate that *Org_MIMOS_Berhad* is related to *F1050600*. This is because new inferred relations can be discovered through the transitive property of "*rdfs:subClassOf*". Fig. 3b depicts the new relations which were inferred based on the MRDCS hierarchical taxonomy. We can now ascertain that *Org_MIMOS_Berhad* is also related to *F1050600, F1050000, F1000000* and *F0000000, owl:Thing*.

## 3.2   Handling Heterogeneity of Data Type and Format

Information about researchers, publications, patents, projects, facilities, equipments, products and organizations are stored in various formats (such as word documents, HTML, PDF and relational databases), thus making retrieval of desired information a big challenge. Metadata is a systematic method for describing resources and thereby improving access to them. Consistent use of language with metadata descriptions can aid in the consistent discovery of resources. The primary tool for ensuring consistent language usage is via controlled vocabulary [10].

Fig. 4 depicts an example of a publication published in PDF format. A clipping tool is used to extract the metadata of the publication and then represent it in RDF by using the FOAF and DC vocabularies. Given the semantics of each term in FOAF and DC, there will be no ambiguity when we use these vocabularies in describing the title, author, organization, emails, etc. For HTML format, the object for the *dc:source* is therefore pointing to a URL from which the resource is derived, in this example it is "www.springerlink.com/content/72707k5078656173/fulltext.pdf".

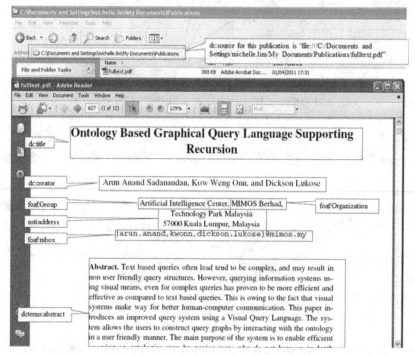

**Fig. 4.** Modelling Publication in PDF Format[2]

The implementation of standard vocabulary will not only address disambiguation but also enables interoperability as it helps both humans and machines understand the terms used.

NST information for KRSTE is also available in various databases schemas. Fig. 5 depicts an overview of mapping multiples tables from various databases into respective concepts in the NST ontology. The database schemas are first analyzed based on relations, attributes and functional and inclusion dependencies. A conversion tool written in Lisp maps the content in the databases into NST ontology. FOAF concepts such as *Group*, *Organization*, *Person*, *Document* and *Project* were used to describe the relevant contents, thus enhancing reusability and extendibility. In

---

[2] Document retrieved from the following URL:
www.springerlink.com/content/72707k5078656173/fulltext.pdf

addition, there were also additional contents about facility, equipment and products which were mapped to NST domain specific concepts (*Equipment, Facility* and *Product*).

Fig. 5 further illustrates the mapping of database tables into NST ontology. From the analysis of the database schemas, the conversion tool developed will extract and construct RDF triples based on the following definition:

*Content in "Organizations" table will be populated as instances of foaf:Organization.*
*Content in "Publications" table will be populated as instances of foaf:Document.*
*Content in "Patents" table will be populated as instances of foaf:Document.*
*Content in "Researchers" table will be populated as instances of foaf:Person.*
*Content in "Projects" table will be populated as instances of foaf:Project.*
*Content in "Products" table will be populated as instances of nst:Product.*
*Content in "Facilities" table will be populated as instances of nst:Facility.*
*Content in "Equipments" table will be populated as instances of nst:Equipment.*

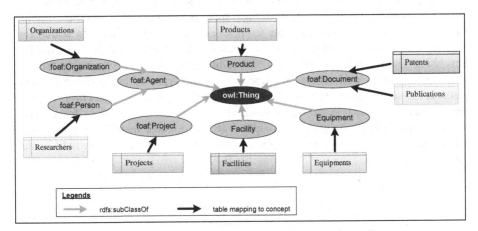

**Fig. 5.** Mapping Tables to Concepts in Ontology

### 3.3    Modelling for Reusability

The semantic web is envisioned as an evolving set of local ontologies that are gradually linked together into a global knowledge network. Many such local "application" ontologies are being built, but it is difficult to link them together because of incompatibilities and lack of adherence to ontology standards [11].

Reusing standard vocabularies from FOAF, DC and SIOC will enhance reusability and interoperability since systems are able to understand the terms used as the same URI [13] is used.

Fig. 6 is an example of how we model the organization, MIMOS Berhad instance, by reusing standard vocabularies, such as *foaf:Organization, foaf:Person, dcterms:isPartOf, dc:source, foaf:member, foaf:name.*

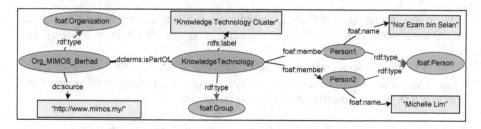

**Fig. 6.** Modelling for Ontology Reusability

## 3.4    Modelling for Performance

It is important to model the ontology for performance to ensure that the user experience will not be dismayed by the response time during information searching. Fig. 7 depicts how we model the NST Ontology for search performance. In Malaysia, there are multiple types of organizations, such as Government Agencies, Public Institutes of Higher Learning, Private Companies, etc. There is a need to search for organizations based on their organization types. We used the DC property, *dc:type* for describing the types of organization. In addition, we make an improvisation to enhance the search performance by representing the types of organization as URI resources instead of string literals. By implementing *dc:type* as an object property, the search operation time is significantly reduced.[3] This is because searching for a literal value involves many queries as it compares all the possible matches of data types, while searching for a resource looks for exactly one value, which greatly reduces the query response time.

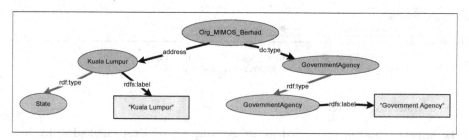

**Fig. 7.** Modelling of KRSTE NST Ontology for Performance

## 4    Comparison to Other Ontology Model Based System

The KRSTE Ontology Model was designed for describing NST activities specific to Malaysia. Table 1 lists the comparisons between KRSTE and other ontology based

---

[3] The response time to query for an organization represented as literal type is 1.9s in comparison to 0.2s when it is represented as a URI resource.

systems, such as euroCRIS [14] and RKBExplorer [15, 16]. From the table, we can conclude that the features for KRSTE are comparable to the others. This creates the opportunity for ontology matching and alignment [17], thus enabling the resources in these ontologies to be shared and reused.

**Table 1.** Features Comparison

|  | KRSTE | euroCRIS | RKBExplorer |
|---|---|---|---|
| Research Areas Classification | Based on MRDCS | Based on Research Classification Scheme [18] | Based on ACM Computing Classifica tion System [19] |
| Concepts | Person, Documents (Patents, Publications, Reports), Organization, Projects and Funds, Facilities, Equipments, Survey. | People, projects, Organizations, products of R&D (patents, products, publications) equipment used for R&D and R&D funding. | People, Publications, Organizations, Research Projects, Research Areas, Resilience – Explicit Mechanisms, Resilience related course |
| Vocabularies | Uses vocabularies from Dublin Core , FOAF and SIOC | Uses Dublin Core vocabulary [20] | The RDF is presented in accordance with the AKT Reference Ontology [21]. |
| Access | Free for public | Subscription fee required | Free for public |

## 5    Future Works and Enhancements

This paper described how we modeled the ontology for Malaysia's NST that uses MRDCS classification as a central terminological reference in which all the artifacts are linked. We also described how we handled the resources which come in various data types and formats by representing the resources metadata with standard vocabulary. We also ensure the reuse of standard vocabularies from SIOC, Dublin Core and FOAF to remove disambiguates and to enable the ontology to be further reused by other system. In addition, we also showed an example on how the ontology was modeled to improve the performance for the search by annotating the statement object as URI resource rather than character string.

We acknowledge that there are still a number of improvements that can be made to the ontology model for the NST in KRSTE. There are currently duplication of references in the knowledge base as our current method of determining that the person is the same is based on similar name and email address. We could improve this by using entity resolution techniques to enable the elimination of duplicate projects, documents and person.

Currently the instances URIs in the knowledge base are populated based on respective concepts (such as *Organization, Project, Facilities, Equipment*, etc). For example, we represent an organization instance, *Org_MIMOS_Berhad*, whereby semantics are binded to the URI. Since URI should not carry any meaning, therefore future works will include renaming of all the instances, removing semantics to the URI. In addition, the current MRDCS classification is developed based on English language terms only. Future enhancements will include Malay language terms to allow for contextual search based on Malaysia's national language. This can be accomplished by extending the language tag on the *rdfs:label*, which is not only limited to Malay language terms but also other terms such as Arabic, Chinese, Latin, etc.

Additional enhancements in the future will include public access to the KRSTE SPARQL endpoint. We are also in the process of linking resources in KRSTE knowledge base to resources that are published as Linked Open Data.

**Acknowledgements.** We acknowledge the efforts of all the members of the KRSTE project team, both from MIMOS Berhad and the Ministry of Science, Technology and Innovation (MOSTI).

# References

1. Science, Technology, Engineering and Innovation for Development: A Vision for the Americans in the Twenty First Century. Organization of American States Executive Secretariat for Integral Development Office of Education, Science and Technology (2004)
2. Sadeh, T., Walker, J.: Library portals: Toward the semantic Web. In: New Library World, vol. 104(1184/1185) (2003)
3. Franz. Inc. AllegroGraph RDFStore Web 3.0's Database, http://www.franz.com/agraph/allegrograph/ (last visited: May 2011)
4. Malaysian Research & Development Classification System, http://mrdcs.mastic.gov.my/mrdcs/index2.php?menu=mapping_cen tre&menu_dyn=classification (last visited: May 2011)
5. SIOC Core Ontology Specification (2006), http://sioc-project.org/ontology (last visited: May 2011)
6. Dublin Core Metadata Element Set, http://dublincore.org/documents/dces/ (2010) (last visited: May 2011)
7. FOAF Vocabulary Specification 0.98, http://xmlns.com/foaf/spec/ (last visited: May 2011)
8. OWL Web Ontology Language Reference, W3C Recommendation (2004), http://www.w3.org/TR/owl-ref/ (last visited: May 2011)
9. W3C Inference (2010), http://www.w3.org/standards/ semanticweb/inference (last visited: May 2011)
10. Taylor, C.: Information Access Service, An Introduction to Metadata, The University Of Queensland, Australia (2003)

11. Brinkley, J.F., Suciu, D., Detwiler, L.T., Gennari, J.H., Rosse, C.: Framework for using reference ontologies as a foundation for the semantic web. In: Proceedings, American Medical Informatics Association Fall 2006 Symposium, Bethesda, MD, pp. 96–100 (2006)

12. Bontas, E.P., Mochol, M., Tolksdorf, R.: Case Studies on Ontology Reuse. In: Proceedings of I-KNOW 2005, Graz, Austria, June 29-July 1 (2005)

13. URIs, URLs, and URNs: Clarifications and Recommendations 1.0 Report from the joint W3C/IETF URI Planning Interest Group, W3C Note, September 21 (2001)

14. The European Organization for International Research Information, euroCRIS (2010), http://www.eurocris.org/Index.php?page=homepage&t=1 (last visited: May 2011)

15. ReSIST RKB Explorer, http://www.rkbexplorer.com (last visited: May 2011)

16. Glaser, H., Millard, I., Anderson, T., Randell, B.: ReSIST Project Deliverable D10: Prototype knowledge base (2006), http://eprints.ecs.soton.ac.uk/14304/1/D10-TR.pdf (last visited May 2011)

17. Granitzer, M., Sabol, V., Kow, W.O., Lukose, D., Tochtermann, K.: Ontology Alignment—A Survey with Focus on Visually Supported Semi-Automatic Techniques. In: Future Internet (2010) ISSN 1999-5903

18. CERIF. euroCRIS Research Classification Scheme (1991), ftp://ftp.cordis.europa.eu/pub/cerif/docs/cerif1991.htm#sectionC (last visited: May 2011)

19. The ACM Computing Classification System (1998), http://www.acm.org/about/class/1998/ (last visited: May 2011)

20. CERIF Ontology (2004), http://www.eurocris.org/Uploads/Web%20pages/CERIF2004/CERIF2004_ontology.xml (last visited: May 2011)

21. AKT Reference Ontology, http://www.aktors.org/publications/ontology/ (last visited: May 2011)

# Norms Detection and Assimilation
# in Multi-agent Systems: A Conceptual Approach

Moamin A. Mahmoud[1], Mohd Sharifuddin Ahmad[1], Azhana Ahmad[1],
Mohd Zaliman Mohd Yusoff[1], and Aida Mustapha[2]

[1] College of Information Technology,
Universiti Tenaga Nasional,
Jalan IKRAM-UNITEN, 43009 Kajang, Selangor, Malaysia
moamin84@gmail.com, {sharif,azhana zaliman}@uniten.edu.my
[2] Faculty of Computer Science & Information Technology,
Universiti Putra Malaysia,
43400 UPM Serdang, Selangor Malaysia
aida@fsktm.upm.edu.my

**Abstract.** In this paper, we propose a technique for a software agent to detect the norms of a community of agents and assimilate its behavior to comply with the local normative protocol, failing which, the agent is refused services and resources. In this technique, the software agent is equipped with an algorithm, which detects and analyzes the normative interactions between local agents. When the detection is successful, it launches another algorithm to request for its assimilation to the local normative protocol, indicating its acceptance by the group of local agents

**Keywords:** Norms Detection, Normative Environment, Multi-agent Systems.

## 1    1 Introduction

In normative agent environments, norms are soft constraints that regulate the behavior of agents and their interactions with other agents in achieving a shared goal. In such circumstance, agents are said to be loosely constrained by some normative protocol with which agents interact according to a set of norms. Such imposed restriction could manifest an orderly social structure in multi-agent systems. However, in an open cyber world in which agents roam in cyberspace to achieve some intended goal, a visitor agent, which is not normally constrained by some normative protocol could undermine the integrity of a host system by posing requests that are not socially congruent with the norms of the local agents. With autonomous capability and being refused services, the visitor agent could launch hostile actions for services and resources, which could be construed as threats to the host system [6], [7]. While such requests could be totally ignored, a more graceful approach would provide an amicable solution to this problem.

In this paper, we present the preliminary findings of our research in norms detection and assimilation. We propose a novel technique for a visitor agent to detect the norms of

D. Lukose, A.R. Ahmad, and A. Suliman (Eds.): KTW 2011, CCIS 295, pp. 226–233, 2012.

a group of local agents and assimilate its behavior to comply with the local norms. In this technique, the visitor agent is equipped with an algorithm, which detects and analyzes the normative interactions between the local agents. When the detection is successful, it launches an algorithm to request for its assimilation with the local agents. When the request is granted only then the visitor agent's actions is governed by the host's normative protocol indicating its acceptance by the group of local agents.

Savarimuthu et al. [10], [11] claim that norms recognition benefits an agent in that it alerts the agent on the expectation of a society that it wishes to join. An agent that is equipped with this capability would be able to modify its expectations of behavior as it joins and leaves different agent societies. In other words, as it experiences new environments, its ability to detect and identify new norms could help the agent to review its plans to achieve its intended goals.

The problem of norms conflict could be modeled in a multi-agent environment by detecting the normative protocol of a group local agents and discovering an emerging pattern of interaction between them. Such pattern could be deciphered as a norms model by a visitor agent. The findings of this research would offer a novel approach to designing software agents which can interoperate effectively across different normative agent platforms. Consequently, this research attempts to answer the following questions:

1. What are the advantages of imposing a local normative protocol on visiting agents?
2. How should a visiting agent detect the norms of a host's agents?
3. How should the assimilation of a visiting agent with the host norms be implemented?

This research is inspired by two main convictions. Firstly, the phenomenon of norms conflict between two people of different culture, which could be simulated in agent-based systems. For example, an English expatriate is requested to work with his/her counterpart in Japan. The expatriate would find it difficult to assimilate and practice the norms of the Japanese people if he/she does not attempt to know, understand, and practice the norms of the Japanese people. A serious problem could emerge if such norms conflict causes misunderstanding in communication, coordination, and social interactions resulting in poor work performance or complete failure. Secondly, in many organizations, work systems are built upon some specific norms, which are known only by the organizational members, but are ambiguous to a visitor. It is not possible to create an agent-based system that contains and covers the entire existing norms of the world, let alone to predict the emergence of new norms in the future. To solve these problems, we initiate a detailed study in norm detection and assimilation to create an intelligent agent system that can autonomously detect the norms in any domain or environment and subsequently assimilate with the norms.

## 2    Related Work

A review of the literature revealed that norms and normative systems have received much attention in recent years due to their capability to coordinate agents' interactions [5], [8]. Generally norms are incorporated in an agent system to guide an agent to behave in a socially congruent manner in a multi-agent environment. An agent is a normative agent when its behavior is determined by the obligations it must comply with, prohibitions that limit the kind of goals it can pursue, social commitments that have been created during its social life and social codes that may not carry punishment, but whose fulfillment could represent social satisfaction for the agent [8]. The normative agent is governed by the norms and is able to adopt or deliberate the norms on its own decision or violate a norm in case of conflicts with other norms or other goals [5].

While a multitude of research work addresses norms regulation in normative frameworks, very few have been found to focus on norms detection and identification, in which an agent attempts to decipher the normative protocol of a group of local agents for eventual assimilation with the local agent society. However, similar works in norms identification have been reported in [2], [4], [10], [11].

Savarimuthu et al. [10], [11] propose a norm identification architecture in which an agent infers the norms of a society without the norms being explicitly given to the agent. Their technique exploits an association rule mining approach to identify the tipping norms in an agent-based simulation of a virtual restaurant. An agent in the system is able to add, remove, and modify norms dynamically and modify the parameters of the system based on whether it is successful in identifying a norm.

The EMIL Project [2] attempts to understand and develop design strategies in coping with particular types of complex entities, i.e. social systems. These strategies are characterized by a two-way dynamics, consisting of emergent and immergent processes: emergence from interaction among individual agents, and immergence of entities (norms) at the aggregate level into the agents' minds.

Campos et al. [4] uses a norm adaptation mechanism based on social power, which consists of providing support to the coordination of agents. They use generic level assisted architecture to develop systems that self-adapt their organization depending on their evolution. They model peer-to-peer sharing network case study as a multi-agent system with two levels of organization. Both levels share the same goal, that all participant agents obtain the data by using minimum time. They add an assistance layer to first level, model the set of computers that share some data as agents within a domain-level. In order to support the coordination of agents, this support consists of adapting domain-level's organization to changing circumstances. They use a learning technique to decide how to adapt domain-level norms depending on current system status. They apply a CBR learning approach, which is based on a heuristic that tries to align the amount of serving/receiving capacity, and this heuristic itself is used by the CBR to suggest a solution when no similar cases are found.

# 3    A Framework for Norms Detection and Assimilation

In this framework, we introduce the concept of norms model matching, in which a visitor agent attempts to detect the norms (1) (see Figure 1) of a group of host agents by analyzing the agents' interaction pattern and matching that pattern with those that are available in its norms model base (2). The model base consists of a repository of norms models of all existing normative frameworks. If there is a match, the visitor agent launches an assimilation request algorithm (3). If the request is granted (4), the visitor agent is allowed to assimilate with the local agents' norms (5) sharing resources and achieving its own goal. If no match is found in the norms model base, either the visitor agent is unable to access the host's services or non-normative access is assumed. In the latter case, the host system is not considered to have a normative protocol installed.

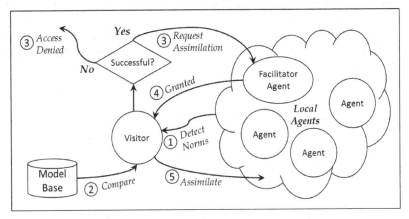

**Fig. 1.** A Conceptual Framework of Norms Detection and Assimilation

## 3.1 Agent Registration and Trust Verification

Figure 2 shows a process for agent registration and trust verification prior to its permission to access the required resources for norms detection.

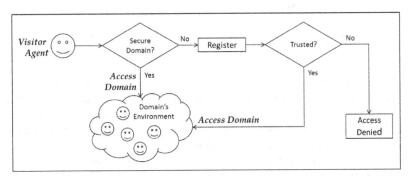

**Fig. 2.** Agent Registration and Trust Verification

When the visitor agent needs to access a specific domain in a host system, it needs to check if this domain is secured. If the domain is not secured then the agent can access the domain to initiate the norms detection and assimilation process. Otherwise it needs to provide enough information to build its trust with the host system. If the agent is trusted, it would be granted the permission to access limited resources from the host system which it needs to detect the norms and assimilate with the local agents.

## 3.2    Norms Detection

When the visitor agent has been granted permission to access the domain, it can then initiate the process of norms detection. In Savarimuthu et al. [10], [11], the agent detects the local norms based on observation of interactions between agents and also recognizing signalling actions of sanctions and rewards, i.e. it is assumed that the agent has access to such interactions between agents and signalling actions. In our technique, the agent has the option to request permission to access the records of interactions between agents (or event history). Figure 3 details out the technique of norms detection.

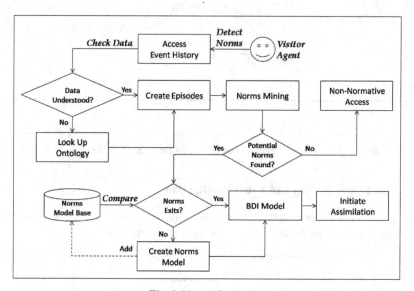

**Fig. 3.** Norms Detection

The visitor agent access the event history log, which contains historical records of interactions between the agents. Being a visitor, it may not be able to understand the meaning of data in the history log. To resolve this problem, the agent must request and gain access to the local ontology as another resource to interpret the meaning of data in the records. Having successfully done so, it then creates a number of event episodes that could represent potential norms. With these episodes, it uses a data mining technique to derive a potential pattern of interaction i.e., the norms model. If a potential norm is not found, the non-normative access is assumed, in which case the host system is not considered to have a normative system installed. But if a potential norms is found, the agent compares it with a set of pre-compiled norms in the norms model base. If the norms models match, then the agent's BDI logic initiates the assimilation process. However, if a matching model is not found, the agent creates a new norms model and adds it to the norms model base.

## 3.3    Norms Model Base and Norms Mining

The Norms Model Base is a compilation of all normative models of agent-based systems that have been developed by researchers in norms. Currently, we have identified a few norms model such as BOID [3], OP-RND [9], and ONI [10]. The model will be analysed to develop a logical model of the norms in prolog clauses. Using these models, the agent would be able to compare the detected norms with the norms in the model base.

In norms mining, the history log which could be of any data structure is read by the agent. All repeated events are filtered leaving unique events that would contribute to a potential norm model. These are then formatted to a specific data structure that matches with the structure of the norms from the model base. Figure 4 highlights the process of norms mining and matching.

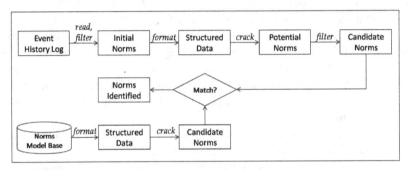

**Fig. 4.** Norms Mining and Matching

In Savarimuthu et al. [8] an event is represented by a predicate with three or four arguments to indicate a series of actions that happen, e.g. *happen(1, arrive, A)*, *happen(2, order, A, W)*, *happen(3, eat, A)*, etc. represent an episode of an agent A arriving at a restaurant, ordering a meal from a waiter, and eating the meal. In our system [1], an event is represented by a record of message exchanged data of the form *(Date, Time, Agent1, Message, Agent2)* or action performed such as *(Date, Time, Agent1, Action, Agent2, Merit/Demerit)*. A sequence of message exchanges and actions could represent a complete episode from which a model of norms could be detected. From such data structure, a norms model is developed by using prolog facts such as *event/5* and *event/6*. For example, the data *(Date, Time, Agent1, Message, Agent2)* is represented as:

event('Wed 01 Jul 2009', '10:00:26', 'Committee', prepare, lecturer).

Similarly, *(Date, Time, Agent1, Action, Agent2, Merit/Demerit Point)* is represented as:

event('Mon 06 Jul 2009', '09:01:46', lecturer, review, moderator, 2).

However, before the agent could use the data structure, the arguments would need to be understood by the agent using some local ontology.

### 3.4   Agent Assimilation

To assimilate with the local agents, the visitor agent sends a request for assimilation message to the facilitator agent of the host system. The agent's request is an indication of its agreement to comply with the norms of the society and the sanctions imposed if it disregards or violates the norms. When the request is granted, only then the visitor agent is allowed to assimilate with the group. However, due to space limitation, we shall not deliberate this process here.

## 4   Conclusions and Further Work

Norms detection could offer an alternative technique in low-level security for normative systems. The development of numerous software agent capabilities, such as agent cooperation, negotiation, trust, and mobility demands additional approach to agents' verification and authentication. At the society level, norms could provide another line of defence by constraining visiting agents to comply with the society's norms, failing which the agent is refused services. Savarimuthu et al. [10] argue that identifying and knowing the norms would help an agent to maintain its utility as the agent could apply the norms when situations warrant it, or subject to sanctions for not complying with the norms.

Specific research issues identified to manifest a comprehensive architecture for norms detection and assimilation includes agent authentication and trust, ontology sharing, norms mining, norms model structure, and norms model matching issues. These will be dealt with in our future work.

## References

1. Ahmed, M., Ahmad, M.S., Yusoff, M.Z.M.: Modeling Agent-Based Collaborative Process. In: Pan, J.-S., Chen, S.-M., Nguyen, N.T. (eds.) ICCCI 2010. LNCS, vol. 6421, pp. 296–305. Springer, Heidelberg (2010)
2. Andrighetto, G., Conte, R., Turrini, P., Paolucci, M.: Emergence in the loop: Simulating the two way dynamics of norm innovation. In: Boella, G., van der Torre, L., Verhagen, H. (eds.) Normative Multi-agent Systems. Dagstuhl Seminar Proceedings, vol. 07122. Internationales Begegnungs-und Forschungszentrum für Informatik (IBFI), Schloss Dagstuhl, Germany (2007)
3. Broersen, J., Dastani, M., Hulstijn, J., Huang, Z., van der Torre, L.: The BOID Architecture: Conflicts Between Beliefs, Obligations, Intentions And Desires. In: AGENT 2001. ACM, Canada (2001)
4. Campos, J., López-Sánchez, M., Esteva, M.: A Case-Based Reasoning Approach for Norm Adaptation. In: Corchado, E., Graña Romay, M., Manhaes Savio, A. (eds.) HAIS 2010. LNCS, vol. 6077, pp. 168–176. Springer, Heidelberg (2010)
5. Castelfranchi, C., Dignum, F., Jonker, M.C., Treur, J.: Deliberative Normative Agents: Principles and Architecture (1999)

6. Chopinaud, C., Seghrouchni, A.E.F., Taillibert, P.: Prevention of harmful behaviors within cognitive and autonomous agents. Paper Presented at the Proceeding of the 2006 Conference on ECAI 2006: 17th European Conference on Artificial Intelligence, Riva del Garda, Italy, August 29 – September 1 (2006)

7. Governatori, G., Rotolo, A.: How do agents comply with norms? In: IEEE/WIC/ACM International Joint Conferences on Web Intelligence (WI) and Intelligent Agent Technology (IAT), vol. 3, pp. 488–491 (2009)

8. López, F., Marquez, A.: An Architecture for Autonomous Normative Agents. In: Proceedings of the Fifth Mexican International Conference in Computer Science, ENC 2004 (2004)

9. Mohd Yusof, M.Z., Ahmed, M., Ahmad, A., Ahmad, M.S., Mustapha, A.: Sacrifice and Diligence in Normative Agent Environments. In: International Conference on Computer and Software Modeling (ICCSM 2010), Manila, Philippines, December 4-5, Indexed by EI (Compendex) and ISI Proceeding, ISTP (2010)

10. Savarimuthu, B.T.R., Cranefield, S., Purvis, M., Purvis, M.: Obligation Norm Identification in Agent Societies. Journal of Artificial Societies and Social Simulation 13(4), 3 (2010)

11. Savarimuthu, B.T.R., Cranefield, S., Purvis, M., Purvis, M.: Internal agent architecture for norm identification. In: Proceeding of the International Workshop on Coordination, Organization, Institutions and Norms in Agent Systems (COIN@AAMAS), pp. 156-172 (2009)

# Model Based Human Pose Estimation in MultiCamera Using Weighted Particle Filters

Mohammad Behrouzifar, Hamidreza Shayegh Boroujeni,
Nasrollah Moghadam Charkari, and Kourosh Mozafari

ippp Lab, Electrical and Computer Engineering Department,
Tarbiat Modares University, Tehran, Iran
behroozi2010@gmail.com,
{shayegh,moghadam,kourosh.mozafari}@modares.ac.ir

**Abstract.** Human motion capture and human pose estimation is one of the most active research area in computer vision. Pose estimation refers to the process of estimating the configuration of the kinematic or skeletal articulation structure of a person. Recent research area refers to multi-camera 3D pose estimation that is very useful and more robust to self-occluding and ambiguous. These approaches use fusion of features that extracted from each camera in different views. In this paper we proposed a novel fusion algorithm based on multiple particle filters in decision level fusion. We use particle filter to tracking the body parts on each camera. Then we use decision fusion by weighting estimation results from each camera. Experimental results in HumanEva dataset and comparison of our work to one of the recent works in this area show the better accuracy especially in some difficult situations.

**Keywords:** Human motion capture, Human pose estimation, Human 3D tracking, Particle filter, Kalman filter, Multicamera fusion, 3D pose estimation.

## 1 Introduction

Human motion capture and analysis of human motion is one of the most active research areas in computer vision due both to the number of potential application and its inherent complexity. It can be defined as the ability to estimate, at each frame of a video sequence, the position of each joint of a human figure, which is represented by an articulated model. Also pose estimation is the most important task in this field. Pose estimation refers to the process of estimating the configuration of the kinematic or skeletal articulation structure of a person. This process may be an integral part of the tracking process as in model-based analysis-by-synthesis approaches or may be performed directly from observations on a per-frame basis [1].

Applications can roughly be grouped under five titles: Advanced user interface, Model-based encoding, Motion Analysis, Smart surveillance systems and Virtual reality. *Advanced user interfaces* are very important applications in HCI such as social interfaces, sign language interpretation and gesture driven application

D. Lukose, A.R. Ahmad, and A. Suliman (Eds.): KTW 2011, CCIS 295, pp. 234–243, 2012.

interfaces. *Model-based encoding* where we use for low bit-rate video compression and low variable animation control. *Motion Analysis* or Analysis applications such as automatic diagnostics of orthopedic patients [2,3] or analysis and optimization of an athletes' performances. Newer applications are annotation of video as well as content-based retrieval [4] and compression of video for compact data storage or efficient data transmission, e.g., for video conferences and indexing. *Smart surveillance systems* cover some of the more classical types of problems related to automatically monitoring and understanding locations where a large number of people pass through such as airports and subways [5]. Applications could for example be: people counting or crowd flux, flow, and congestion analysis. Newer types of surveillance applications perhaps inspired by the increased awareness of security issue are analysis of actions, activities, and behaviors both for crowds and individuals for example in queue and shopping behavior analysis, detection of unusual activities, and human identification. Another branch of applications is within the car industry where much vision research is currently going on in applications such as automatic control of airbags, sleeping detection, pedestrian detection, lane following, etc[1].

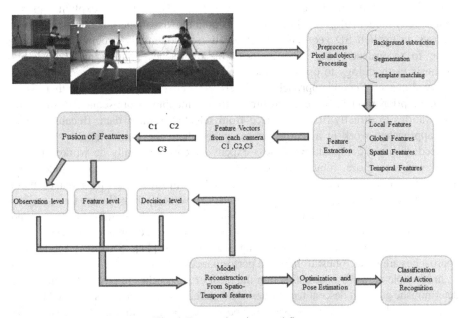

**Fig. 1.** Pose estimation workflow

The research area contains a number of hard and often ill-posed problems such as inferring the pose and motion of a highly articulated and self-occluding non-rigid 3D object from observed images. This complexity makes the research area challenging from a purely academic point of view. Recent algorithms for resolving self-occluding and ambiguous in pose estimation, use multi-camera approaches that very useful for robust pose estimations. These approaches use fusion of features that extracted from each camera in different view. Fusion process can be applied in three approaches:

1) Fusion in observation level
2) Fusion in feature level
3) Fusion in decision level

In the first approaches we compound observations from each camera and we make one observation from mixture of observations.

In fusion in the feature level we compound feature vectors that extracted from each camera. Then we produce one feature vector from mixture of vectors.

In this paper we proposed a novel fusion algorithm based on multiple particle filter in decision level approaches algorithms, we concentrate our effort on tracking the body parts of a human. Our results are evaluated against the Human Eva (HE) data set, which is becoming the standard for assessing human pose estimation based on multi camera algorithms.

## 2     Related Work

Approaches to vision-based human motion capture can widely be divided into two categories: 1) Direct model use and 2) Model free. The first class explicitly uses a human body model [6], [7], [8] that describes both the visual and kinematic properties by uses an explicit 3D geometric representation of kinematic structure and human shape to reconstruct pose. The majority of approaches employ an analysis-by-synthesis methodology to optimize the similarity between the model projection and observed images. Second approaches [9], [10] class covers methods where there is no explicit a priori model. These algorithms learn the mapping from sequences of image observations to directly 3D pose space from selected training data. Because model free approaches work in a learned pose space where the dimensionality has been reduced, they are computationally less expensive, can potentially be worked in real time, and are more robust to occlusions or noises. Furthermore, model free approaches allow the recovery of poses with less information, which makes them more suitable for monocular applications. However, they have two serious drawbacks: their accuracy relies on the similarity between the gesture to recover and the data used in the training data set. And their performance tends to decrease when the variety of poses used in a training set increases [11]. So direct model use approaches are more robust to increase performance.

In Multiple view 3D pose estimation the most of approaches to pose estimation used deterministic gradient descent techniques to iteratively estimate changes in pose [12].

The extended Kalman filter was applied to tracking and human pose estimation with low-order dynamics used to estimate a person pose. Recent work using model-based analysis by-synthesis has extended deterministic gradient descent based approach to more complex actions. For example Plankers and Fua [13] applied upper body tracking of arm movements with self-occlusion using stereo and silhouette cues. A common limitation of gradient descent approaches is the use of a single pose or state estimate which is updated at each time step. In practice if there is a fast movement or visual ambiguities pose estimation may fail. To achieve more robust pose estimation, techniques which employ a deterministic or stochastic search of the pose state space have been investigated. Stochastic tracking techniques, such as the particle filter, that we used in our method were introduced for robust visual tracking of objects where sudden changes in movement or cluttered scenes can result in failure. The principal difficulty with their application to human pose estimation is the dimensionality of the state space. The number of samples or particles required increases exponentially with dimensionality. In our method we reduce number of particles to each body part region. Typically whole-body human models use more than 20 degrees-of-freedom making direct application of particle filters computationally prohibitive. Mac Cormick and Isard [14] proposed partitioned sampling of the state space for efficient 2D pose estimation of articulated objects such as the hand. However, this approach does not extend directly to the dimensionality required for whole-body pose estimation. Deutscher et al. [15] introduced the annealed particle filter which combines a deterministic annealing approach with stochastic sampling to reduce the number of samples required. At each time step the particle set is refined through a series of annealing cycles with decreasing temperature to approximate the local maxima in the fitness function. Results [15, 16] demonstrate reconstruction of complex motion such as a hand-stand. A hierarchal stochastic sampling scheme to efficiently estimate the 3D pose for complex movements or multiple people is presented in [17]. This approach estimates the torso pose for each person and propagates samples with high fitness to estimate the pose of adjacent body parts. Recent work has combined deterministic or stochastic search with gradient descent for local pose refinement to recover complex whole-body motion. Carranza et al. [18] applied whole-body human motion estimation from multiple views combining a deterministic grid search with gradient descent. Pose estimation is performed hierarchically starting with the torso. For each body part a grid search first finds the set of valid poses for which the joint positions project inside the observed silhouettes. A fitness function is then evaluated for all valid poses to determine the best pose estimate. Finally gradient descent optimization is applied to refine the pose that estimated. This search process is made feasible by the use of graphics hardware to evaluate the fitness function which is based on the overlap between the projected model and observed silhouette across all views. In related work Kehl et al. [19] propose stochastic Meta descent for whole-body pose estimation with 24 degrees-of-freedom from multiple views. Stochastic Meta descent combines a stochastic sampling of the set of model points used at each iteration of a gradient descent algorithm. This introduces a stochastic search element to the optimization which allows the approach to avoid convergence to local minima. In summary, the introduction of stochastic sampling and search approaches has achieved whole-body pose estimation of complex motions from multiple views. Current approaches are limited to gross-body pose estimation of torso, arms, and legs and do not capture detailed movement such as hand-orientation or axial arm rotation. Multiple hypothesis sampling achieves robust tracking but does not provide a single temporally consistent motion estimate resulting in jitter which must be smoothed to obtain visually acceptable results.

## 3    Proposed Framework

First we use Kalman filter for position tracking of human body and global location of the person [20] then we find the body parts of human and relative pose of the limbs then we track them using a set of particle filters from each camera. This work flow can be applied in each camera in parallel .So we have estimation results of particle filters from each camera. Then we use decision level weighting strategy for estimation results of particle filters of each camera. The flow diagram of proposed method has been shown in figure 2.

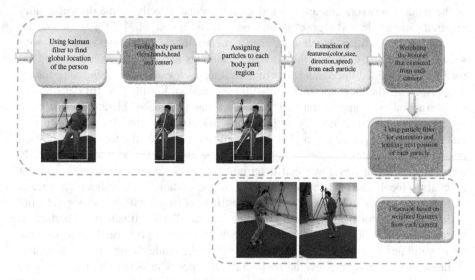

**Fig. 2.** Proposed framework

### 3.1    Position Extraction Based on Kalman Filter

Using a Kalman filter, we track the bounding box of the person under observation. The state vector $X_{Whole} = \{ X_1, X_2, X_3, X_4, X_5, X_6 \}$ where  is the global location in the (x, y)-coordinates. The likelihood function is based on a motion detector that extracts the blob that corresponds to the subject. Details completely reported in [20].

### 3.2    Finding Body Parts and Assigning to Each Particle

After finding location of the person we find relatively center of each body part. First we find center of global location and we calculate the distance of each contour to center, then we marked the local maxima of distances as a convex point. Then we assign each convex point to one body part.

### 3.3    Tracking Body Parts Using Particle Filter

Particle filters are usually used to estimate Bayesian models in which the latent variables are connected in a Markov chain similar to a hidden Markov model (HMM), but typically where the state space of the latent variables is continuous rather than discrete, and not sufficiently restricted to make exact inference tractable [22].

For tracking each body part and estimating the motion of each part, we use a particle filter method with one particle for each body part. For each view, recorded by each camera, we use independent particle filter. The probabilistic results of them, fused in a weighting method. Particle filter estimate the probability of each candidate observation based on similarity measurement of features, and one of the observations is selected as the tracked body part. Bhattacharyya coefficient is used for similarity measurement:

$$p(z_i^t \mid x_{t-1}) = \prod_{\forall r \in R(x_t)} \frac{1}{N} (\sum_{i=1}^{N} p^e (I_t, i)^{\alpha_e} \tag{1}$$

Where $I_t$ is the original image in RGB, $r$ represents each of the Regions that compose the articulated model, N are all the pixels that compose the region, and $\alpha e >$ 0 is an empirical factor. For each observation $z_i$ in the time t, the similarity is calculated and the one of the observations with the most probability, are selected based on the particle filter decision:

$$q = p(z_t \mid E[x_t]).f_{rel}(t) \tag{2}$$

Where $p(z_t \mid E[x_t])$ are, the measurement values (based on color and edge likelihoods) of the, $h$ is the length of the step (in frames), $xt$ is the state vectors of the tracking, and $f_{rel}$ is the decreasing function that takes into account the reduction of reliability over time:

$$f_{rel}(t) = e^{-\beta(1-t)} \tag{3}$$

The state vector of each body part is described by the following equation

$$X_{bodypart} = [x_{hip}, y_{hip}, \dot{x}_{hip}, \dot{y}_{hip}, \Theta_{thigh}, \Theta_{bodypart}, \dot{\Theta}_{thigh}, \dot{\Theta}_{bodypart}, l_{thigh}, l_{bodypart}, \dot{l}_{thigh}, \dot{l}_{bodypart}] \tag{4}$$

Where $x$ and $y$ are the coordinates in pixels, $\theta$ is the angle between a limb and the $x$ axis and $l$ is the length of the limb.

### 3.4    Fusion of the Tracking Results of Each Camera

After tracking body parts using particle filter for each camera recorded sequence independently, a weighting combination is done on the results of each body part.

Before starting the pose estimation process, a learning phase is done based on HMM that learns the changing sequence of body parts. For example moving the right hand in the right direction based on last motion sequence might be more probable than a left direction moving. Thus if the particle filter for a camera detect a right direction change, this estimation is a high weight one and if other particle filter for another camera with different view, estimate a down direction in the motion, the estimation weight will be low. The HMM learns in the training phase by consideration of body parts as the states and are weighted based on the frequency of occurrences.

Then all of estimations with different weighting are fused and the final estimation decision is gained.

The fusion idea that is used for motion estimation shows in figure 3.

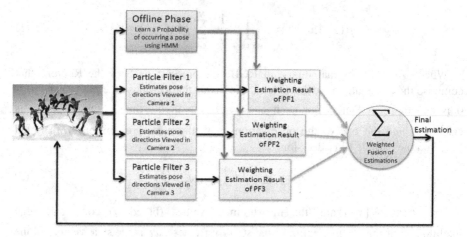

**Fig. 3.** The fusion idea for combination of particle filters results based on a HMM offline phase

The fusion idea is based on the change direction comparison:

$$E = vector - sum(w_1[\Delta x, \Delta y], w_2[\Delta x, \Delta y], w_3[\Delta x, \Delta y]) \qquad (5)$$

Where $w_i = P(z_t \mid x_{t-1}, x_{t-2}, \dots)$

## 4    Experiments

The proposed method has been tested on well-known Human Eva Data set [21] and has been compared with one state of the arts methods in this scope [20]. four different action sequences, two from human Eva 1 and two from human Eva 2 have been tested: *S2 Walking 1 , S2 Combo 1* from HE I and *S2 Combo 1, S2 walking 2* from HE. For testing we use two metrics: accuracy rate of estimation of body parts, and error rate of estimation of them. The first measurement has been calculated in frame sequences by step 5:

$$Accuracy\,rate = \cfrac{\displaystyle\sum_{\forall frame+5} \cfrac{\displaystyle\sum_{\forall body\_part} (number\_of\_matched\_pixels)/(number\_of\_all\_pixels)}{number\_of\_body\_parts}}{(number\_of\_frames)/5} \tag{6}$$

The error measurement is calculated based on unmatched and false-matched pixels in the selected pixels for each body part:

$Error\_rate\_of\_body\_part\_det\,ecton =$

$$\sum_{\forall frame+5} \frac{(number\_of\_unmatched\_pixels)+(number\_of\_false\_matched\_pixels)}{number\_of\_really\_true\_pixels+number\_of\_really\_false\_pixels} \tag{7}$$

The results show that proposed method is better than the last method presented by Martinez et al [20] only a few. But in some situations it works much better. If the human motions lead to hiding some parts in one or two camera views but these parts have been shown in other cameras, our method, can estimates the body part more accurate than the Martinez method. One of such motions shows in HumanEva that the left hand of human can't be observable in c1 and c2 but it is in c3.

Figure 4 shows the accuracy results in frame sequences *S2 Walking 1 , S2 Combo 1 S2 Combo 1, S2 walking 2 respectively,  about 40 frame from each sequence* for two methods: proposed and Martinez methods.

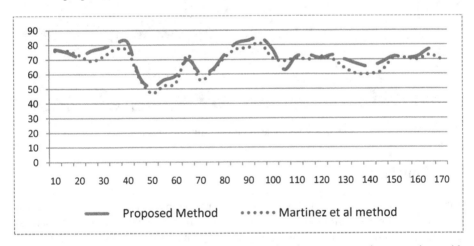

**Fig. 4.** The accuracy rate of pose estimation in the 4 action sequences in comparison with Martinez method [20].

In general the accuracy of martinez method is about 70% and the accuracy of proposed method is about 71.5%. but in the frames with body part hidings in two camera, the accuracy rates are 65% and 70% respectively. In the other hand we can say that our proposed method is more robust than another in error rate such as self-occlusion situations and it is more accurate than using single camera.

Figure 5 shows the error rate of frames. This chart confirms such conclusion.

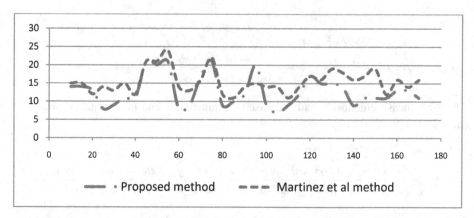

**Fig. 5.** The error rate of pose estimation for 4 action sequences in comprison with Martinez method

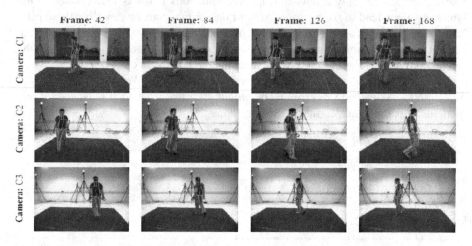

**Fig. 6.** Shows some of output estimation results

## 5    Conclusion

In this paper we proposed a method for pose estimation in video sequences based on a novel weighted particle filter approach that cause to better results in error rate such as self-occlusion situations and it is more accurate than using single camera. In this method, the result of each particle filter in each camera view, weights with a probability of observation based on HMM that is learned in an offline phase for all of sequences. We test our method on some of Human Eva sequences. A big problem in this method is that the human body part detection phase can't detect the body parts accurately and thus the estimation phase that is affected from this phase can't be accurate more. The manually detection of body parts makes better results but they are unreal results and they aren't suitable for applications. In the future we work on the body part detection.

# References

1. Moeslund, T.B., Hilton, A., Krüger, V.: A survey of advances in vision-based human motion capture and analysis. Computer Vision and Image Understanding 104, 90–126 (2006)
2. Zhou, H., Hu, H.: Human motion tracking for rehabilitation—A survey. Biomedical Signal Processing and Control 3, 1–18 (2008)
3. Popata, H., Richmonda, S., Benediktb, L., Marshallb, D., Rosinb, P.L.: Quantitative analysis of facial movement A review of three-dimensional imaging techniques. Computerized Medical Imaging and Graphics 33, 377–383 (2009)
4. Pradhan, G.N., Prabhakaran, B.: Indexing 3-D Human Motion Repositories for Content-Based Retrieval. IEEE Transactions On Information Technolog. Biomedicine 13(5) (September 2009)
5. Yao, J., Odobez, J.-M.: Multi-Camera 3D Person Tracking With Particle Filter In A Surveillance Environment. In: 16th European Signal Processing Conference, EUSIPCO (2008)
6. O'Rourke, J., Badler, N.I.: Model-based image analysis of humanmotion using constraint propagation. IEEE Trans. Pattern Anal. Mach. Intell. PAMI-2(6), 522–536 (1980)
7. Kakadiaris, I.A., Metaxas, D.: Model-based estimation of 3-D human motion with occlusion based on active multiviewpoint selection. In: Proc. IEEE Conf. Comput. Vis. Pattern Recog., pp. 81–87 (1996)
8. Li, B., Meng, Q., Holstein, H.: Articulated pose identification withsparse point features. IEEE Trans. Syst., Man, Cybern. B: Cybern. 34(3), 1412–1422 (2004)
9. Hou, S.B., Galata, A., Caillette, F., Thacker, N., Bromiley, P.: Real-time body tracking using a Gaussian process latent variable model. In: Proc. ICCV, pp. 1–8 (2007)
10. Lee, C.S., Elgammal, A.: Body pose tracking from uncalibrated camera using supervised manifold learning. In: Proc. NIPS Workshop Eval. Articulated Human Motion Pose Estimation(EHuM), Whistler, BC, Canada (2006)
11. Darby, J., Li, B., Costen, N., Fleet, D., Lawrence, N.: Backing off: Hierarchical decomposition of activity for 3-D novel pose recovery. In: Proc. BMVC, London, U.K. (2009)
12. Plänkers, R., Fua, P.: Articulated soft objects for multiview shape and motion capture. IEEE Transactions on Pattern Analysis and Machine Intelligence 25(9), 1182–1187 (2003)
13. Plänkers, R., Fua, P.: Articulated soft objects for multiview shape and motion capture. IEEE Transactions on Pattern Analysis and Machine Intelligence 25(9), 1182–1187 (2003)
14. MacCormick, J., Isard, M.: Partitioned Sampling, Articulated Objects, and Interface-Quality Hand Tracking. In: Vernon, D. (ed.) ECCV 2000. LNCS, vol. 1843, pp. 3–19. Springer, Heidelberg (2000)
15. Deutscher, J., Blake, A., Reid, I.: Articulated body motion capture by annealed particle filtering. In: Computer Vision and Pattern Recognition, Hilton Head Island, South Carolina, June 13–15 (2000)
16. Davison, A.J., Deutscher, J., Reid, I.D.: Markerless motion capture of complex full-body movement for character animation. In: Eurographics Workshop on Computer Animation and Simulation, Manchester, UK (September 2001)
17. Mitchelson, J., Hilton, A.: Hierarchical tracking of multiple people. In: British Machine Vision Conference, Norwich, UK (September 2003)
18. Carranza, J., Theobalt, C., Magnor, M., Seidel, H.-.P.: Free-viewpoint video of human actors. In: ACM SIGGRAPH, pp. 565—577 (2003)
19. Kehl, R., Bray, M., VanGool, L.: Full body tracking from multiple views using stochastic sampling. In: Computer Vision and Pattern Recognition, San Diego, California, USA, June 20-25 (2005)
20. Martínez del Rincón, J., Makris, D., Uruñuela, C., Nebel, J.: Tracking Human Position and Lower Body Parts Using Kalman and Particle Filters Constrained by Human Biomechanics. IEEE Transactions on Systems (2010)
21. Human Eva Data Set, http://vision.cs.brown.edu/humaneva/
22. Ristic, B., Arulampalam, S., Gordon, N.: Beyond the Kalman Filter: Particle Filters for Tracking Applications. Artech House (2004)

# Route Guidance System Based on Self-Adaptive Algorithm

Mortaza Zolfpour-Arokhlo, Ali Selamat,
Siti Zaiton Mohd Hashim, and Md Hafiz Selamat

Faculty of Computer Science & Information Systems,
Universiti Teknologi Malaysia, 81310 UTM Skudai, Johor, Malaysia
zolfpour@gmail.com, {aselamat,sitizaiton,mhafiz}@utm.my

**Abstract.** Self-adaptive systems are applied in a variety of ways, including transportation, telecommunications, etc. The main challenge in route guidance system is to direct vehicles to their destination in a dynamic traffic situation, with the aim of reducing the motoring time and to ensure an efficient use of available road resources. In this paper, we propose a self-adaptive algorithm for managing the shortest paths in route guidance system. This is poised to minimize costs between the origin and destination nodes. The proposed algorithm was compared with the Dijkstra algorithm in order to find the best and shortest paths using a sample simplified real sample of Kuala-Lumpur (KL) road network map. Four cases were tested to verify the efficiency of our approach through simulation using the proposed algorithm. The results show that the proposed algorithm could reduce the cost of vehicle routing and associated problems.

**Keywords:** Traffic control, route guidance system (RGS), self-adaptive, shortest path problem (SPP), Dijkstra's algorithm, Urban road network(URN).

## 1 Introduction

Due to the variety of road network and environment the corresponding route guidance systems are becoming more complex and present many new characteristics such as traffic congestion, time to travel, etc. Therefore, the vehicle driver needs to use a route guidance system in order to reduce the traffic delays. Nowadays, there are various successful algorithms for finding the shortest path in the route guidance system. One of the most popular of these types of algorithms is Dijkstra algorithm. The vehicle drivers with different available solutions need a quick response to the change in road network conditions. The main challenge faced by the route guidance system is directing the vehicles to their destination in the dynamic traffic situation, with the aim of reducing travel times and efficient use of available network capacity. Therefore, the vehicle driver needs to use a route guidance system based on self-adaptive algorithms. Any Change in the environment can be perceived by the self-adaptive algorithm because the self-adaptive system is a complex system as it has the ability to change structure and behavior itself [1], [2]. Generally, to solve the problems, using a fast path planning method seems to be an effective approach to improve the route guidance system. The main thrust of this is to compare the

D. Lukose, A.R. Ahmad, and A. Suliman (Eds.): KTW 2011, CCIS 295, pp. 244–253, 2012.
© Springer-Verlag Berlin Heidelberg 2012

shortest path algorithms with Dijkstra algorithm. The shortest path (i.e. lowest cost) is calculated using the proposed algorithm, and it will be suggested to vehicle drivers in each intersection (or node). Dijkstra algorithm is a search graph for routing problems and producing a graph of the shortest path in a graph [3]. This algorithm finds the shortest path (i.e. lowest cost) between origin and destination nodes in a directed graph. For example, Dijkstra algorithm can find the shortest path (or lowest cost) between one points of the city to another point of city in a trip [4]. The shortest path problem has been investigated extensively, and many modern algorithms have been proposed to solve this problem. Yet, many of the existing algorithms fail to meet some of the requirements of the solutions applicable in the real world. Therefore, this paper proposes a method to solve the path planning problem in route guidance systems in terms of accuracy and speed. The results obtained from the approach have been compared with Dijkstra algorithm [3]. The rest of the paper is organized as follows. In Section 2, we start with an overview of path planning problem, and some basic concepts related to the study. In Section 3, we propose a self-adaptive algorithm for route guidance system, a formal problem formulation and dynamic shortest path algorithm in route guidance system. In section 4, experimental comparison and simulation, results are analyzed with some samples. The conclusion and some future works are discussed in the last section.

## 2    Related Works

In this section, we present the essential background for our contribution. A brief review of related studies is given with particular dynamic shortest path algorithms focused on a route guidance system. A new approach in the urban road network, which can reserve time and space for cars at an intersection or junction has been studied in [6]. A shortest path problem is used on finding the path with minimum travel distance, time or cost from one or more origins to the one or more goals or destinations through a connected network [4]. It is an important issue because of its wide range of applications in transportation, robotics and computer networks. The shortest path from one node to every other node can be found by using a number of various algorithms [5]. Dijkstra algorithm is probably the best-known and most important, popular, and successfully implemented shortest path algorithms for both lecturers and students in computer science [7]. Therefore, road traffic control flow through the use of the self-adaptive algorithm for route guidance system in dynamic transportation is one of the reasons for this paper. However, most route guidance system based applications are concentrated on modeling and simulation. Hence, new efforts and researchers should be done for real-world applications. In this research, we study a route guidance system framework for solving urban road traffic problems.

### 2.1    Review of the Self-Adaptive System

Over the past decade, numerous frameworks have been done for the successful implementation of self-adaptive systems [8]. Several systems properties can be controlled using self-adaptive systems [9] like self-healing system, which can automatically diagnose, correct faults, and discover. In the alternative, a self-optimizing system can automatically monitor and adapt resource usage to ensure optimal functioning relative to defined requirements [10].

## 2.2    Review of the Dijkstra algorithm

Dijkstra algorithm is widely used and is a successful algorithm for routing problems to find the shortest path in a weighted directed graph network graph [3]. It is often used in routing. As a result, the shortest path first is widely used in network routing protocols.

# 3    Proposed Route Guidance System

The proposed algorithm can also be used for shortest path and finding costs in the urban road network. A route guidance system (RGS) will calculate the shortest path from current vehicle position to destination locations, so that the traveling cost or time for drivers will be minimized. We have used shortest path computation (SPC) to calculate the nearest distance of the destinations with the support of vehicles in the route guidance system. The updated information of the vehicle will be used by the system to find the shortest path from current vehicle position to destination zone for drivers based on current conditions. However, one of the requirements of intelligent transportation system dynamic is real and current information about travel time for a continuous path that can be acquired by several detectors such as magnet sensors, video cameras, GPS, GSM (global system for mobile communication) and also other network traffic sensors on the routes to transport [12]. We have applied the roles of path planning algorithm in route guidance system as follows:

- Transferring information acquired through the sensors;
- Receiving a route request from vehicles;
- The shortest path computation and sending it to vehicle's drivers;

## 3.1    Problem Formulation and Properties

A road directed graph presented by G=(V,A) is a directed dynamic route guidance system based on an electronic map, a set of '$N$' nodes (V) and '$M$' directed edges (A). Each $R_{(s,d)}$ edge is a nonnegative number which stands for the cost while '$s$' is a start node and '$d$' is a finish node connected to the $R_{(s,d)}$. Consider a directed graph, G, which is composed of a set of '$s$' nodes and a set of '$r$' directed edges. Set a number of edges as cost(C) table. Furthermore, G file is filled in data file. If  S= { $s_1$, $s_2$, ....., $s_n$ },

D= { $d_1$, $d_2$, ....., $d_n$ } then $R_{(s,d)}$ = { $R_{(s_1,d_1)}$, $R_{(s_2,d_2)}$, ....., $R_{(s_n,d_n)}$ }

$R_{(s,d)}$ consists of the sum of all edge distances (costs) participating in the network path. Therefore, according to the trip origin (s) node and destination (d) node, this issue can be solved as the optimization problem based on real transportation network that is defined as follows:

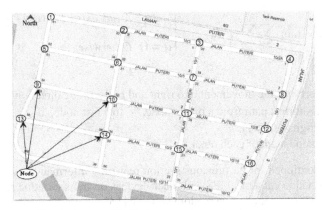

**Fig. 1.** A part of Kuala Lumpur city road network map

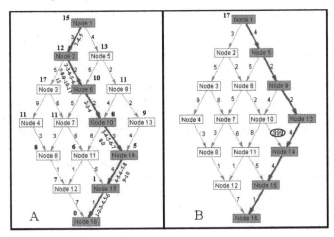

**Fig. 2.** G (16,24) is the optimal path in ideal (A) and blocked(B) routes status in Case 1, respectively

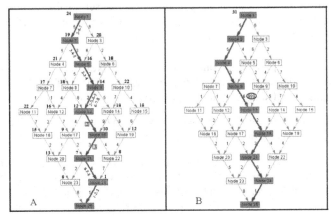

**Fig. 3.** G (25,40) is the optimal path in ideal (A) and blocked(B) routes status in Case 2, respectively

$$R^*_{(s,d)} = \min \sum_{i=0}^{d} \sum_{j=0}^{d} a \; R_{(s_j,d_j)} \begin{cases} a = 1 & \text{A link exists between } i \text{ and } j \\ a = 0 & \text{Otherwise} \end{cases} \qquad (2)$$

Where:

— $i$ and $j$ are current state movement into right and bottom side directions.
— $R_{(s,d)}$ is the shortest path from a origin node, 's' to a last node, 'd'.
— $a$ is binary digits ( 0 or 1).
— Links are independent of each other.

However, as mentioned in the previous section real-time information systems can be acquired using video cameras, GPS and other devices on transportation routes. In this paper, travel cost is assumed to be random variable [13].

### 3.2    Route Guidance System Based on Self-Adaptive Algorithm (RGSBSAA)

Figure 1 shows the road network map of a part of Kuala Lumpur (KL), the capital of Malaysia. A grid network of Figure 1 has been drawn in Figure 2A. Proposed algorithm (RGSBSAA) can also be used for finding shortest path (i.e. lowest cost) from origin node (i.e. node 1) to destination node (i.e. node 16) in a road network graph. The input of the algorithm consists of distances directed between nodes as $r$ and a source data from origin and destination paths as $s$, $d$ in the network graph G. If the distance between two nodes can be defined as path cost, the total cost of a network graph is the sum of all route distances between nodes in a graph. This algorithm is expressed based on node weight. Indeed, in this study, a "*node weight*" is defined as the shortest distance from each node to last network node. If "$n$" is the last node number, "$n$-$1$" and "$k$" are two node numbers connected to "node n" in the network. And also, regarding equation (2) and node weight definition, the procedure is presented as follows: All node's weights in the road graph are calculated in two steps as follows:

- **Step 1**
  - Distance between node $n$-$1$ and node $n$ are set to the weight of node $n$-$1$ and also distance between node $k$ and node $n$ is set to the weight of node $k$ (expect for first and last nodes, each node has two input and one output routes or at least one input and one output route).
- **Step 2**
  - Similarly, the procedure continues to calculate weights of all next nodes until the first node weights is calculated. Finally, the weight of the first node will be the minimum distance paths between first points to last nodes in the network.

Now, consider calculated node weights in RGSBSAA algorithm as the input data file, the shortest path computation procedure is presented as follows:

- Transportation route is started from first node (node 1), then amount of each distance route connecting the first node minus the weight of first node. If the result number is equal to weights of the each node, then the second node is connected to first node and it will be a priority route of network or next optimal route.
- Procedure will continue along the route connected to the last node then the shortest path network is determined.

The above algorithm uses the listed assumptions as follows:

- The road network is one-way, and it is from left to right and towards bottom direction.
- There is the one-way directional connection between all nodes.
- The information of distance routes between all nodes exists before the start of the travel.
- The travel is based on electronic map.

RGSBSAA Algorithm: Rote Guidance System based on Self-adaptive Algorithm.

| | |
|---|---|
| 1. | **INPUT:** |
| 2. | G( V, E) is a data file which composed a set of V nodes and set of E directed edges. |
| 3. | **PARAMETERS:** |
| 4. | $R_{(s,d)}$ , a nonnegative number stands for the cost where "s" is start node and "d" is last node. |
| 5. | i, j , k; loop index, G( 1, i) is array of vertexes source; G( 2, i) is array of vertexes destination. |
| 6. | G( 3, i) is array of edge distance(or cost); W(i) is array of node costs(node weights) table |
| 7. | for each node and P(i) is array of shortest path from the origin to final node. |
| 8. | **OUPUT:** |
| 9. | W(k) is a cost data table for all nodes. P(k), a shortest path data table for a graph. |
| 10. | S(j), cost limit s of shortest path routes. |
| 11. | **INITIALIZATION:** |
| 12. | // All nodes from last to first nodes are examined for the routes connected nodes. |
| 13. | // For each edge do operation in two steps as follows: |
| 14. | **set** W[1... n-1] = 999 , W(n) = 0 , P(i) = 0 ; |
| 15. | **BEGIN** // step 1: Node weight computation |
| 16. | **for all nodes** // for each node and edge pick the costs in W(k). |
| 17. | **for** j = first to last edges // j is set to the destination node of edges. |
| 18. | **if** (G( 2 , j) = i ) // k is set to the source node of edges. |
| 19. | W(k) = W(i) + G( 3 , j); |
| 20. | **end if** |
| 21. | **end for** |
| 22. | **end for** |
| 23. | // step 2: Shortest Path computation |
| 24. | **for** i = first to last edges |
| 25. | **while** (the origin (k) is the same in graph, G ) |
| 26. | **if** (G( 3 , i) = W(k) - W(j)) |
| 27. | P(k) = G( 2 , i); |
| 28. | **else** |
| 29. | i = i + 1; |
| 30. | k = G( 1 , i); |
| 31. | **end if** |
| 32. | **end while** |
| 33. | **end for** |
| 34. | // step 3: Cost limit computation for self-adaptive |
| 35. | **for** j = first to last edges // j is set to the destination node of edges. |
| 36. | **while** (edge belong to p(k) & defined shortest path is true ) |
| 37. | C(j) = G( 3 , j) + 1; |
| 38. | S(j) = { G( 3 , j), C(j) }; cost limit s of shortest path routes. |
| 39. | **end while** |
| 40. | **end for** |
| 41. | **END** |

**Table 1.** Performance measure for weighted directed graphs on sparse graphs

| Criterion | Description |
|---|---|
| BPC | Best path cost among all the RGSBSAA and Dijkstra runs |
| AvePC | Average path cost obtained for all RGSBSAA and Dijkstra |
| AvePCGap | Difference between the average path cost found (AvePC) and the best found path cost (BPC) of the test problem. AvePCGap = (AvePC - BPC)/BPC*100% |

**Table 2.** Simulation results of RGSBSAA and Dijkstra algorithms for weighted directed graphs on sparse graphs

| ( V , E ) | Method | Elapsed time | BPC | AvePC | AvePCGap |
|---|---|---|---|---|---|
| $G(16, 24)$ | RGSBSAA | 0.001124 | 17 | 17.00 | 0.00% |
| | Dijkstra | 0.001211 | 19 | 19.00 | 0.00% |
| $G(20, 31)$ | RGSBSAA | 0.001210 | 27 | 27.00 | 0.00% |
| | Dijkstra | 0.001215 | 26 | 26.00 | 0.00% |
| $G(20, 31)$ | RGSBSAA | 0.001214 | 25 | 25.10 | 0.40% |
| | Dijkstra | 0.001281 | 27 | 27.35 | 1.30% |
| $G(20, 37)^*$ | RGSBSAA | 0.001282 | 131 | 131.50 | 0.38% |
| | Dijkstra | 0.001315 | 135 | 136.80 | 1.33% |
| $G(20, 49)$ | RGSBSAA | 0.001294 | 145 | 145.00 | 0.00% |
| | Dijkstra | 0.001341 | 151 | 152.00 | 0.66% |
| $G(46, 69)$ | RGSBSAA | 0.001232 | 162 | 163.35 | 0.83% |
| | Dijkstra | 0.001377 | 155 | 156.20 | 0.77% |
| $G(46, 71)$ | RGSBSAA | 0.001263 | 164 | 165.30 | 0.79% |
| | Dijkstra | 0.001370 | 158 | 159.50 | 0.95% |
| $G(100, 197)^*$ | RGSBSAA | 0.001422 | 850 | 852.00 | 0.24% |
| | Dijkstra | 0.001534 | 810 | 815.20 | 0.64% |
| $G(100, 203)^*$ | RGSBSAA | 0.001411 | 862 | 865.00 | 0.35% |
| | Dijkstra | 0.001538 | 843 | 860.30 | 2.05% |
| $G(100, 216)^*$ | RGSBSAA | 0.001446 | 1145 | 1164.00 | 1.66% |
| | Dijkstra | 0.001548 | 1162 | 1188.46 | 2.28% |
| $G(100, 217)$ | RGSBSAA | 0.001433 | 1125 | 1141.30 | 1.45% |
| | Dijkstra | 0.001559 | 1210 | 1231.00 | 1.73% |

* In these types of graphs, there are more than two connected edges for some of nodes ( i.e. 3 or 4 edges).

# 4    Comparison of Experimental Results

In this section, simulation experiments have been carried out on different network topologies for road networks consisting of 6 to 100 nodes with different edges (between 10 to 220 edges). In addition, Table 2 is simulation results of RGSBSAA algorithm and Dijkstra algorithm for weighted directed graphs on sparse graphs (i.e. road network graphs). BPC and AvePC report the shortest and average path cost found in the mentioned graphs in RGSBSAA algorithm runs respectively. Also, AvePCGap reports the measures of the gap between shortest path and average path cost. To summarize, for all graphs proposed in the Table, the average path cost gaps performance in RGSBSAA algorithm were obtained less than the Dijkstra algorithm performance.

## 4.1    RGSBSAA Algorithm - Case 1

Figure 2A shows the weight of node 1 is 15, and the weight of node 2 is 12... Finally the cost (or weight) of node 16 is 0 that is calculated with first part of proposed algorithm procedure. The cost (or weight) of 15 is the shortest distance from node 1 to last node (destination node). Figure 2A presents the optimal routes (the red color path) in ideal route status as follows: node 1 → node 2 → node 8 → node 10 → node 14 → node 15 → node 16. Figure 2B shows the route between node 10 and node 14 which is blocked (let cost= ∞) because of traffic congestion; therefore, some connected routes are re- moved automatically in the new optimal route computation and the next route which is connected node 1 (i.e. node 2) is selected as first choice route in this figure. Furthermore, Figure 2A shows the self-adaptive algorithm performance measures of the shortest path routes. Self-adaptive algorithm will provide specifications as to what costs exist for a particular route within a given time frame, after which an alternative route must be taken. The RGSBSAA algorithm uses the received traffic congestion information of shortest path routes from the traffic-control devices in each intersection with acceptable route critical costs suggested by the proposed algorithm in Algorithm RGSBSAA. For example, in first route of Figure 2A, the costs limit of node 8 to node 10,  2,3,4 show that the cost's limit encountered that vehicle driver can use the costs of 2,3 and 4 for taking this shortest path. The out of 2,3,4 range which causes are the change in the recommended shortest path. Therefore, consider the above-mentioned constraints and proposed RGS- BSAA algorithm, the optimal path in blocked route status (Figure 2B) is the red color path between node 1 and node 16. Last, consider that RGSBSAA algorithm and Figure 2B routes of node 10 and node 14 is congested (or blocked) therefore, its cost tends to ∞. Figure 2B shows the minimum cost (or distance) between node 1 and node 16 is changed from first minimum cost (in Figure 2A), 15 to 17 in Figure 2B.

## 4.2    RGSBSAA Algorithm - Case 2

Consider the previous section discussion; Figure 3A presents a transportation network that has 40 routes and 25 nodes as intersection. The distance between node 1 and node

25 (or last node) is 24 i.e weight of node 1. Figure 3A presents the optimal routes in ideal route status as follows: node 1→node 2→ node 5→ node 9→ node 13→ node 18→node 21→node 24→ node 25. Also, Figure 3A shows the self-adaptive algorithm performance measures of the shortest path routes. For example, in first route of Figure 3A, the costs limit of node 1 to node 2, 5,6,7 show that the cost's limit encountered that vehicle driver can use the costs of 5,6 and 7 for taking this shortest path. The out of 5,6,7 range which causes are the change in the recommended shortest path. Therefore, considering the above-mentioned constraints the proposed RGSBSAA algorithm, the optimal path in blocked route status (Figure 3B) is the red color path between node 1 and node 25. Lastly, consider that RGSBSAA algorithm routes of node 9 → Node 13 is congested (or blocked) therefore, their costs tend to ∞. Figure 3B shows the minimum cost (or distance) between node 1 and node 25 is changed from first minimum cost in Figure 3A, 24 to 31 in Figure 3B. However, the simulation program is tested with network graph information. The acquired results affirmed the proposed algorithm results were the convergence to acquire the shortest path in each network case. In order to assess the performance of self-adaptive algorithm, Table 2 shows the comparison of some of our cases experimental study with Dijkstra algorithm. We can see that the result of our proposed algorithm achieves better results than Dijkstra algorithm. This comparison in a real transportation network shows the evaluation of proposed method that is route guidance system based on a proposed path planning algorithm. Finally, in all experimental cases, average path cost gaps (AvePCGap) of RGSBSAA method are less than average path cost gaps of Dijkstra algorithm and has a better performance result.

# 5    Conclusion and Future Work

This paper proposes a new self-adaptive system algorithm in route guidance system for finding the critical route costs and dynamically responds to the environmental conditions for any stated path. The behavior of the route guidance system change based on the proposed algorithm. The performance of the proposed algorithm was evaluated using adaptive routing agents that dynamically and adaptively find the minimum delay and maximize time between the routes in the route guidance system (see Table 2). Through simulation, the proposed algorithm (RGSBSAA) was applied for managing the shortest path routes for many network graphs with a number of edges ranging from 4 to 200(like Figures 2A and 3A ). In each node (intersection), the shortest path (or lowest cost) was determined by the Dijkstra algorithm. Whenever the agent reports traffic congestion on in a specific route using the initial shortest path, the drivers take next shortest path based on the proposed algorithm. Given the above results, our contributions are made as follows:

–   We have demonstrated a new self-adaptive algorithm to find the shortest path problem in the road network graph.
–   It develops a novel approach to the route guidance system in the underlying network since the vehicle driver receives information from next road status he/she follows to the next shortest path.

- The study provides empirical grounds for the route guidance system based on self-adaptive algorithm that could perform well on network graphs.

Future works are suggested as follows: Vehicle route selection could be adapted online, i.e. vehicles could coordinate their behavior based on real-time traffic conditions in the road network. In the future study, many real-world factors in the environment should be considered such as vehicle accidents, weather, illegal parking, etc.

**Acknowledgement.** The authors wish to thank Ministry of Higher Education Malaysia (MOHE) under Fundamental Research Grant Scheme (FRGS) Vot 4F031 and Universiti Teknologi Malaysia under the Research University Funding Scheme (Q.J130000.7110.02H47) for supporting the related research.

# References

1. Sloman, M., Lupu, E.: Engineering policy-based ubiquitous systems. The Computer Journal, 1113–1127 (2010)
2. Che, K., Li, L.-L., Niu, X.-T., Xing, S.-T.: Research of software development methodology based on self-adaptive multi-agent systems. In: 2009 IEEE International Symposium on IT in Medicine & Education (ITME 2009), vol. 1, pp. 235–240 (2009)
3. Nagib, G., Ali, W.G.: Network routing protocol using Genetic Algorithms. International Journal of Electrical & Computer Sciences 10, 40–44 (2010)
4. Dijkstra, E.W.: A Note on Two Problems in Connection with Graphs. Journal of Numerical Mathematics 1, 269–271 (1959)
5. Arokhlo, M.Z., Selamat, A., Hashim, S.Z.M., Selamat, M.H.: Multi-agent Reinforcement Learning for Route Guidance System. IJACT: InternationalJournal of Advancements in Computing Technology 3(6), 224–232 (2011)
6. Vasirani, M., Ossowski, S.: A market-inspired approach to reservation-based ur- ban road traffic management. In: AAMAS 2009: Proceedings of The 8th International Conference on Autonomous Agents and Multiagent Systems, Budapest, Hungary, pp. 617–624 (2009)
7. Schulz, F., Wagner, D., Weihe, K.: Dijkstra's algorithm on-line: an empirical case study from public railroad transport. J. Exp. Algorithmics 5, 12 (2000)
8. Andersson, J., de Lemos, R., Malek, S., Weyns, D.: Reflecting on self-adaptive software systems. In: ICSE Workshop on Software Engineering for Adaptive and Self-Managing Systems, Vancouver, Canada, pp. 38–47 (2009)
9. Zhang, J., Goldsby, H.J., Cheng, B.H.: Modular verification of dynamically adaptive systems. In: Proceedings of the 8th ACM International Conference on Aspect-oriented Software Development, AOSD 2009, pp. 161–172. ACM, New York (2009)
10. Herrmann, K., Gero, M., Geihs, K.: Self-management: The solution to complexity or just an- other problem. IEEE Distributed Systems Online 6 (2005)
11. Chen, B., Cheng, H.H., Palen, J.: Integrating mobile agent technology with multi-agent systems for distributed traffic detection and management systems (2009)
12. Zhang, Z., Xu, J.-M.: A dynamic route guidance arithmetic based on reinforcement learning. In: Proceedings of 2005 International Conference on Machine Learning and Cybernetics, China, vol. 6, pp. 3607–3611 (2005)
13. Zou, L., Xu, J.-M., Zhu, L.: Application of genetic algorithm in dynamic route guidance system. Journal of Transportation Systems Engineering and Information Technology 7, 45–48 (2007)

# Location Recognition with Fuzzy Grammar

Nurfadhlina Mohd Sharef

Intelligent Computing Group,
Faculty of Computer Science and Information Technology,
University Putra Malaysia,
UPM Serdang, Malaysia
fadhlina@fsktm.upm.edu.my

**Abstract.** Fuzzy grammar has been introduced as an approach to represent and learn text fragments where the set of learned patterns are represented by combining similar segments to represent regularities and marking interchangeable segments. This paper is dedicated to present a procedural scheme towards learning text fragment in text categorization task facilitated by fuzzy grammars. A few issues are involved in developing fuzzy grammars which are (i) determination of the number of text classes to develop (ii) the selection of text fragments, $F$ relevant to each text class (iii) determining frequent and important terms or keywords, $V$ to develop the set of terminal, $T$ and compound grammars, $N$. (iv) Conversion of text fragments into grammars, (v) Combination of grammars into a compact form. Comparison between fuzzy grammar and other location entity identifier such as LbjTagger, LingPipe, Newswire and OpenCalais is observed where results have shown that this method outperforms other standard machine learning and statistical-based approach.

**Keywords:** Fuzzy grammar, text categorization, evolving fuzzy grammar, text mining.

## 1 Introduction

Named entity (NE) involves identification of *proper names* in texts, and classification into a set of predefined categories of interest. However, in order to develop a classifier and reveal interesting patterns in a text, one must first impose some structure to the text before attempting any mining task. This involves identifying the feature, or text constructs related to the actual word usage in the text in order to capture the underlying structure of the text. Nonetheless, the number and type of the features are based and are dependent on the text content; therefore, they cannot be anticipated in advance (e.g, ambiguity of NE types; 'Washington' could be both a person name and a location).

Since the enumeration of all described phenomena to be encountered in any text is infinite, it is not possible to use it as the set of features. All these render difficult the mapping of textual information into manageable datasets in standard form.

D. Lukose, A.R. Ahmad, and A. Suliman (Eds.): KTW 2011, CCIS 295, pp. 254–261, 2012.

This scenario is made more critical by the knowledge that linguistic and domain knowledge bases are very expensive to build, and almost impossible to generate "on the fly" for any corpus of potentially interesting texts selected or retrieved by some process. Therefore, we must find an easy and systematic approach to impose a simple structure, yet one rich and descriptive enough to allow us to use discovery techniques and reveal interesting patterns in such texts. This relates to the text representation method which often depends on the task in hand and it allows for easier and more efficient data manipulation.

A widely-used approach to facilitate this task for NER application is to identify the words distribution in the texts that represent a text class; such as statistical approach (e.g, Hidden Markov Model [1,2], maximum entropy [3], conditional random field [4]) but this method can neglect the semantic properties in the texts. Another common approach in NER is the rule-based method, which includes (i) lookup list, similar to gazetteer approach [5,6], which recognizes entity only based on its list, (ii) shallow parsing approach such as regular expression [7,8] which utilizes entities based on a training of a strict pattern.

On a slightly different view, this paper presents the fuzzy grammar method which utilizes the local degree of structure within text fragments as the means of representation. The underlying structure is then called fuzzy grammars, which can be combined for more compact presentation using the evolving fuzzy grammar method. One of the approaches for building fuzzy grammars is by utilizing the power of regular expression. Instead of the tedious, manual generation of regular expression rules which can grow complicated as more patterns are trained, the fuzzy grammar method produces an automated, compact representation of rules for location expression.

The contribution of the paper is three-fold; (i) to introduce fuzzy grammar as a text fragment learning model, (ii) to present a procedural schema on the development of fuzzy grammars as a tool for building text categoriser, (iii) to present the strength of fuzzy grammar as location entity classifier, as one of the common task under named entity recognition. The first part of the paper gives a brief introduction on the properties of fuzzy grammars, the second part focuses on the notation used in fuzzy grammar, the third section provides steps involved in building text categoriser with fuzzy grammar, section four provides the implementation setting to perform text categorization using fuzzy grammar and section five concludes the paper.

## 2    Learning Location Expression with Fuzzy Grammar

Previous related papers [9-12] on fuzzy grammars have focused on the evolving aspects of fuzzy grammars learning and application of fuzzy grammars in textual event detections (e.g, terrorism incidents, and events attached with numerical values). This paper presents a procedural mechanism for performing text categorization using fuzzy grammars. A fuzzy grammar is represented by 5-tuples, $G=<V,f,F,T,N>$ where

1. the vocabulary, $V$, is the set of symbols appearing in the text fragments. The text is split into meaningful symbols prior to processing; the symbols may be words, numbers, separators, etc.

2. $f$: a text fragment, i.e. a sequence of adjacent symbols (words, numbers, etc) from a text source. We write $f = v_1 v_2 \ldots v_n$ where each $v_i \in V$. For example, the common indicators to identify location are district name, town, and country. Each attribute can be further represented by a closed set of patterns such as (i) word followed by district, (ii) country name indicated by words beginning with capital letters and (iii) closed set vocabulary such as 'Republic of' and 'United'.

3. $F$: a set of text fragments $\{ f_i, i=1, \ldots, m \}$. For example, a few text fragments can then be prepared such as shown in Fig. 1. One approach towards this is by crafting a set of main keywords regarding the text class to be built.

- Brussels, Belgium
- Barakaldo, Spain
- Manchester, City and Borough of Manchester England
- Quetta, Balochistan Province, Pakistan
- Mafraq, outside of Ba'qubah, Iraq

**Fig. 1.** Examples of location expression

4. $T$: a set of terminal classes. Each class is named by a grammar token $T_i$ and represents a fuzzy or crisp subset of the vocabulary $V$ (and possibly punctuation), with membership function $\mu_{Ti}$ based on enumeration of the class members or on regular expressions. The extension of T, $Ext(T)$ is the definition or parse-able strings by each token, $T_i$. For example, if T:=<number>, $Ext$(number)= $<number>$ := { 1, 2, 3, . . . }

```
<conjunction>:={and,of,between}
<whereabout>:={island,city,village,district,borough}
locPrefix:={in,close,near,between,east,west,south,north}
<loc>:=^[A-Z][a-z]+\D*
```

**Fig. 2.** Terminal Grammars

5. $N$: a set of non-terminal grammar tokens, where the definition of each non-terminal grammar token $N_i$ is a sequence of one or more symbols from the vocabulary and grammar tokens (these may be terminal or non-terminal tokens). In general, $N_i := E^*_i$ where $E^*_i$ represents a sequence of elements (vocabulary symbols, terminal and non-terminal grammar tokens). At most one token in the definition may be enclosed in square brackets, meaning that it is optional. Using the terminal grammars depicted in Fig. 2, the text fragments as in Fig. 1 will be transformed into grammars as shown in Fig. 3. Each derived grammar has a perfect parsing/matching (membership 1.0) on its source string. The grammar parsing is the opposite of grammar derivation where parsing score with membership 0.0 indicates no matching and 1.0 indicates perfect matching. The algorithm for grammar derivation was presented in [9,11,13]

**Table 1.** Conversion of text fragments into grammars

| Text Fragments | Grammars |
|---|---|
| Peureulak, Aceh Province, Indonesia | locPrefix-provinceName-Country |
| in Kirkuk, Iraq | locPrefix-loc-Country |
| near Chaplingehera, Tripura, India | locPrefix-loc-loc-Country |
| in Parang, Mindanao Island, Philippines | locPrefix-loc-loc-whereabout-Country |
| in Narathiwat and Yala Provinces, Thailand | locPrefix-loc-conjunction-provinceName-Country |
| in Mangalapur, near Bhairahawa, Nepal | locPrefix-loc-locPrefix-loc-Country |
| west of Kirkuk, Iraq | locPrefix-conjunction-loc-Country |

```
Location-0:=whereabout-province
Location:=locPrefix-loc-[Location-0]-Country
Location-1:=conjunction-loc
Location:=locPrefix-loc-loc-[Location-1]-Country
Location:=locPrefix-loc-loc-[conjunction]-loc-loc-Country
Location-2:=aw
Location-2:=whereabout
Location:=in the loc Location-2 loc loc Country
```

**Fig. 3.** Compact Grammars

The grammars are the representation of the learned text fragments. Once they are obtained, they can be used for further text learning task, such as text extraction, text classification and text matching. However, a compact representation of the grammars provides a more efficient management and utilization of the computational resource. The evolving fuzzy grammar (EFG) algorithm [12] combine fuzzy grammars to produce a more compact representation of the set of learned text by identifying similar and interchangeable segments of the learned text as illustrated in Fig. 3. These problems are fuzzy because it involves grammar optimization and minimal change in producing the compact rules. As shown in Fig. 3, the produced result is in different syntax but the semantic is preserved, and this is made feasible by the soft computing concept.

The developed fuzzy grammars can then be used as the rules in a text categorizer. Fuzzy grammar parsing can be utilized to measure the degree of changes needed by the grammar to parse the text. The text chunk with the highest parsing score is then returned as the portion of recognized text.

## 3    Applying Learned Grammars to Perform Text Categorization

General machine learning setting for performing text categorization is guided by (i) training the text classifier; for example by following the steps in section III, (ii) testing the text classifier; which validates the degree of recognisability by the developed classifier. Setting (ii) indicates that there is possibility of limited coverage by the learned classifier. Typically machine learning setting requires thorough re-development of the text learner but the incremental evolving learning technique presented in [11,13,14] achieve this goal by slight modification of the learned patterns without needing to regenerate the classifier. An empirical observation has shown advantageous of EFG as incremental learner in time consumption versus batch learning method at 1:3 ratio for learning additional (re-training) data [14].

An experiment on identifying location in a summary of terrorism incidents was performed. For the purpose of the experiment, evaluation is done by comparing the location tags by EFG, benchmark dataset from Worldwide Incidents Tracking System (WITS)[1] and other NER programs namely Newswire, LbjTagger, LingPipe and OpenCalais. The WITS data are tagged in XML-format according to subject, summary, incident date, weapon type list, location and victim list. Multi or single event types are coded in the database for example as armed attack, arson/firebombing, assassination and assault. Weapons used are also tagged.

The Newswire NER[2] uses an HMM-based chunk tagger is based on simple deterministic internal feature of the words, such as capitalization and digitalization. It also creates an internal semantic feature of important triggers; internal gazetteer feature; and external macro context feature. The system can classify seven named entity categories including person, location, organization, date, time, percent and money.

LingPipe[3] is a suite of Java-based natural language processing tools for the linguistic analysis of human language distributed with source code by Alias-i. The LingPipe application used in this experiment is the Named Entity Demo for News English which was trained on the MUC 6 Corpus. It is developed based on probabilistic n-gram based character language models. It is also founded on hidden Markov model interface with several decoders: first-best (Viterbi), n-best (Viterbi forward, A* backward with exact Viterbi estimates), and confidence-based (forwardbackward). A chunking implementation that codes a chunking problem as an HMM tagging problem using a refinement of the standard BIO coding is also introduced.

The Illinois Named Entity Tagger (LBJ based)[4] project at Cognitive Computation Group, University of Illinois provides a publicly available NER. The method used includes non-local features, external knowledge such as gazetteers derived from Wikipedia, and features based on word cluster hierarchy derived using distributional similarity over a larger text corpus. The NER baseline is based on regularized averaged perceptron. The tagging classifier is based on BILOU (Beginning, the Inside and the Last tokens of multi-token chunks as well as Unit-length chunks) scheme.

---

[1] Available at http://wits-classic.nctc.gov/
[2] Available at http://nlp.i2r.a-star.edu.sg/demo_mucner.html
[3] Available at http://lingpipe-demos.com:8080/
  lingpipe-demos/ne_en_news_muc6/response.xml
[4] Available at http://l2r.cs.uiuc.edu/~cogcomp/LbjNer.php

Open Calais[5] is a strategic initiative from Thomson Reuters to support the interoperability of content across the digital landscape. It is a rapidly growing toolkit that incorporates state-of-the-art semantic functionality. It provides a fast way to tag the people, places, facts and events in text content. OpenCalais is also called Web 3.0 which is a concept related to the Semantic Web[6], Linked Data[7] , etc. OpenCalais analyzes a document and finds the entities (such as people, geographies and companies), events (such as position, alliance, and person-education) and facts (such as sports and management) within it using natural language processing (NLP), machine learning and other methods.

**Table 2.** Evaluation of location recognition

| Evaluation | Data | Newswire | LingPipe | LBJ | EFG | OpenCalais |
|------------|------|----------|----------|------|------|------------|
| Precision | JAN | 0.75 | 0.67 | 0.80 | 0.96 | 1.00 |
|  | FEB | 0.60 | 0.48 | 0.53 | 0.96 | 1.00 |
|  | MAR | 0.89 | 0.86 | 0.85 | 0.79 | 0.98 |
|  | APR | 0.82 | 0.43 | 0.64 | 0.91 | 1.00 |
| Accuracy | JAN | 0.65 | 0.62 | 0.77 | 0.96 | 0.74 |
|  | FEB | 0.51 | 0.38 | 0.43 | 0.96 | 0.70 |
|  | MAR | 0.81 | 0.77 | 0.77 | 0.76 | 0.86 |
|  | APR | 0.80 | 0.42 | 0.60 | 0.91 | 0.69 |

The result for the location entity recognition is evaluated based on precision (the proportion of true results) and recall (degree of closeness between the classified locations with the benchmark data) and shown in TABLE 2. Gathered results show in terms of precision OpenCalais is leading, followed by EFG, Newswire, LBJ, and LingPipe. OpenCalais has higher precision compared to EFG because OpenCalais has incorporated other named entity recognition such as facility, organization and person, which has tagged wrongly by the rest of the applications. However, EFG has the highest accuracy. This is because the sum of false negative and false positive by EFG is not as much as OpenCalais because despite able to recognize other entities and events, OpenCalais is not trained on uncommon location patterns such and has leaded into false positive such as 'Al Fallujah, <person>Al Anbar</person>, <Country>Iraq<Country>'. EFG has also demonstrated advantage over the statistical method in LingPipe and Newswire and gazetteer method in LBJTagger. This is because the statistical method depends highly on the statistical distributions of words, thus many important words and patterns to be learned can be ignored, compared to

---

[5] Available at http://viewer.opencalais.com/
[6] The Semantic Web is a "web of data" that facilitates machines to understand the semantics, or meaning, of information on the World Wide Web.
[7] Linked data describes a method of publishing structured data so that it can be interlinked and become more useful.

wider coverage in EFG. Although EFG demands more human involvement in the creation of terminal grammars, this method typically obtains better precision. The high accuracy obtained by EFG show the ability of the model to learn both syntactic and semantic properties of the text and the fuzzy grammar method provides a comprehensive measure to determine the degree of closeness between a grammar and the text. The grammar consists of combination of terminal and non-terminal grammars where each of these can have multiple definitions, thus during matching more possibility of matched string-grammar is provided.

## 4    Conclusions

The fuzzy grammar approach discussed in this paper utilizes the underlying structure within the text, by focus given on the text fragment unit. This method is distinguished from other previous research in the field because this method allows human-like representation of the segment of the texts. These are then converted into the symbolic representation called fuzzy grammars and are then combined as compact fuzzy grammar rules. The evolving fuzzy grammar method discussed in this paper is a complementary approach to traditional machine learning setting application which is typically constrained by the scalability of the training set. Results on location recognition have shown that EFG is at least comparable with the other method and has also outperformed other statistical based applications. This imposed fuzzy grammar as a potential an alternative text classifier in addition to the common statistical and manual based tagger approach. Besides, the syntactical representation of the semantic/lexical representation of the text being focused provides a higher level of human understanding and representation alike compared to other existing methods.

**Acknowledgements.** The research on fuzzy grammar was carried on during the PhD study of the author under the supervision of Professor Trevor Martin in University of Bristol, United Kingdom. The author wishes to express her appreciation to Professor Trevor Martin for his guidance in performing the research. The PhD work was co-sponsored by Ministry of Higher Education, Malaysia and Universiti Putra Malaysia, Malaysia from May 2007 until November 2010.

## References

[1] Todorovic, B.T., Rancic, S.R., Markovic, I.M., Mulalic, E.H., Ilic, V.M.: Named entity recognition and classification using context Hidden Markov Model. In: 2008 9th Symposium on Neural Network Applications in Electrical Engineering, pp. 43–46 (2008)

[2] Xu, R., Supekar, K., Huang, Y., Das, A., Garber, A.: Combining Text Classification and Hidden Markov Model Techniques for Structuring Randomized Clinical Trial Abstracts. In: Annual Symposium of AMIA, pp. 824–828 (2006)

[3] Chieu, H.L.: Named Entity Recognition: A Maximum Entropy Approach Using Global Information. Language

[4] McCallum, A., Pereira, F., Lafferty, J.: Conditional random fields: Probabilistic models for segmenting and labeling sequence data. In: 18th International Conference on Machine Learning (ICML 2001), pp. 282–289 (2001)

[5] Nadeau, D., Turney, P.D., Matwin, S.: Unsupervised Named-Entity Recognition: Generating Gazetteers and Resolving Ambiguity, pp. 266–277 (2006)

[6] Mikheev, A., Moens, M., Grover, C.: Named Entity Recognition without Gazetteers. Technology (1998)

[7] Watt, A.: Beginning Regular Expressions. Wiley (2005)

[8] Brazma, A.: Learning of Regular Expressions by Pattern Matching. In: Computational Learning Theory, pp. 392–403 (1995)

[9] Martin, T., Shen, Y., Azvine, B.: Incremental Evolution of Fuzzy Grammar Fragments to Enhance Instance Matching and Text Mining. IEEE Transactions on Fuzzy Systems 16, 1425–1438 (2008)

[10] Martin, T., Shen, Y., Azvine, B.: Automated semantic tagging using fuzzy grammar fragments. In: 2008 IEEE International Conference on Fuzzy Systems (IEEE World Congress on Computational Intelligence), pp. 2224–2229 (June 2008)

[11] Sharef, N.M., Martin, T., Shen, Y.: Order Independent Incremental Evolving Fuzzy Grammar Fragment Learner. In: Ninth International Conference on Intelligent Systems Design and Applications, pp. 1221–1226 (2009)

[12] Sharef, N.M., Shen, Y.: Text Fragment Extraction using Incremental Evolving Fuzzy Grammar Fragments Learner. In: World Congress on Computational Intelligence, pp. 18–23 (2010)

[13] Sharef, N.M., Martin, T., Shen, Y.: Minimal Combination for Incremental Grammar Fragment Learning (2009)

[14] Sharef, N.M.: Text Fragment Identification with Evolving Fuzzy Grammars (2010)

# High Level Semantic Concept Retrieval Using a Hybrid Similarity Method

Sara Memar Kouchehbagh, Lilly Suriani Affendey, Norwati Mustapha,
Shyamala C. Doraisamy, and Mohammadreza Ektefa

Faculty of Computer Science and Inforamtion Technology,
Universiti Putra Malaysia, 43400 Serdang, Malaysia
{sr.memar,mrektefa}@gmail.com,
{suriani,norwati,shyamala}@fsktm.upm.edu.my
http://www.fsktm.upm.edu.my

**Abstract.** In video search and retrieval, user's need is expressed in terms of query. Early video retrieval systems usually matched video clips with such low-level features as color, shape, texture, and motion. In spite of the fact that retrieval is done accurately and automatically with such low-level features, the semantic meaning of the query cannot be expressed in this way. Moreover, the limitation of retrieval using desirable concept detectors is providing annotations for each concept. However, providing annotation for every concept in real world is very challenging and time consuming, and it is not possible to provide annotation for every concept in the real world. In this study, in order to improve the effectiveness of the retrieval, a method for similarity computation is proposed and experimented for mapping concepts whose annotations are not available onto the annotated and known concepts. The TRECVID 2005 dataset is used to evaluate the effectiveness of the concept-based video retrieval model by applying the proposed similarity method. Results are also compared with previous similarity measures used in the same domain. The proposed similarity measure approach outperforms other methods with the Mean Average Precision (MAP) of 26.84% in concept retrieval.

**Keywords:** Video Retrieval, Video Analysis, Semantic Knowledge, Similarity Measures.

## 1 Introduction

In video retrieval, users can easily and intuitively express their needs in natural language in terms of semantic concepts. However, some earlier techniques for retrieval are not intuitive enough for users. So, the inconsistency between the user's perception and the way video is retrieved leads to semantic gap which is considered as a vital and important problem and issue in multimedia computer society. In order to retrieve video shots in a more intuitive manner, several studies [3,7,10] utilized semantic similarity methods rather than using low-level features and concept detectors. The limitation with low-level features is failure in retrieving queries expressed in terms of semantic concept. Hence, lately, a number

D. Lukose, A.R. Ahmad, and A. Suliman (Eds.): KTW 2011, CCIS 295, pp. 262–271, 2012.
© Springer-Verlag Berlin Heidelberg 2012

of concept detectors have been developed by different researchers to help with the semantic video retrieval. Among them, Large Scale Concept Ontology for Multimedia (LSCOM) with the collection of more than 400 concept annotations [16], Media Mill which is provided with a set of 101 concept annotations [17], Columbia374 [20], and Vireo374 [11] are considered as the largest and the most popular concept detectors. The drawback with such concept detectors is that the annotations of concepts should be provided for retrieval. However, providing annotations for every concept in the real world is not feasible.

Recently, several studies in [3,7,10] utilized semantic word similarity measures [15] for automatic concept-based video retrieval. Authors in [3] tried to retrieve new concepts - concepts whose annotations are not provided- by utilizing semantic word similarity measures through concept detectors. However, since knowledge-based semantic word similarities are based on WordNet, they face some limitations. For example, some similarity measures like Resnik, Jcn, and Lin are defined based on information content (IC). Information content in the context of WordNet is drawn from the Brown University Standard Corpus of Present-Day American English (a.k.a the brown Corpus). Although there are more than 155,000 unique words and phrases in WordNet, only about 37,000 of them are available in the Brown Corpus. Therefore, there is the gap of losing concepts in the Brown Corpus, and this matter causes empty information content for approximately 75% of words and phrases from WordNet. Furthermore, some studies [3,7,10] did not take into account string similarity which is claimed to have some impacts on the success of query retrieval [9].

To address the mentioned problems, Latent Semantic Analysis (LSA), which is independent of WordNet, is utilized as a semantic similarity measure. In addition, due to the essence and importance of string similarity, Longest Common Consequence (LCS) is also adopted as a string similarity measure in concept-based video retrieval model. This is unlike [3] in which string similarity was not taken into account. Both LCS and LSA are common similarity measures which have been applied in a variety of domains by different researchers as well [18,19,12,9]. In concept-based video retrieval model, query which can be a concept or a group of concepts (concepts whose annotations are not available for the system) is given to the model. Then, the similarity of query is calculated with the 374 annotated concepts which are trained by Columbia374 concept detector for retrieving video shots. For mapping query terms with the annotated concepts of Columbia374, the proposed similarity method which is the linear combination of semantic similarity (LSA) and string similarity (LCS) measures is applied for improvement purpose in concept-based video retrieval.

## 2   Related Work

For retrieving video shots more intuitively, a method for similarity computation is proposed and experimented in this study. Therefore, several similarity techniques are reviewed in this section.

Recently, text similarity or semantic similarity of words has been utilized in the area of video retrieval. There have been a large number of studies on word-to-word similarity metrics ranging from distance-oriented measures computed on semantic networks, to metrics based on models of distributional similarity derived from large text collections [15]. Word-to-word similarity metrics are mainly divided into two groups as follows: knowledge-based and corpus-based. Corpus-based measures attempt to introduce the similarity between two words using the information which come from large corpora.

Corpus-based measures fall into two metrics, namely Pointwise Mutual information (PMI) and Latent Semantic analysis (LSA) [15]. PMI-IR is a simple method for computing corpus-based similarity of words. It uses Pointwise Mutual Information which was proposed as a semantic word similarity in [18]. In PMI-IR measure, similar concepts have this tendency to be viewed together in the documents more than dissimilar ones. LSA is another metric for computing the similarity semantically, and it was suggested by [13]. LSA was defined as "a fully automatic mathematical/statistical technique for extracting and inferring relations of words in passages" [13]. LSA has a variety of applications. In the context of information retrieval, it is also called Latent semantic Indexing (LSI). The application of the LSA word similarity in text similarity was explained by [5]. Another application of LSA is measuring the textual coherence [6]. [19] applied LSA in predicting physiological phenomena as well. LSA was also used to recognize the synonyms on TOEFL [18]. As mentioned above, LSA has many applications in the context of text similarity, TOEFL exam and so on, but in the context of concept based video retrieval, only a few studies have been experimented yet. Therefore, in this study, the strength of LSA in the context of concept based video retrieval is also shown by experimental results.

Knowledge-based semantic word similarity measures were developed to quantify the degree in which two words are semantically related by using information drawn from semantic network like WordNet hierarchy. In [3], seven of knowledge-based semantic word similarities were utilized in the context of concept based video retrieval.

In most concept based video retrieval models, semantic similarity measures have been applied. On the other hand, the essence and importance of string similarity is inevitable. [9] indicated the importance of string similarity by giving an example. For instance, two words "Maradona" and "Maradena" correspond to one entity, but the second one "Maradena" is misspelled. In such cases, for computing similarity value between these two proper names, dictionary-based similarity measures cannot be effective. Moreover, corpus-based similarity measures are very weak and not acceptable enough. Hence, string similarity is considered as a better and more proper alternative for such examples. Longest Common Subsequence is the employed string similarity measure in this paper. LCS finds the longest common subsequence which is common in two input strings [1]. Authors in [4] presented a comprehensive comparison of well-known LCS algorithms for two input strings. There are many studies contributing to the modified and normalized version of LCS [12,14]. In this study, the modified and normalized version of the LCS which was suggested by [9] is utilized as a string similarity metric.

# 3  Proposed Similarity Method

In order to map concepts of query to pre-determined concepts of Columbia374, a method for similarity computation is proposed. Hence, in this study, one of the corpus-based semantic word similarity measures named LSA is combined with LCS string similarity measure. So, here, the details of each similarity measure are presented in this section, and then the proposed similarity method is explained as well.

## 3.1  LCS String Similarity Algorithm

For string similarity, [9] normalized and modified version of LCS string matching algorithm which was proposed by [1] is utilized. In this paper, modified and normalized version of LCS which was proposed by [9] is applied for computing the string similarity between two terms. In this version, three different versions of LCS were normalized and given a weight. Afterwards, all three normalized and weighted values were summed together, and the result is the score which shows the string similarity between two concepts or terms.

## 3.2  LSA Semantic Similarity Algorithm

LSA was defined as a fully automatic mathematical/statistical technique for extracting and inferring relations of words in passages [13]. It does not depend on the usage of humanly constructed dictionaries, knowledge bases, semantic networks, grammars, syntactic parsers, or morphologies like a traditional natural language processing or artificial intelligence program. LSA as a corpus-based semantic word similarity measure is utilized in the proposed similarity method for computing the similarity between two words or concepts. LSA [13] computes the similarity of words or sentences within a large corpus of texts or semantic space by determining the occurrence of each word or term. Semantic information can be understood in raw text, and the meanings of words are more tangible and realizable in a text or corpus. LSA is a mathematical technique. In other words, in LSA, a high-dimensional semantic space or a term-document matrix is generated. In this matrix, occurrence of each word or term (concept) in a document or corpus is determined. In our concept based video retrieval model, the similarity between concept(s) of query and annotated concepts of Columbia374 should be computed. So, if a term-document matrix is assumed as M, rows of M are considered as $C_i$ which shows the occurrence of concept, and columns of M are regarded as Pj which present a unit of corpus (paragraph). Unit of corpus is paragraph. This corpus was constructed from the reading materials of an average high school student in the United States. It consists of approximately 11 million words which yield a co-occurrence matrix of more than 92,000 word types and more than 37,000 documents.

### 3.3   Final Hybrid Similarity Method

In this study, for computing the similarity between concept(s) of query and annotated concepts of Columbia374, both string and semantic similarity measures are
taken into account. In string similarity, Longest Common Subsequence (LCS)
string similarity algorithm is utilized. For semantic purpose, Latent Semantic
Analysis (LSA) is applied as well. In proposed hybrid method, both LSA and
LCS are combined together as follows: if LCS string similarity function is denoted as $S_1$ and LSA semantic similarity function is denoted as $S_2$, the proposed
hybrid similarity method is denoted as $S$ and it is defined as below:

$$S(c_j, c_n) = w_1 * S_1(c_j, c_n) + w_2 * S_2(c_j, c_n) \tag{1}$$

In this equation, $S_1(c_j, c_n)$ is the string similarity between concept(s) of query
($c_n$) and annotated concepts of Columbia374 ($c_j$) by LCS, and $w_1$ is the weight of
function $S_1$. $S_2(c_j, c_n)$ is the semantic similarity between concept(s) of query and
annotated concepts of Columbia374 by LSA, and $w_2$ is its weight. In the proposed
hybrid similarity method, combination is done using a linear combination with
an equal weight for each similarity measure [2].

## 4   Overview of Retrieval System

In video retrieval system, firstly, a query is expressed in terms of a new semantic
concept (new concept is a concept whose annotations are not provided by any
concept detector). Then, this query is given to the system and mapped onto the
374 concepts of Columbia374. In mapping, the similarity of concepts of query
and 374 concepts of Columbia374 is computed to see how relevant and close
they are. This similarity score is used for computing the confidence score for all
video shots. The confidence score shows the degree of relevancy between query
and shot. Therefore, for each query, the confidence score of all shots from the
test set should be computed. The dataset used in this study is TRECVID 2005.
TRECVID 2005 consists of development set and test set which will be explained
in details later.

The similarity of each new concept ($c_n$) - concept whose annotations are
not available - is measured with annotated concepts which are provided by
Columbia374 concept detector. Columbia374 acts as a learner for new concepts,
and it is presented as $C = \{c_i\}_{i=1}^{R}$, where R is the number of all annotated concepts which is assumed at most 374 due to Columbia374; $c_i$ is the $ith$ concept
in the corpus. Video shot database is presented as $D = \{S_k\}_{k=1}^{N}$ , where N is
the number of all video shots which is at most N=64256 for test set. Then, the
similarity measure is utilized in (6). Authors in [3] computed similarity measure
between two concepts by utilizing seven of knowledge-based semantic word similarity, four of corpus-based semantic word similarity and visual co-occurrence.
In the video retrieval system, for mapping, the proposed similarity method is
applied and used in (6) as well.

$$Score_{c_n}(S_k) = \frac{\sum_{i=1}^{R} sim(c_i, c_n) \times Score_{c_i}(S_k)}{\sum_{i=1}^{R} Score(c_i)} \tag{2}$$

Since the number of video shots for test set of TRECVID 2005 is 64256, there will be 64256 confidence scores for each query. Then, all confidence scores should be sorted, and the first 1000 ranked shots are selected and evaluated in terms of MAP.

## 5  Experimental Results

Experiments are done on the TRECVID 2005 (TV05) search dataset for evaluating the effectiveness of video retrieval system by applying the proposed similarity method. This dataset consists of 160 hours of multilingual television news. The first 80 hours is considered as development set and the remaining ones in related to test set. A set of 374 concept detectors, namely Columbia374 [20], is also used. These detectors are trained based on the development set of TRECVID 2005. In retrieving video shots for new concepts, seven concepts which were defined by NIST are used as queries. The retrieval system should return the ranked list of up to 1000 shots for each query. The ground truth for all queries is provided by NIST as well.

A variety of measures for evaluating the performance of information retrieval systems have been presented. A common measure of retrieval effectiveness called "non-interpolated average precision" is also defined by NIST over a set of retrieved documents (shots in our case are considered as the unit of testing and performance assessment). Average precision emphasizes on ranking relevant documents higher. It is the average of precision of the relevant documents in the ranked order. Let R be the number of true relevant documents in a set of size S; L is the ranked list of documents returned. At any given index j, let be the number of relevant documents in the top j documents. Let if the jth document is relevant and 0 otherwise. Assuming the non-interpolated average precision (AP) is then defined as [8]:

$$\frac{1}{R} \sum_{j=1}^{S} \frac{R_j}{j} \times I_j \tag{3}$$

Finally, the Mean of Average Precision (MAP) is calculated for all queries which are considered as an overall performance measure.

For comparison purpose, the proposed similarity method is compared to corpus-based semantic word similarity measures [3] which were applied in concept-based video retrieval. Table 1 indicates the performance of concept based video retrieval methods which are experimented in terms of AP and MAP. As can be observed from Table 1 "Car" and "Map" with APs of 0.53 and 0.5 are top two best-retrieved concepts, respectively by applying hybrid similarity method. "Prisoner" is the worst-retrieved concept, and AP for this concept using all concept based retrieval methods is close to 0. In overall, LSA with MAP of 18.33%

outperforms the other four corpus-based similarity measures used in [3]. The reason may be due to the genre of corpus, the genre of query or concepts. Since LSA performed very well among corpus-based semantic measures, it is combined with LCS string similarity measure. The Hybrid method performs very well and it has the best performance on many concepts and evaluations with MAP of 26.84%. As the results show, besides semantic similarity, string similarity can have beneficial impacts in query retrieval. That is why in this study, both semantic and string similarity measures are taken into account, and the essence and importance of considering both semantic and string measures together has been supported by experimental results.

**Table 1.** The performance of concept based video retrieval methods on TRECVID 2005 with Columbia374 in terms of AP and MAP

| Concept Name | Explosion or fire | Map | Waterscape | Mountain | Prisoner | Sports | Car | **MAP** |
|---|---|---|---|---|---|---|---|---|
| Topic ID | 1039 | 1040 | 1043 | 1044 | 1045 | 1046 | 1047 | |
| **AP of Corpus-based semantic similarity measures** | | | | | | | | |
| PMI-WebNEAR | 0.00335 | 0.0005 | 0.1175 | 0.157 | 0.000016 | 0.148 | 0.23 | 9.47% |
| PMI-WebAND | 0.0052 | 0.007 | 0.00256 | 0.0065 | 0 | 0.12 | 0.09 | 3.30% |
| PMI-WebImage | 0.0097 | 0.0001 | 0.042 | 0.226 | 0.000047 | 0.3 | 0.15 | 9.45% |
| PMI-WebFLKR | 0.0036 | 0.0001 | 0.0064 | 0.107 | 0.00002 | 0.153 | 0.07 | 4.85% |
| LSA | 0.045 | 0.142 | 0.0386 | 0.248 | 0.00001 | 0.305 | 0.5 | 18.33% |
| **AP of proposed hybrid similarity method** | | | | | | | | |
| LSA+LCS | 0.0556 | 0.503 | 0.1748 | 0.286 | 0.000004 | 0.325 | 0.53 | 26.84% |

Precision-recall curves for different methods are also plotted. As can be seen from Figure 1, the hybrid method and LSA are the top two concept-based video retrieval methods according to areas under their precision-recall curves. Among other corpus-based semantic similarity measures, PMI-IR-WebNEAR measure performs slightly better, compared to the others.

In order to show the success and capability of our proposed similarity method, we also apply it to some concepts different from concepts of Columbia374 and also seven test concepts or queries which were defined by NIST. Therefore, we define four concepts which are different from concepts of NIST and Columbia374. Moreover, ground truth for these four concepts is not available, and it is not possible to compute AP and MAP for such concepts by applying the proposed similarity method. Therefore, only top 100 video shots, which are retrieved for each concept by using the proposed similarity method, are demonstrated. These four concepts are as follow: "doctor", "sea", "fighting", and "politic". The retrieved video shots for these concepts are shown in Figure 2. Approximately all shots are semantically in very close relation to the defined concepts.

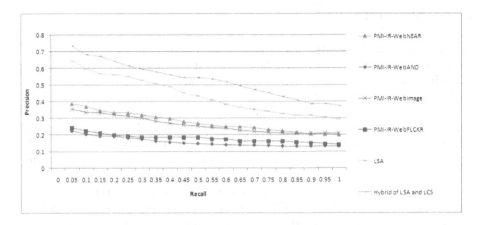

**Fig. 1.** Precision-recall curves for different concept based video retrieval methods on TRECVID 2005 with Columbia374

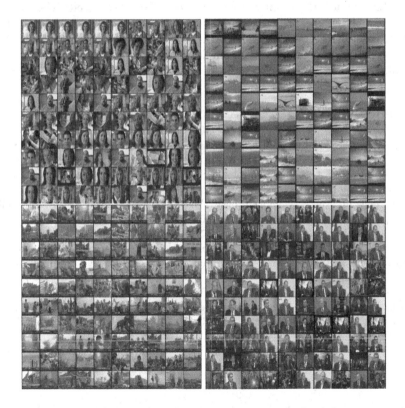

**Fig. 2.** Top 100 Retrieved Video Shots for Four Defined Concepts

# 6    Conclusion

In video search and retrieval, user's need is expressed in terms of query. Early video retrieval systems usually matched video clips with such low-level features. The difference between the user level and the low-level features of video leads to the one of the challenging issues named "Semantic Gap". Moreover, the limitation of retrieval using desired concepts detector is providing annotation. However, providing annotation for every concept in real world is very challenging and time consuming. Several previous studies [3,7,10] utilized semantic similarity measures for retrieving video shots based on semantic concept without relying on video annotations. However, due to some limitations with some semantic similarity measures in the previous studies and ignoring the string similarity measures, in this study, a method for similarity computation is proposed. The proposed similarity method is based on the linear combination of the both semantic and string similarity measures for semantic query retrieval in spite of unavailability of annotations for video shots. a series of experiments were done on TRECVID 2005 for evaluating the effectiveness of concept-based video retrieval model by applying the proposed similarity method in terms of AP and MAP. As experimental results shows, the proposed similarity method with MAP of 26.845 outperforms previously used similarity measures in concept-based video retrieval.

# References

1. Allison, L., Dix, T.I.: A bit-string longest-common-subsequence algorithm. Inf. Process. Lett. 23(6), 305–310 (1986)
2. Awang Iskandar, D.N.F., Thom, J.A., Tahaghoghi, S.M.M.: Content-based image retrieval using image regions as query examples. In: Proceedings of the Nineteenth Conference on Australasian Database, ADC 2008, vol. 75, pp. 38–46. Australian Computer Society, Inc., Darlinghurst (2007),
http://portal.acm.org/citation.cfm?id=1378307.1378319
3. Aytar, Y., Shah, M., Luo, J.: Utilizing semantic word similarity measures for video retrieval. In: IEEE Conference on Computer Vision and Pattern Recognition, Anchorage, Alaska (2008)
4. Bergroth, L., Hakonen, H., Raita, T.: A survey of longest common subsequence algorithms. In: Proceedings of the Seventh International Symposium on String Processing Information Retrieval (SPIRE 2000), p. 39. IEEE Computer Society (2000)
5. Berry, M.W.: Large scale sparse singular value computations. International Journal of Supercomputer Applications 6, 13–49 (1992)
6. Foltz, P.W., Kintsch, W., Landauer, T.K.: The measurement of textual coherence with latent semantic analysis (1998)
7. Haubold, A., Natsev, A.: Proceedings of the 2008 International Conference on Content-based Image and Video Retrieval, pp. 437–446. ACM, Niagara Falls (2008)
8. Hauptmann, A., Yan, R., Lin, W., Christel, M., Wactlar, H.: Can High-Level concepts fill the semantic gap in video retrieval? a case study with broadcast news. IEEE Transactions on Multimedia 9(5), 958–966 (2007)

9. Islam, A., Inkpen, D.: Semantic text similarity using corpus-based word similarity and string similarity. ACM Trans. Knowl. Discov. Data 2(2), 1–25 (2008)
10. Jiang, Y.G., Ngo, C.W., Chang, S.F.: Semantic context transfer across heterogeneous sources for domain adaptive video search. In: Proceedings of the Seventeen ACM International Conference on Multimedia, MM 2009, pp. 155–164. ACM, New York (2009)
11. Jiang, Y., Ngo, C., Yang, J.: Towards optimal bag-of-features for object categorization and semantic video retrieval. In: Proceedings of the 6th ACM International Conference on Image and Video Retrieval, pp. 494–501. ACM, Amsterdam (2007)
12. Kondrak, G.: N-Gram Similarity and Distance. In: Consens, M.P., Navarro, G. (eds.) SPIRE 2005. LNCS, vol. 3772, pp. 115–126. Springer, Heidelberg (2005)
13. Landauer, T., Foltz, P., Laham, D.: An introduction to latent semantic analysis. Discourse Processes (25), 259–284 (1998)
14. Melamed, I.D.: Bitext maps and alignment via pattern recognition. Comput. Linguist. 25(1), 107–130 (1999)
15. Mihalcea, R., Corley, C.: Corpus-based and knowledge-based measures of text semantic similarity. In: AAAI 2006, pp. 775—780 (2006)
16. Naphade, M., Smith, J.R., Tesic, J., Chang, S., Hsu, W., Kennedy, L., Hauptmann, A., Curtis, J.: Large-Scale concept ontology for multimedia. IEEE Multimedia 13(3), 86–91 (2006)
17. Snoek, C., Koelma, D., van Rest, J., Schipper, N., Seinstra, F., Thean, A., Worring, M.: Mediamill: Searching multimedia archives based on learned semantics. In: IEEE International Conference on Multimedia and Expo, ICME 2005, pp. 1575–1577 (2005)
18. Turney, P.: Mining the web for synonyms: PMI-IR versus LSA on TOEFL (2001)
19. Wolfe, M.B.W., Goldman, S.R.: Use of latent semantic analysis for predicting psychological phenomena: two issues and proposed solutions. Behavior Research Methods, Instruments, & Computers: A Journal of the Psychonomic Society, Inc. 35(1), 22–31 (2003); PMID: 12723777
20. Yanagawa, A., Chang, S.F., Kennedy, L., Hsu, W.: Columbia university's baseline detectors for 374 lscom semantic visual concepts. Columbia University ADVENT technical report (2007)

# A Conceptualisation of an Agent-Oriented Triage Decision Support System

Shamimi Halim[1,*], Muthukaruppan Annamalai[1],
Mohd Sharifuddin Ahmad[2], and Rashidi Ahmad[3],

[1] Faculty of Computer and Mathematical Sciences, Universiti Teknologi MARA, Malaysia
[2] College of Information Technology, Universiti Tenaga Nasional, Malaysia
[3] Department of Emergency Medicine, School of Medical Sciences,
Universiti Sains Malaysia Health Campus, Malaysia
{shamimi,mk}@tmsk.uitm.edu.my, sharif@uniten.edu.my,
shidee_ahmad@yahoo.com

**Abstract.** Triage is a complex process in ED that necessitates accurate, consistent and timely decision for assessment and management of incoming patients. With triage officers having different levels of expertise and experience, the triage scales are applied inconsistently, as evident from the lack of uniformity in the existing triaging system in Malaysian hospitals. In many hospitals, the triaging largely occurs in an ad hoc basis, placing patients at risk and contributing to inefficient resource utilisation. A computer system can be used to overcome these problems. Consequently, the paper proposes the development of a knowledge-based, agent-oriented decision support system for triage assessment and management. The ability to implement goal-directed behaviour, communicative processes and schedule-oriented, but concurrently executed tasks among others, make the agent technology suitable to be applied. We intend to adopt a modelling view to knowledge engineering, and plan to use the CommonKADS methodology as a basis for knowledge engineering, modelling and knowledge base development.

**Keywords:** Multi-agent system, BDI, Clinical Decision Support System, Triage Assessment, Emergency department.

## 1    Introduction

The Emergency Department (ED) is a primary care unit that provides initial treatment with a broad spectrum of illnesses and injuries that could be life threatening and require immediate attention and treatment [1]. The ED staffs consists of doctors, physician assistants, paramedic and nurses, emergency medicine technicians, respiratory therapists and healthcare assistants who work as a team to give the best emergency services around the clock. The ED staffs consistently face challenging decision making presentations under pressing time and resource constraints.

---

[*] Corresponding author.

D. Lukose, A.R. Ahmad, and A. Suliman (Eds.): KTW 2011, CCIS 295, pp. 272–282, 2012.

Triage is the first point of access to ED. Triage officers are responsible for quick assessment and management of the incoming patients. Decision making in Triage is not easy because it is made under conditions of uncertainty and also time and resource constraints. Triage decision making includes the ability to: i) receive information, ii) process and understand the information, iii) deliberate on the information, iv) make and defend the resulting choices [5].

Triage acuity scales or emergency severity index guidelines are available to help triage officers to make their triage decisions. The more recent and robust Triage scales (e.g. Canadian Triage and Acuity Scale) that consist of five or more categories and are capable of supporting a triage officer to make precise triage decision. These scales consider various entities and variables, and multiple sets of rules when triaging the patient, are however, quite complex. In practice, many years of clinical experience are required before a triage officer acquires the necessary skills to apply such scales.

In contrast, the Malaysian hospitals use a three-tier triage scale, except for Hospital Universiti Kebangsaan Malaysia and Pusat Perubatan Universiti Malaya that use a four-tier triage scale, and have remained unchanged for a long time. The ED at Hospital Universiti Sains Malaysia for example, is still using the 3-tier triage scale since it was established in 1982 [29]. The acquisition of a more robust Triage scale has been held back because of high staff retraining and implementation costs.

While the simple Triage scales contain fewer grey areas, it faces the stark reality of placing patients at risk due to under-triage or inadvertently depletes ED resources due to over-triage. An under-triage occurs when a severe case is triaged with lower acuity than it should be whereas an over-triage happens when a less severe case is triaged with high acuity.

In spite of the simplification, the officers having different levels of expertise and experience tend to apply the triage scale inconsistently. This is evident from the non-uniformity of the Triage scales across the Malaysian hospitals [20], [21], [22], [23], causing variable patients waiting times.

On one hand, the increased in the arriving patients combined with the shortage of proficient triage officers prolongs triage assessment. On the other hand, when rapid assessment is demanded of the officers, it gives rise to decision making errors. Either way, it may have detrimental effects on the patients' conditions. Under-triage jeopardises patients' life and prolong their pain and suffering. Over-triage delays giving treatment to waiting patients.

Therefore, an accurate and immediate triage decision is essential to ensure patients survival and maintain a good quality of service in ED. Hence, it is a good reason to have a computer system that can aid to triage the patients.

The usage of a decision support system (DSS) enables adaption of a robust Triage scale that can ensure correct assessment and precise allocation of triage category, i.e., overcome the problems of over- and under- triage. It has the potentials to maximize resources sufficiently and avoid prolong waiting times of patients. Unlike the manual triage, the result of the Triage DSS will be more objective, could be validated and retrieved for future reference. The Triage DSS will also promote the application of a uniform Triage scale across the local hospitals.

We regard agent technology as a promising platform to model, design and develop the Triage DSS. Agents are said to be suitable for problems where independent tasks can be easily distinguished [24], as apparent in the Triage process. In this respect, the complexity of triage decision making can be handled by decomposed intelligent elements owned by collaborating autonomous agents.

This paper provides motivation for, and a conceptualisation of a Triage DSS and outlines an agent-oriented model to this effect. The rest of the paper is organised as follows. In section 2, we briefly discuss the decision making in ED, but paying attention to the Triage process. The clinical DSS is discussed in section 3, and the motivation for using software agents to enact such DSS is given in the end. In section 4, we describe the conceptualisation of a proposed agent-oriented Triage DSS. Finally, in section 5, we summarise the contribution of this paper.

## 2    Decision Making in ED

There are three main sequential work processes in ED; beginning with Triage, then Treatment, and ending with Disposition. In Triage, the triage officers need to provide timely triage to patients requiring immediate care and must ensure those whose conditions are not likely to deteriorate could wait safely for care. Subsequently, the physicians need to provision the patients with the right treatment and transfer them to appropriate ward for continuous treatment if necessary. In the ensuing sub-sections, we will discuss the decision making in each of these work processes, but will elaborate on the Triage, the focus of this paper.

### 2.1    Triage

Upon arrival to ED, a patient will undergo a triage assessment to determine the order in which the patient will receive medical diagnosis and treatment based on the nature and severity of the illness condition. The assessment is both subjective and objective in nature, and is carried out by triage officers (e.g. paramedics, staff nurses and healthcare assistants).

The subjective assessment begins with patient's chief complaint, i.e., patient's statement of the problem (e.g. headache, nausea and pain), followed by oral history interview (e.g. *when did the condition start? how long did it last? does it come and go?* etc.) [3]. The general observation (e.g. physical appearance, degree of distress and emotional response) and the ensuing physical examination through objective assessments such as measuring vital signs as body temperature, heart beat rate, respiratory rate and blood pressure and so on, validate the patient's chief complaint, confirm the symptom severity and the clinical urgency, providing justification of the patient's tag category (e.g. red, yellow and green).

The tags indicate the waiting period before the patient is seen by a physician for treatment. For example, a patient with life threatening conditions will be issued a red tag and will be seen immediately by a physician while a patient with less severity symptoms will be issued a yellow or a green tag, which stipulates a waiting time up to thirty and hundred and twenty minutes, respectively [22].

Decision making during assessment involve the interpretation, discrimination and evaluation of information elicited from the patient and the physical examination to gauge the severity and acuity of the patient's condition, which results in the tagging of the patient [6]. While a red tagged patient is deemed to be unstable or potentially unstable and is diagnosed and treated immediately by physicians, the triage officers must take full responsibility over the patients who are issued with yellow or green tag, whose conditions are deemed not likely to deteriorate while waiting to be treated. Making a triage decision for the patients who fall in grey area (e.g. yellow and green category at the same time) can be tricky. Inability to make quick decisions can have detrimental effects on patient's condition and prolong a patient's waiting time and suffering.

## Triage Scales

A Triage scale is an emergency patient prioritisation tool that provides some guidance to triage patients in ED. The scale assists triage officers to evaluate the acuity level and resource needs. The Triage scales that are currently in wide use are The Canadian Triage and Acuity Scale (CTAS), Manchester Triage Scale (MTS) and Australian Triage Scale (ATS).

CTAS utilises five levels of classification: Level I – resuscitation (immediate treatment), Level II – emergent (treatment within fifteen minutes), Level III – urgent (treatment within thirty minutes), Level IV – less urgent (treatment within sixty minutes) and Level V – non urgent (treatment within hundred and twenty minutes). This scale has the ability to evaluate the acuity level and resource needs based on several factors such as respiratory, neurological, musculo-skeletal, shock, gastrointestinal, code and many others [4].

MTS also categorises a patient according to five clinical priorities: Level I – immediate (Red – immediate treatment), Level II – very urgent (Orange - treatment within ten minutes), Level III – urgent (Yellow - treatment within sixty minutes), Level IV – (Green - treatment within hundred and twenty minutes) and Level V – non-urgent (Blue - treatment within two hundred and forty minutes). This scale triage a patient based on the following six discriminators: i) life threat, ii) pain, iii) haemorrhage, iv) consciousness, v) temperature and vi) acuteness [2]

ATS assigns a patient into one of the five categories: Category 1 – resuscitation (immediate simultaneous assessment and treatment), Category 2 – emergency (treatment within ten minutes), Category 3 – urgent (treatment within thirty minutes), Category 4 – semi-urgent (treatment within sixty minutes) and Category 5 – non-urgent (treatment within one hundred and twenty minutes). Clinical descriptors used to triage a patient such as cardiac, respiratory, airway risk, circulation and many others [25].

The three-tier Triage scales applied in most of the Malaysia hospitals are developed based on Malaysian Triage Category (MTC) approved by the Ministry of Health [26]. The local Triage scale is founded on the following principles: save life, save organ or limb, relieve patient's suffering, prevent further deterioration, and do no harm.

## 2.2    Treatment

In general the outcome of sick patients depends on two main factors, which are disease management and response time. Continuum of care from the site of incident, accident or injury to the ED and eventually into the respective ward improves patient's survival. Management of sick patients can be divided into three treatment categories – i) Domestic aid, ii) First aid, and iii) Medical aid.

Domestic aid and first aid is provided at the scene of incident or casualty by herself or with the help of others. It is mainly for non urgent cases. First aid on the other hand is carried out by trained personnel whereby they capable to assess the victim's condition, providing temporary treatment and immediately refer them to the hospital. First aid saves life and comforts the victims.

In ED, both first aid and medical aid are carried out. During triage process, the triage officer provides first aid treatment to the patient based on the condition at resuscitation bay. Each bay may equipped with defibrillator, airway equipment, oxygen, intravenous lines and fluids and emergency drugs. This area also has ECG machines, limited x-ray facilities, non-invasive ventilation and portable ultrasound devices.

The duty physician institute a medical treatment after suspected diagnosis is made. Making a treatment plan and determine the prognosis are the tasks of a duty physician. The medical treatment is planned based on the result of the diagnosis, and by taking into consideration the patient's status and needs.

## 2.3    Disposition

The next process in ED is disposition. In many cases, once the patient's condition has been recognised as requiring hospitalisation has begun, much of the emergency physician's clinical work is accomplished. Still there are other reasons for continued care. The duty physician has to decide where the appropriate and safe place is for patient's continuing their treatment. Therefore, a decision to discharge patient from ED whether to Intensive Care Unit (ICU), ward for inpatient care or to other hospital that provides facility needed based on the prognosis must be done correctly. For example, in general those patients with mortality of 30% and above require intensive care. Those who require subspecialty management may need to be referred to respective hospital.

The most difficult patients cared for the emergency physician are the ones who are discharged. This is doubly true for those who leave without a clear diagnosis. Unfortunately, in a limited and often first time assessment, the data available may be insufficient for a disease process to be identified. To ensure optimal care, an appropriate discharge position should include the patient's basic understanding of 1) the underlying problem that caused them to seek emergency care, 2) the evaluation and treatment given in the ED, 3) follow up date, 4) criteria for immediate readmission.

As the nature of ED, which includes increases of ED attendances, outbreaks, staff shortage and obviously the need for rapid assessment and decision making, errors at disposition decision may place patient at risk. Note: The decision making in the Treatment and Disposition processes are outside the scope of this paper.

# 3    Clinical Decision Support System

Decision support systems (DSS) have emerged in healthcare industry over the decades [8]. They are designed to aid clinical decision making in which the characteristics of individual patients are matched to a computerised clinical knowledge base and recommendations are then presented to the clinician for a decision [14]. There are three categories of clinical DSS [19]:

i)   System for information management that provide data and knowledge for clinicians. This type of DSS includes medical information retrieval systems used to manage and extract medical knowledge and patient records.
ii)  System for focusing attention that prompt clinicians about actions that might require attention. This type of DSS can be used in ICU or laboratories to flag abnormal readings.
iii) Systems for providing patient-specific recommendations that provide assessments or advice based on patient-specific clinical data. This type of DSS should help clinician to make decisions for diagnostic and patient management.

The third category of invention of clinical DSS can help to address some of the complex decision making issues, which has resulted in unexpected consequences in ED. Dealing with an environment that cannot afford to dispense with human expertise, there is a need to design a DSS that has socio-technical elements. A socio-technical system embeds social structures such as the work processes, clinical roles, culture adaptation and new technologies in their scope of its design [11].

Based on the characteristics of clinical DSS and the decision making involved in Triage, an agent technology is suitable to be applied since it is a promising platform to model, design and develop a complex system. An agent is a computer system that is situated in some environment, and that is capable of autonomous action in this environment in order to meet its delegated objectives [28].

Agents could be deployed to share tasks between other agents by implementing goal-directed behaviour, communicative processes and other mundane and schedule-oriented tasks. In addition, a multi-agent system uses distributed problem solving techniques by decomposing a complex task into sub-tasks, and solving it in a decentralised fashion. The use of agents in healthcare has been shown to improve the performance of a computerised system in terms of interoperability, scalability and re-configurability [16].

# 4    The Proposed Agent-Oriented Triage DSS

The envisaged agent-oriented Triage DSS will support the tasks performed by ED staffs as the registration of the incoming patients, the Triage assessment, the scheduling of the treatment and the monitoring of the turns of the waiting patient. The agent oriented environment will work on the basis of "more severe, first serve" principle.

## 4.1     Agent Roles

We propose to employ the following six agent roles to enact the Triage system:

i)   **Registrar** agent registers the incoming patients at ED and sends the registered information to the Assessor agent.
ii)  **Vital_Sign_Reader** agent records the vital sign readings from the sensors and sends the data to the Assessor agent for triage assessment.
iii) **Assessor agent** assesses the severity and acuity of a patient's condition based on triaging knowledge and patient data.  The knowledge sources will be based on the Triage scale guidelines and physicians' knowledge. The data sources are the recorded vital signs, triage officers observation and the patient's history.  The result of the assessment is a triage tag (e.g. Red, Yellow and Green, indicating the ordered severity level from the highest to the lowest).

   The Assessor agent sends the information of a red tagged patient to the Treatment_Caller agent to facilitate the patient's diagnosis and treatment.  It sends the information of the yellow and green tagged patients to the Scheduler agent for determining the patients' treatment turns.
iv)  **Scheduler** agent performs a number of tasks.  Its main task is to schedules and manages the yellow and green tagged patients' treatment turns by maintaining a set of priority queues.  It creates a Monitor agent for each patient to keep track of the patient's waiting time, reschedules the patient if the waiting time expires, informs the Monitor agent to alert the its assigned patient to seek treatment when the patient's turn has arrived and kills the Monitor agent once the patient reports for treatment.
v)   **Monitor** agent notifies the Registrar agent and its assigned patient to register for re-triage once the allotted waiting time runs out.  It also alerts the patient to seek treatment when the patient's turn has arrived before reporting to the Treatment Caller agent.
vi)  **Treatment_Caller** agent is responsible for calling a patient for treatment through the Scheduler agent when the resources for treatment are available.  It also instructs the Scheduler agent to kill the Monitor agent assigned to a patient, once the patient reports for treatment.

## 4.2     Agent Interactions

These agents interact among themselves through information exchange.  Figure 1 delineates the agents' interaction model. Communications between agents are identified by arrows, where the direction of arrows identifies the initiating and receiving ends. The labels on the arrows are descriptors identifying interactions and information flow between the agent-agent and agent-human.  Note that the integrated system still requires human in the loop, particularly to input certain data (e.g. chief complaint, general observation and physical examination).  For the purpose of illustration, we consider below four basic scenarios and list the flow of information for each case at the end.

i)   The registered patient is issued a red tag (highest severity level) based on the triage assessment. The patient is immediately seen for diagnosis and treatment. (Information flow: $1 - 2 - 3 - 4$)

ii)  The registered patient is issued a yellow (medium severity level) or a green (lowest severity level) tag. While waiting for his turn, the patient gets called for treatment. (Information flow: $1 - 2 - 3 - 5 - 6 - 7 - 8 - 9 - 10 - 11$)

iii) The registered patient is issued a yellow or a green tag. While waiting, the waiting time allotted to the patient expires, and so he needs to be re-triaged. (Information flow: $1 - 2 - 3 - 5 - 6 - 12$)

iv)  The registered patient is issued a yellow or a green tag. While waiting, the patient develops new condition, and so needs to be re-triaged. (Information flow: $1 - 2 - 3 - 5 - 6 - 13$)

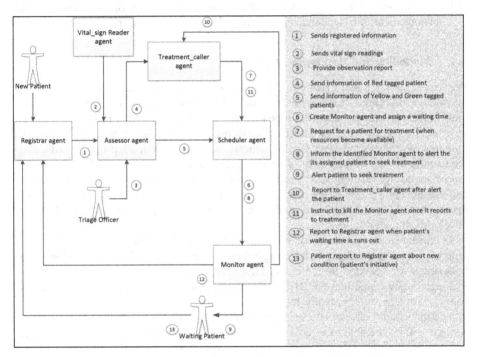

**Fig. 1.** Triage DSS agents' interaction model

## 4.3    Agent Features

The objectives of the agent are to implement the functionalities by exhibiting goal-directed behaviours that are triggered signals from physical entities or messages from other agents. Based on definition of agent, there are several features that are exhibit by agents [17], which demonstrate their decision-making ability in Triage:

i)   **Autonomy** describes the ability of agent to do a decision making task with minimal or no interventions from humans or other entities. The agents have a control over their actions and internal state.

ii)   **Reactivity** describes the ability to react or perceive the ED triage environment and respond to changes that occur in it in timely manner.

iii)  **Pro-activeness** describes the ability to do an accurate triage, schedules the patients, monitor them and take initiative when appropriate.

iv)   **Social ability** describes the ability to interaction between agent (and human) when there is necessity.

v)    **Modularity** describes the ability to perform triage decision making by dividing and distributing the complex task among multiple agents.

vi)   **Efficiency** describes the ability to enhance the overall decision making process by parallel and concurrent execution.

vii)  **Flexibility** describes the ability to dynamically add or remove agents at runtime based on execution requirements.

viii) **Event management** describes the ability to continuously monitor the management of triage since agents operate in a natural way and react to changes.

### 4.4    Modelling and Development Plan

We envisage an agent-oriented Triage DSS that captures the decision making knowledge structured in a knowledge base. We intend to adopt a modelling view to knowledge engineering, and so, propose to use the CommonKADS methodology [30] as a basis for knowledge engineering, modelling and knowledge base development. Due to strong interaction problem hypothesis, we see value in adopting and adapting the CommonKAD's assessment, scheduling and monitoring generic task templates to aid in acquiring, refining, and evaluating the requisite decision making knowledge, and subsequently guide in the representation of the knowledge. In order to model the agent system, we intend to apply an extension of the GAIA [31,32] agent development methodology.

In current implementations of many agent-based systems, the agents are endowed with humanistic behaviours like having belief, desires and intentions [33]. In fact such behaviours are encoded as the Belief-Desire-Intention (BDI) agent architecture. In this regards, we opine that the BDI architecture could be implemented in the triage as follows: An agent's beliefs are its knowledge about the world, its internal state, and other agents and their mental states. Its desires correspond to the tasks allocated to it. Since it will not be able to achieve all its desires, it must decide on some subset of its desires and commit resources to achieving them. The chosen desires represent its intentions which correspond to its goal [28].

Consequently, we will develop the system using a BDI-based agent programming language. And, we plan to check the consistency of the system with itself and with expert using the simulation approach.

## 5    Conclusions

In this paper, we have conceptualised a Triage DSS for ED in Malaysian hospitals. We discussed the main ED processes and focussed on the decision making involved in Triage, which is a challenging process associated with complex clinical assessments and management of patients. Given the nature of triage and the problems identified, we propose an agent-oriented DSS for triage. The conceived system consists of six agent roles, namely Registrar agent, Assessor agent, Vital_sign Reader agent, Treatment_caller agent, Scheduler agent and Monitor agent. The tasks of each of these agent roles have been defined, and a model of interaction between the agents is delineated. We also sketched out few possible scenarios to illustrate how the agents could interact to achieve their design. We hope to show that the planned DSS can help to expedite triage assessment in a consistent manner and at the same time increase the control over the management of the patients in ED.

# References

1. Daknou, A., Zgaya, H., Hammidi, S., Hubert, H.: Agent Based Optimization and Management of Healthcare Processes at The Emergency Department. International Journal of Mathematics and Computers in Simulation 2, 285–294 (2008)
2. Jones, K.M.: Emergency Triage – Manchester Triage Group. BMJ Publishing Group, London (1997)
3. Beveridge, R., Clarke, B., Janes, L., Savage, N., Thompson, J., Dodd, G., Murray, M., Jordan, C.N., Warren, D., Vadeboncoeur, A.: Implementation Guidelines for the Canadian Emergency Department Triage & Acuity Scale (CTAS). The Canadian Association of Emergency Physicians (CAEP), The National Emergency Nurses Affiliation of Canada (NENA) and L'association des medecins d'urgence du Quebec, AMUQ (1998)
4. Steel, I.R.: Evolution of Triage Systems. Emergency Medicine Journal 23, 154–155 (2006)
5. Larkin, G.L., Marco, C.A., Abbott, J.T.: Emergency Determination of Decision-making Capacity: Balancing Autonomy and Beneficence in The Emergency Department. Academic Emergency Medicine 8, 282–284 (2001)
6. Cronin, J.G.: The Introduction of the Manchester Triage Scale to an Emergency Department in The republic of Ireland. Accident and Emergency Nursing 11, 121–125 (2003)
7. Franklin, A., Liu, Y., Li, Z., Nguyen, V., Johnson, T.R., Robinson, D., Okafor, N., King, B., Patel, V.L., Zhang, J.: Opportunistic Decision Making and Complexity in Emergency Care. Journal of Biomedical Informatics 44, 469–476 (2011)
8. Dhar, V., Stein, R.: Intelligent Decision Support Methods: The Science of Knowledge Work. Prentice Hall, Upper Saddle River (1997)
9. Kawamoto, K., Houlihan, C.A., Balas, E.A., Lobach, D.F.: Improving Clinical Practice using Clinical Decision Support Systems: A Systematic Review of Trials to Identify Features Critical to Success. BML 330, 765–772 (2005)
10. Moreno, A.: On the Evolution of Applying Agent Technology to Healthcare. IEEE Intel. Syst. 21, 8–10 (2006)
11. Coiera, E.: Four Rules for the Reinvention of Health Care. BMJ 328, 1197–1199 (2004)
12. Bradley, V.M.: Placing Emergency Department Crowding on the Decision Agenda. Nursing Economics. Journal Emergency Nurse. 31, 247–258 (2005)
13. Geary, U., Kennedy, U.: Clinical Decision-Making in Emergency Medicine. Emergencias 22, 56–60 (2010)
14. Sim, I., Gorman, P., Breenes, R., Haynes, R., Kaplan, B., Lehmann, H., et al.: Clinical Decision Support Systems for the Practice of Evidence-based Medicine. Journal Med. Inform. Assoc. 8, 527–534 (2001)
15. Berlin, M., Sorani, M., Sim, I.: A Taxonomy Description of Computer-Based Clinical Decision Support Systems. Journal of Biomedical Informatics 39, 656–667 (2006)

16. Isern, D., Sanchez, D., Moreno, A.: Agent Applied in Health Care: A Review. International Journal of Medical Informatics 79, 145–166 (2010)
17. Sanchez, D., Isern, D., Moreno, A.: Agent Technology and Healthcare: Possibilities, Challenges and Examples of Application. In: Bichindaritz, A., et al. (eds.) Computational Intelligence in Healthcare, vol. 4, pp. 25–48. Springer, Heidelberg (2010)
18. Schreiber, G., Wielinga, B., Hoog, R.: CommonKADS: A Comprehensive Methodology for KBS Development. Journal IEEE Expert: Intelligent Systems and Their Applications 9, 28–37 (1994)
19. Wojtek, M., Slowinski, R., Wilk, S., Faridon, K.J., Pike, J., Rubin, S.: Design and Development of a Mobile System for Supporting Emergency Triage. Methods of Information in Medicine 44, 14–24 (2005)
20. Azhar, A.A., Ismail, A.S., Ham, F.L.: Patient Attendance at a Major Accident and Emergency Department: Are Public Emergency Services Being Abused? Med. J. Malaysia 55, 164–168 (2000)
21. Guidelines for Triaging at Hospital Universiti Sains Malaysia. Hospital Universiti Sains Malaysia
22. Guidelines for Triaging at Hospital of Kementerian Kesihatan Malaysia. Kementerian Kesihatan Malaysia
23. Laporan Tahunan Pusat Perubatan Universiti Malaya 2009. Pusat Perubatan Universiti Malaya
24. Aylett, R., Brazier, F., Jennings, N., Luck, M., Nwana, H., Preist, C.: Agent Systems and Applications. In: Second UK Workshop on Foundations of Multi-Agent Systems (1997)
25. Guidelines on the Implementation of the Australasian Triage Scale in Emergency Departments. Australasian College for Emergency Medicine (2005)
26. National Patient Safety Indicators: Technical Specifications. Ministry of Health, Malaysia (2009)
27. Wooldridge, M.: An Introduction to MultiAgent Systems. Wiley, West Sussex (2009)
28. Wooldridge, M.J., Jennings, N.R.: Intelligent Agents: Theory and Practice. Knowledge Engineering Review 10, 115–152 (1995)
29. Kamari, Z.: Hospital Universiti Sains Malaysia (HUSM): 25 Years of Excellent Service. Malaysian Journal of Medical Sciences 16, 16–24 (2009)
30. Schreiber, G., Akkermans, H., Anjewierden, A., Hoog, R., Shadbolt, N., Velde, W.V., Wielinga, B.: Knowledge Engineering and Management: the CommonKADS Methodology. Massachussettes Institute of Technology (2000)
31. Akbari, Z., Faraahi, A.: Evaluation Framework for Agent-Oriented Methodologies. World Academy of Science, Engineering and Technology 45, 418–423 (2008)
32. Wooldridge, M., Jennings, N.R., Kinny, D.: The Gaia Methodology for Agent-Oriented Analysis and Design. Autonomous Agents and Multi-Agent Systems 3, 285–312 (2000)
33. Rao, A.S., Georgeff, M.P.: BDI Agents: From Theory to Practice. In: Proceedings of the First International Conference on Multi-Agent Systems (ICMAS 1995) (1995)

# Evaluating Multiple Choice Question Generator[*]

Sieow-Yeek Tan, Ching-Chieh Kiu, and Dickson Lukose

Artificial Intelligence Center, Mimos Berhad,
Technology Park Malaysia, Bukit Jalil, Kuala Lumpur, Malaysia
{tan.sy,cc.kiu,dickson.lukose}@mimos.my

**Abstract.** Semantic-based computer-assisted automated question generator has been increasingly popular as a tool for creating personalized assessment questions. Various question generator tools have been proposed, such as those which generate structured question from a text file, generate Multiple Choice Question (MCQ) from a text file or from ontology-based knowledge representation. A comparison framework and evaluation methodology is required to evaluate different question generator tools. This paper discusses the requirement and criteria to carry out performance comparison of different MCQ generator. A feature comparison of Question Generation (QG) tool, namely Mimos-QG with the existing QG tools is presented. We have evaluated our QG tool based on several standard criteria such as the correctness of (a) distractor generation, (b) answer choice grouping strategy and (c) syntactical and pedagogical quality on three different domain ontologies. The experimental result indicated that Mimos-QG is capable of producing good quality direct type and grouping type multiple choice questions.

**Keywords:** Question Generator, Multiple Choice Question Generator, Ontology, Semantic Web, Education Technology.

## 1 Introduction

In the last decade, various computer-assisted educational tools have been introduced for different teaching domains. Question-answering (Q&A) is an important method to evaluate students' understanding and performance in either electronic (online) or traditional (offline) learning process. Studies in [1], [2] and [3] show that students have substantially benefited from automated question-answering approach offered via e-learning. Studies [4] and [5] have also shown that high number of questions from the same subject drives an in-depth understanding of the topic and subject. Multiple choice question (MCQ) is one popular mean of self-assessment in learning evaluation. MCQ comprises of a short text describing the question, a number of alternative choices where one of the choices is the correct answer, and the others are wrong answers (a.k.a distractors [6]).

---

[*] The authors developed this system under the Semantic Technology Research & Development initiative funded by the Ministry of Science, Technology & Innovation (MOSTI), Malaysia.

D. Lukose, A.R. Ahmad, and A. Suliman (Eds.): KTW 2011, CCIS 295, pp. 283–292, 2012.

The use of ontology-based knowledge representation in semantic-based question generation is becoming common nowadays. Domain ontologies can be considered as a proper formalism which serves the basis of automatic assessment. It contains the domain knowledge in the form of definitions of terms (concepts), individuals (instances) belong to these terms and the relationship between these terms and individuals. The formalism is required to conform to Semantic Web Technology standard in Web Ontology Language (OWL) representation [9]. The trend from the literature survey showing work on ontologies in e-learning domains has been primarily focused on ontological formalization. Studies in [7, 8] have proposed methods in constructing the learning objects, instructional processes and learning design by using ontology formalism. Hence, an ontology-driven QG is needed to be researched and developed to utilize the benefit of ontology in the domain subject in the e-learning system [11].

By leveraging on the pre-defined OWL standard, automatic generation of questionnaires based on existing ontological descriptions have been used to formulate multiple choice questions from MCQ generator. Papasalouros et. al. [6] proposed various strategies to automatically generate MCQ from domain ontologies. It takes OWL ontology as inputs and generates multiple choice questions. The MCQs are generated based on the basic meta-ontology relations and 'binary role' between class and instances. Their strategies were further implemented as a plugin for Protégé ontology editor by Tosic and Cubric [10]. Cubric et. al. [11] enhanced the abovementioned strategies by generating assessment for different knowledge domains. They proposed a proprietary MCQ ontology and used it as a pre-defined question template. Semantic interpretation is then applied to perform mapping between an arbitrary domain and the pre-defined MCQ ontology, in order to generate the MCQ. Bloom's taxonomy standard is used for supporting the MCQ generation process; by grouping the questions into several categories which required different comprehension levels to answer.

Inevitably, good quality question is the key to perform effective assessment over a learning process. Many criteria of evaluation have been used in evaluating MCQ generation strategies. For example, in [12], questions are evaluated by checking whether it is correctly written, meaningful, clear, appropriate to the context and etc. However, most of the evaluation processes are still done manually and are subjective to expert domain. For instance, the ontology used as an input dataset for a QG is normally constructed by the domain experts that are also going to evaluate the questionnaires generated by the QG [6]. Thus, a comparison framework and evaluation methodology is required to evaluate different MCQ generator.

The aim of the paper is to discuss the requirement and criteria to carry out performance comparison for each MCQ generator. We evaluate our QG tool, namely Mimos-QG based on several standard criteria such as distractor generation correctness, answer choice grouping strategy correctness, syntactical correctness and pedagogical quality on three different domain ontologies. In addition, we present a feature comparison of Mimos-QG with existing QGs tools.

The structure of the paper is outlined as follow: Section 2 discusses the requirement for evaluating MCQ QG tool. A literature review and comparison between Mimos-QG and the other MCQ generators are explicated in Section 3. Section 4 presents the experimental evaluation results of Mimos-QG. Conclusion and further work are concluded in the last section.

## 2     Requirement of Evaluating MCQ Generator

MCQ question formulation includes generating multiple distractors against a correct answer in a list of choices. While generating MCQ from ontology, the distractors generation can be generally categorized into three techniques, these are: (1) Class-based instance strategy, (2) Property-based strategy and (3) Terminology-based strategy. Ontology consists of concepts or classes, properties and instances. The classes are connected with 'is-a' relationship to denote the super-class and sub-class hierarchy. In class-based instance strategy, distractors are generated based on classes and instances by choosing the proper classes different from those that appear in the correct answer. Meanwhile in property-based strategy, the question components and distractors are generated based on properties which denote relationship between classes in the ontology. Lastly, in terminology-based strategy, the question component and distractors are generated based on concept or sub-concept relationships.

In order to effectively evaluate an ontology independent MCQ generator tool, a set of evaluation requirement and criteria need to be defined. These criteria are listed as below but are not limited to the following:

(1) Input requirements: refer to the content used by QG tool to generate multiple choice questions. The quality of input will definitely influence the quality and type of the questions generated.
(2) Level of user interaction: some QG tools require user intervention in producing a set of MCQ questionnaires. For example, user intervention is required for defining tags or question categories and defining templates for specific tags [12]. Some tools require user intervention in selecting qualify questions to be part of a questionnaire set or in defining question template before generating questions.
(3) Type of output: different type of QG tools will generate different type of multiple choice questions. For example, the MCQ's question and answer pair can be True or False or providing with a list of answer choice for selection purposes. Some output will be in different representation and require further post-processing steps to display them in a proper MCQ format.
(4) Quality of question: pedagogical quality, linguistic or syntactical quality which is normally judged and ranked manually by instructors or domain experts [13].

In order to compare and evaluate ontology independent MCQ generator tool, we need to define and develop the following resources: source of ontologies as the input requirements, benchmarking results and metrics for comparing the performance of the tools.

## 3     Survey of the Multiple Choice Question Generators

### 3.1     Related Works

In year 2009, an international patent [14] had been filed for a method and tool for automatically generating questions for a programming language. The proposed method created a set of pre-defined question templates containing compliable codes.

Some misinterpretation templates were used to generate options for a question which included various correct and incorrect answers. Questions are automatically generated by changing certain variables' value in the compliable code. The incorrect options are served as distractors which were generated by changing the tolerance limit value defined by the user.

Study in [15] defined an innovative approach for generating MCQs by adapting to a student understanding model. A question is generated by applying term information to a pre-defined template stored in question-template database. The questions are generated using "is-a" and "part-of" attributes, which show the relations between concepts. In addition the proposed methodology introduced a method to generate questions of varying levels of difficulty. More difficult questions are generated when other alternatives are conceptually near to the correct alternative. More importantly, the difficulty of the generated questions is adapting to different students based on their understanding level for a studied subject.

In [11], an ontology driven QG for serving the purpose of automated generation of computer-assisted assignment (CAA) from the Semantic Web was proposed. The method defines a proprietary MCQ ontology to act as the basis for development of question types. Different question type consists of its own specification to produce its template. The question template then is used for MCQ generation by mapping to subject knowledge described in the underlying domain ontology. Also, the idea of categorizing the generated questions at multiple level of comprehension on a subject was introduced. Question that is categorized at higher level of comprehension will require higher level of understanding on a subject to answer. The level of comprehension is represented from one of the most influential models of educational competencies, namely Bloom's Taxonomy [16]. These levels are noted as "Knowledge", "Comprehension", "Application" and "Analysis". The abovementioned method successfully links the generated question to different comprehension level by using annotation-based and semantic measurement strategies when generating the question-answer pairs. For example, distractors from a MCQ which are conceptually having "high-similarity" to the correct answer will require higher level of comprehension to answer it correctly.

## 3.2    Similarities and Differences in Question Generation Strategy

In general, the approach from [14] described at previous section holds certain similarities to our Mimos-QG. One of them is that it employed a question template technique in generating the correct answer and distractors. However our QG implementation enables dynamic question template selection which is independent from subject domain. Our approach comply with ontology standard in generating question which indirectly allow conceptual intelligent to be taken into consideration during question generation process. Unlike method proposed by [14], in which the questions are generated by just changing certain programming variable values in a set of pre-defined compliable code templates, our QG strategy that leverages on ontology standard composes question by making use of the additional relationship between the concepts described under OWL formalism.

The QG approach from the study in [15] applies question template technique in MCQ generation. As the QG approach also leverages on ontology formalism, the method allows conceptual space consideration in generating the MCQ options. The evaluation of the difficulty level of the generated question has done indirectly by measuring the conceptual space between the correct answer and the distractors. This strategy is similar to Mimos-QG approach. The difference falls in the techniques used for generating MCQ options. In [15], the generation of question options is based on pre-defined contents. The contents are structurally organized and stored in a terms-information database. Meanwhile, our approach generates question options based on a rule-based conceptual relationship. In other words, the generation of correct answer and the distractors is governed by a set of pre-defined rules baseline to ontology formalism.

In [11], an ontology driven QG was proposed which holds certain similarity to Mimos-QG methodology. The author defined a new MCQ ontology as a basis for the development of the MCQ question format specification, which is similar to the question template database used in the Mimos-QG. In addition, the author proposed rules to govern the semantics mapping between the domain knowledge and MCQ to enable conceptual intelligent consideration when generating questions. The differences between the QG tool when compared to Mimos-QG are: the question template defined in MCQ ontology results in a rigid format specification as compared to Mimos-QG approach which has maintained flexibility by enabling dynamic question template selection in MCQ generation. Also, they use annotation-based strategies in generating question-answer pair as well as the distractors instead of rule-based.

### 3.3    Features Comparison of MCQ Generator Tools

Different QG adopt different strategies and methodologies to automatically generate multiple choice questions. In this section, a comparison of features between the above mentioned QG tools and Mimos-QG are outlined in Table 1. The comparison study is defined based on the knowledge obtained from the reference citations of the tools. Fourteen features have been identified for the comparison as shown in Table 1. If the QG tool contained the specific feature, it will be noted as 'Yes', otherwise it will be noted as 'No'.

As summarized from the features comparison table, a qualitative feature comparison between the three QG tools to Mimos-QG is depicted in Fig. 1. The qualitative feature (QF) comparison can be categorized into several aspects, namely: Various types of question generation using template (QF1), Conceptual space consideration in generating correct answer and question (QF2), Conceptual space consideration in generating distractors (QF3), Dynamic question generation rule independent from question template (QF4) and Evaluate question difficulty (QF5).

**Table 1.** Features Comparison in Multiple Choice Question Generators

| Question Generator Features | Tool-1 [14] | Tool-2 [15] | Tool-3 [11] | Mimos-QG |
|---|---|---|---|---|
| 1. Able to identify topic level ontology for question generation | No | No | No | Yes |
| 2. Enable conceptual intelligent consideration between correct answer and generated question | No | Yes | Yes | Yes |
| 3. Applying question template technique in MCQ question generation | Yes | Yes | Yes | Yes |
| 4. Enable dynamic question template selection in question generation | No | Yes | No | Yes |
| 5. Enable dynamic question template selection independent from target concept | No | No | No | Yes |
| 6. Generating question-answer pair based on conceptual relationship | No | Yes | Yes | Yes |
| 7. Generating question-answer pair based on rule-based conceptual relationship | No | No | No | Yes |
| 8. Generating distractors in question generation | Yes | Yes | Yes | Yes |
| 9. Generating distractors by considering conceptual space relationship | No | Yes | No | Yes |
| 10. Generating distractors govern by rule-based conceptual relationship | No | No | Yes | Yes |
| 11. Implementing method to evaluate question difficulty level | No | Yes | No | Yes |
| 12. Evaluate question difficulty level by considering conceptual space between correct answer and distractors | No | Yes | No | Yes |
| 13. Evaluate question difficulty level by rule-based conceptual relationship calculation on correct answer and distractors | No | No | No | Yes |
| 14. Categorize difficulty level based on Bloom Taxonomy (BT) level using quantified difficulty value | No | No | Yes | Yes |

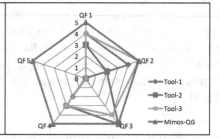

| Qualitative Features | Tool-1 | Tool-2 | Tool-3 | Mimos-QG |
|---|---|---|---|---|
| QF 1 | 4 | 3 | 4 | 5 |
| QF 2 | 4 | 2 | 5 | 5 |
| QF 3 | 0 | 5 | 4 | 5 |
| QF 4 | 2 | 3 | 3 | 5 |
| QF 5 | 0 | 0 | 0 | 5 |

(Rated 1 to 5) where 1 is the lowest and 5 is the highest

**Fig. 1.** Qualitative Feature Comparison

# 4    Mimos-QG Evaluation

Mimos-QG is an ontology independent Multiple Question Generator tool. The class-based instance strategy and terminology-based strategy are adopted to generate MCQ. The class-based strategy is categorized into three approaches known as: direct approach, negative approach and group approach. Fig. 2 shows the direct answer type of MCQ compared to group answer type of MCQ. The algorithm and framework of the Mimos-QG can be referred from [17].

## 4.1    Data Set

We have identified three ontologies from different domain: Accounting, Travel and Conference which were used as input dataset for evaluating the Mimos-QG. The details of ontology are presented in Table 2 and conceptualized in Fig. 3.

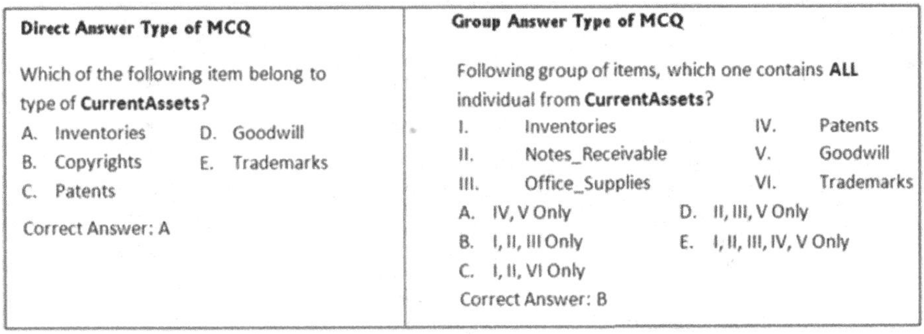

| Direct Answer Type of MCQ | Group Answer Type of MCQ |
|---|---|
| Which of the following item belong to type of **CurrentAssets**?<br><br>A. Inventories    D. Goodwill<br>B. Copyrights    E. Trademarks<br>C. Patents<br><br>Correct Answer: A | Following group of items, which one contains **ALL** individual from **CurrentAssets**?<br><br>I.  Inventories    IV.  Patents<br>II.  Notes_Receivable    V.  Goodwill<br>III.  Office_Supplies    VI.  Trademarks<br>A.  IV, V Only    D.  II, III, V Only<br>B.  I, II, III Only    E.  I, II, III, IV, V Only<br>C.  I, II, VI Only<br>Correct Answer: B |

**Fig. 2.** Direct answer type of MCQ and group answer type of MCQ

**Table 2.** Dataset for Mimos-QG experimental evaluation

| Ontology | Class | Object Property | Datatype Property | Instance |
|---|---|---|---|---|
| Accounting | 21 | 2 | 8 | 59 |
| Conference | 98 | 60 | 32 | 0 |
| Travel | 30 | 6 | 4 | 10 |

## 4.2    Experimental Result

During experimental study, twenty questions were generated from each ontology assigned as input dataset. The questions were evaluated based on the capability of the ontology to generate each type of question from the Mimos-QG's question generation methods, quality of distractors and grouping answer choice, syntactically correctness and pedagogical quality. As shown in the Table 3 and Fig. 4, the MCQ generation from ontology is subjective to the nature of the ontology, whereas by the hierarchy of concepts and number of concepts and instances of the ontology has affected the evaluation criteria of the MCQ questions. As shown in Fig. 4(a), four types of multiple choice questions (Direct, Negative and Group approach of Class-Based Instance Strategy and Terminology-based Strategy) are generated from the Accounting ontology. As depicted by Fig. 4(b), all type of multiple choice questions except Negative approach of Class-Based Instance Strategy are generated from the Travel ontology. On the other hand, Conference ontology was only able to generate Group approach of Class-Based Instance Strategy and Terminology-based Strategy type of multiple choice questions as shown in Fig. 4(c).

The Mimos-QG performance on syntactical correctness and pedagogical skill measurement was good and acceptable as shown in Fig. 4(d). The results have proven that the Mimos-QG is capable of generating quality multiple choice questions to support assessment of different subject learning and understanding. However, the distractors and grouping answer choices generated from the Conference ontology is poor due to the lack of instances in the ontology itself.

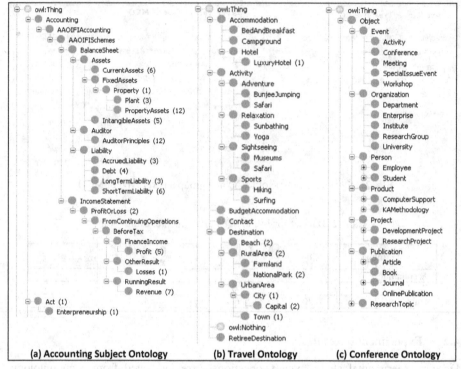

(a) Accounting Subject Ontology     (b) Travel Ontology     (c) Conference Ontology

**Fig. 3.** Sample Ontologies from Different Domains

**Table 3.** Result of the Mimos-QG based on three ontologies

| Question | Accounting | | | | Travel | | | | Conference | | | |
|---|---|---|---|---|---|---|---|---|---|---|---|---|
| | (a) | (b) | (c) | (d) | (a) | (b) | (c) | (d) | (a) | (b) | (c) | (d) |
| Q1 | 3 | Y | N | N | 3 | Y | Y | Y | 3 | N | Y | Y |
| Q2 | 1 | Y | Y | Y | 1 | Y | Y | N | 3 | N | Y | Y |
| Q3 | 3 | N | Y | N | 4 | Y | Y | Y | 1 | Y | Y | Y |
| Q4 | 3 | Y | N | Y | 4 | Y | Y | Y | 1 | Y | N | N |
| Q5 | 3 | Y | N | Y | 4 | Y | Y | N | 3 | Y | Y | Y |
| Q6 | 3 | Y | Y | Y | 3 | Y | Y | Y | 1 | Y | N | Y |
| Q7 | 2 | Y | Y | Y | 3 | Y | Y | Y | 1 | N | N | Y |
| Q8 | 1 | Y | Y | Y | 4 | Y | N | N | 3 | Y | Y | Y |
| Q9 | 4 | Y | N | Y | 3 | Y | Y | Y | 3 | N | Y | Y |
| Q10 | 3 | Y | Y | N | 3 | N | Y | Y | 3 | N | Y | Y |
| Q11 | 3 | Y | Y | Y | 4 | Y | Y | Y | 3 | N | Y | Y |
| Q12 | 3 | Y | N | N | 4 | Y | N | N | 3 | N | Y | Y |
| Q13 | 3 | N | Y | Y | 4 | Y | Y | N | 3 | N | Y | N |
| Q14 | 3 | Y | N | N | 4 | Y | N | Y | 3 | N | Y | Y |
| Q15 | 3 | Y | Y | Y | 3 | N | Y | Y | 3 | Y | Y | Y |
| Q16 | 3 | Y | Y | Y | 4 | Y | N | N | 3 | N | N | Y |
| Q17 | 2 | Y | Y | Y | 3 | Y | Y | Y | 1 | Y | Y | Y |
| Q18 | 3 | Y | Y | Y | 3 | N | Y | Y | 3 | N | Y | N |
| Q19 | 3 | Y | Y | N | 3 | Y | Y | Y | 3 | N | Y | Y |
| Q20 | 1 | Y | Y | Y | 3 | N | Y | Y | 3 | Y | Y | Y |

| Label | Type of question: |
|---|---|
| (a) Type of question | Type 1: Class-based Instance Strategy (Direct Approach) |
| (b) Distractor & Grouping | Type 2: Class-based Instance Strategy (Negative Approach) |
| (c) Syntatically | Type 3: Class-based Instance Strategy (Group Approach) |
| (d) Pedagogical | Type 4: Terminology-based Strategy Methodology |
| Y - Yes | |
| N - No | |

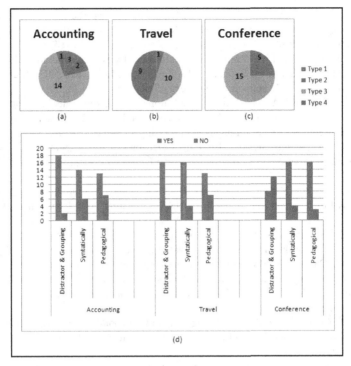

**Fig. 4.** Evaluation result of MCQ questions generated by Mimos-QG

## 5    Conclusion

In this paper, we have discussed the criteria and requirements for evaluating an Independent Ontology Question Generator. We have also presented the experimental result on Mimos-QG. The experimental result have indicated that Mimos-QG is capable of generating quality MCQ questions to support assessment of different subject learning and understanding

For the future direction of Mimos-QG development, we intend to extend it in several prospective directions: (1) extend and enrich the question template to achieve higher level of syntactically correct question generation and (2) extend the MCQ generation strategies to incorporate other ontology components such as rules, axioms, restrictions, events and etc.

## References

1. Andrenucci, A., Sneiders, E.: Automated Question Answering: Review of the Main Approaches. In: 3rd International Conf. on Information Technology and Applications, pp. 514–519 (2005)

2. McGough, J., Mortensen, J., Johnson, J., Fadali, S.: A Web-Based Testing System with Dynamic Question Generation. In: 31st ASEE/IEEE Frontiers in Education Conference, vol. 3, pp. 3–23 (2001)
3. Wang, W.M., Hao, T.Y., Liu, W.Y.: Automatic Question Generation for Learning Evaluation in Medicine. In: Leung, H., Li, F., Lau, R., Li, Q. (eds.) ICWL 2007. LNCS, vol. 4823, pp. 242–251. Springer, Heidelberg (2008)
4. Jonathan, C., Gwen, A., Eskenazi, E.M.: Automatic question generation for vocabulary assessment. In: Proceedings of the Conference on Human Language Technology and Empirical Methods in Natural Language Processing, Canada, pp. 819–826 (2005)
5. Burdescu, D.D., Mihaescu, M.C.: Building a decision support system for students by using concept maps. Proceedings of International Conference on Enterprise Information Systems (ICEIS 2008), Spain (2008)
6. Papasalouros, A., Kotis, K., Kanaris, K.: Automatic generation of multiple-choice questions from domain ontologies. In: IADIS e-Learning 2008 Conference, Amsterdam (2008)
7. Sicilia, M.A., Barriocanal, E.G.: On the convergence of formal ontologies and standardized e-learning. International Journal of Distance Education 3(2), 13–29 (2005)
8. Knight, C., Gasevic, D., Richards, G.: An ontology-based framework for bridging learning design and learning content. Educational Technology & Society 9(1), 23–27 (2006)
9. OWL Web Ontology Language Overview, http://www.w3.org/TR/owl-features/
10. Tosic, M., Cubric, M.: SeMCQ – Protégé Plugin for Automatic Ontology-Driven Multiple Choice Question Tests Generation. In: 11th Intl. Protégé Conference, Poster and Demo Session (2009)
11. Cubric, M.: Tosic. M.: Towards automatic generation of eAssessment using semantic web technologies. In: International Computer Assisted Assessment (CAA) Conference, Southampton (2010)
12. Stanescu, L., Spahiu, C.S., Ion, A., Spahiu, A.: Question Generation for Learning Evaluation. In: International Multi-Conference on Computer Science and Information Technology, pp. 509–513 (2008)
13. Heilman, M., Noah, S.A.: Question Generation via overgenerating transformations and ranking. In: The 14th Annual Conference on Artificial Intelligence in Education Workshop on Question Generation (2009)
14. Ishaq, A., Nanjappa, S.B., Acharya, S.: Method and system for automatically generating questions for a programming. Infosys Technologies Limited, assignee. Patent 12/319566 (2009)
15. Tsumori, S., Kaijiri, K.: System Design for Automatic Generation of Multiple Choice Questions Adapted to Students' Understanding. In: Proceedings of the 8th International Conference on Information Technology Based Higher Education and Training, Japan, pp. 541–546 (2007)
16. Anderson, L.W., Sosniak, L.A.: Bloom's Taxonomy: A forty-year retrospective The National Society For The Study Of Education (1994)
17. Tan, S.Y., Lukose, D.: System and method for ontology-based question generation. Mimos Berhad. (Patent - Pending)

# Ontology Model for Herbal Medicine Knowledge Repository

Supiah Mustaffa, Ros'aleza Zarina Ishak, and Dickson Lukose

Knowledge Engineering Center, MIMOS Berhad
Technology Park Malaysia, 57000 Kuala Lumpur, Malaysia
{supiah,rosaleza.ishak,dickson.lukose}@mimos.my

**Abstract.** In the recent years, the trend of adopting semantic technology to represent knowledge on herbal medicine has increased tremendously. A formal representation of herbal medicine knowledge is crucial for the conservation and utilization of biological resources. Due to the abundance of herbal medicine knowledge, the approaches to integrate diverse sources of knowledge to fulfill needs of various domain experts in multiple fields is a major challenge. The Herbal Medicine Knowledge Repository is modeled to overcome this challenge. The repository consists of the knowledge that interrelates knowledge of herbs (plants), knowledge on traditional medicine, the process involved to produce the medicine  and the associated pharmacology that can be shared and utilized by various interest groups. In addition to presenting the model of the ontology, this paper will also outline the experience in modeling and engineering the Herbal Medicine Knowledge Repository.

**Keywords:** Herbal Medicine, Semantic Technology, Knowledge Representation, Ontology Modeling, Ontology Engineering.

## 1 Introduction

Herbal Medicine (or sometimes also know as traditional medicine) is the knowledge, skills and practice of holistic health care, recognized and accepted for its role in the maintenance of health and the treatment of diseases. It is based on indigenous theories, beliefs and experiences that are handed down from generation to generation [1]. Herbal medicine is widely practiced in many countries including Malaysia. In the past, knowledge about herbal medicines is handed down from generations to generations through informal practices and self experiences. Today, quite often we can obtain this knowledge on the web. Since this knowledge is often found in unstructured format (eg., PDF or MS-WORD) and they are usually scattered on various data sources, the understanding on their usage relied heavily on human interpretations.

Many research and projects have been implemented focusing in the domain of Herbal Medicine, for example, the Chinese government has funded the Unified Traditional Chinese Medical Language System (UTCMLS) which involve a collaboration of 16 research groups to model and engineer the traditional Chinese

D. Lukose, A.R. Ahmad, and A. Suliman (Eds.): KTW 2011, CCIS 295, pp. 293–302, 2012.

medicine knowledge repository, concept-based information retrieval and information integration. The UTCMLS contains 14 ontologies that represent Chinese medicine. In the other hand, Nantiruj et al. [10] developed an e-health advice system to provide user who wants to treat diseases using Thai Herbs. The system provides the recommended Thai herbs to treat some symptoms based on 3 critical elements i.e. patient information, Thai Herb data and inference rules for diagnosis. They are using ontology to express the relations among provinces, symptoms, anamnesis and Thai Herbs. Other related studies on Herbal Medicine ontology development is presented by Zaharudin, I. et.al [22] who proposed a semi-automated approach for acquiring domain knowledge of the medicinal herbs domain from academic literature.

However, these ontologies mainly focus on Herbal Medicine knowledge only. There is not much focus on the integration of other knowledge like herb (plant), herbal medicine, process involved and pharmacology. This paper will focus on how the herbal medicine knowledge and related knowledge like herb, process and pharmacology knowledge is modeled by using well-structured methodology called the Methontology [5].

This paper is structured as follows: In section 2, the overview of Herbal Medicine Knowledge Repository is explained. In Section 3, the different types of ontology development methodologies are reviewed, while Section 4 describes the methodology used to develop Herbal Medicine Knowledge Repository. Finally, Section 5 concludes this paper with the summary of the team experience in developing this repository and future works planned in enhancing the repository.

## 2     Overview of Herbal Medicine Knowledge Repository

Herbal Medicine Knowledge Repository which consists of knowledge on herb, traditional medicine, process and pharmacology can be shared and used by different domain experts. The picture below shows how the Herbal Medicine Knowledge Repository is relevant to various experts, such as botanists, pharmacologists, pharmacists and medical practitioners. The botanists use the herb ontology to study plants and understand its life processes. The pharmacologists are the professionals who responsible to develop, identify and test medicine to cure, treat and prevent certain disease. The pharmacologist will make use of the pharmacology ontology provided by this repository. The practice of mixing the herbal ingredients to form medications is performed by the pharmacists. The process ontology is very useful for the pharmacists in performing these tasks. The medical practitioners are responsible for the diagnosis, care and treatment of the diseases, infections and well-beings of the patients. Thus, they may use the traditional medicine ontology in prescribing treatments and medications suitable to the patients' illness for remedial purpose.

We have identified four major knowledge categories that will be used to develop the Herbal Medicine Knowledge Repository. The categories are knowledge on Herb, knowledge on Traditional Medicine, knowledge on the Process of creating the Herbal Medicine and knowledge on Pharmacology. Each of these different types of knowledge are modeled and engineered as separate ontology such as Herb Ontology, Traditional Medicine Ontology, Process Ontology, and Pharmacology Ontology.

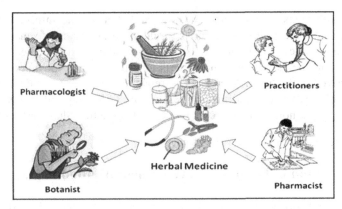

**Fig. 1.** Different Perspective of Herbal Medicine Knowledge Required by Different Experts

The Herb ontology will contain all the knowledge about the herbal plant. For example, the knowledge will consist of information about its habitat, its cultivation management and other related information such as its chemical contents, its demographic location and etc. The Traditional Medicine ontology will contain all the knowledge about traditional medicine. For example, the composition of herbs and its plant part used, the amount of quantity required, the treated diseases and its usage prescription. The Process ontology will contain the information about the process and activity involved to produce the medicine. Finally, the Pharmacology ontology will contain all the information about the possible pharmacology activity and toxicity that can be utilized from the herb, which is important for drug discovery.

Figure 2 depicts an overview of the Herbal Medicine Knowledge Repository. Having separate ontology for different knowledge categories allows changes to be made easily, hence increases the maintainability of the repository in the future. Furthermore, each of the ontology can be shared and reused by other applications, making it extensible and flexible.

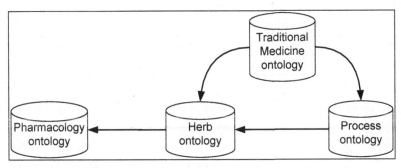

**Fig. 2.** An Overview of Herbal Medicine Knowledge Repository

As we can see in the Figure 2, the Herbal Medicine Knowledge Repository will play an important role as it links together knowledge about Herb, Traditional Medicine, Process and Pharmacology.

HMKR will be used as a comprehensive herbal medicine information system utilizing semantic technology that may be used as an effective informatics tool for the discovery of new drugs and herbal recommendation for medical treatment. Other than that, the researchers and experts may utilize the repository as a single point of reference for herbal recipes. The repository is able to provide a materia medica for Malay/Native Traditional Medicine in TCM Umbrella body, which at the moment is of non-existence compared to Ayurvedic, Shendong and Unanic medicine.

## 3    Evaluation of Ontology Development Methodologies

There are several well known methodologies for modeling ontologies. Among them are Gruniger and Fox [11], Uschold and King [12],   METHONTOLOGY [5] and KACTUS [2]. In our attempt to evaluate these methodologies for our modeling purpose, we conducted a comparison study on the functionalities of these methodologies. Table 1 list the result of the evaluation.

**Table 1.** Comparison of Ontology Development Methodologies

| Criteria | Uschold and King | Gruniger and Fox | METHONTOLOGY | KACTUS |
|---|---|---|---|---|
| Lifecycle | ✗ | To be detailed | Evolving prototypes | ✗ |
| Strategy for building application | Application Independent | Application Semidependent | Application Independent | Application Dependent |
| Strategy for identifying concepts | Middle-out strategy | Middle-out strategy | Middle-out strategy | Top-down |
| Feasibility Study | ✗ | ✗ | ✗ | ✗ |
| Specification | ✓ | ✓ | ✓ | ✓ |
| Conceptualization | ✗ | ✓ | ✓ | ✓ |
| Formalization | ✗ | ✓ | ✓ | ✓ |
| Implementation | ✓ | ✓ | ✓ | ✓ |
| Maintenance | ✗ | ✗ | ✓ | ✗ |
| Evaluation | ✓ | ✓ | ✓ | ✗ |
| Details of Methodology | Very Little | Little | Very detailed | Very little |

From the above evaluation, we concluded that METHONTOLOGY is the most appropriate methodology to be adopted for our modeling purpose, mainly because it provided broader set of activities compared to others, which will enable us to conduct comprehensive and systematic ontology modeling.

# 4    Developing Herbal Medicine Knowledge Repository

The Herbal Medicine Knowledge Repository is modeled and engineered using the METHONTOLOGY that consists the following phases: (1) Specification; (2) Knowledge Acquisition; (3) Conceptualization; (4) Integration; (5) Implementation; (6) Evaluation and (7) Documentation.

Based on our experiences using METHONTOLOGY, it shows that the specification, conceptualization, integration, and implementation can be performed repeatedly as often as required. Thus, this feature of evolving prototype lifecycle allows the knowledge engineers to go back from any state to another if some definition is omitted or erroneous. However the methodology did not indicate the order and depth in which such activities should be done especially on evaluation activity. The specific task of evaluation is self defined and there is a need to strategies on how to evaluate the ontologies wisely. The following subsection will describe activities and outcome of each of these phases.

## 4.1    Specification

In this phase, the domain knowledge is acquired to put together a preliminary version of Requirement Specification that covers the ontology's goal and scope. The goal of Herbal Medicine Knowledge Repository is to present knowledge in the domain of Herbal Medicine to be used by the subject matter experts, researchers, academicians in discovering new medicine which may potentially treat or have positive impact on certain diseases. The scope limits the ontology, specifying what must be included and what must not [6]. The scope definition will guarantee that the relevant terms are not to be missed out and duplication or irrelevant term will be eliminated from the ontology.

## 4.2    Knowledge Acquisition

Knowledge Acquisition is not an independent activity in the ontology development process. Usually, it is done simultaneously with the specification phase. Experts, books, handbooks, figures, tables and even other ontologies are sources of knowledge from which the knowledge can be elucidated using in conjunction techniques such as: Brainstorming, Interviews, Formal and Informal Analysis of Texts, and Knowledge Acquisition Tools [5]. The technique opted for the knowledge acquisition phase of the Herbal Medicine Knowledge Repository project is Text Analysis, where thorough studies are carried out on the given documentation and the main concepts of the ontology are being identified. In order to get specific, detailed knowledge and to verify the correct selection of the concepts and properties, informal interviews with domain experts is conducted.

## 4.3    Conceptualization

This phase organizes and converts an informally perceived view of a domain into a semiformal specification, using a set of intermediate representations that the domain expert and ontologist can understand [8]. Fernandez et. al [5] proposed the use of

Glossary of Terms (GT). Terms include in the GT are concepts, instances, verbs and properties, which will describe the domain knowledge and its definition. After completing the GT, the terms are then grouped into concepts and verbs. Each member or term inside the same group is closely related to each other. The four ontologies residing in the Herbal Medicine Knowledge Repository are graphically modeled using Class and Property Diagram. The Class Diagram uses relation such as *rdfs:subClassOf* to denote sub-class relationship and outlines the taxonomy of each of the ontology. Every instance that exists belongs to either the main class or subclasses of a particular class. On the other hand, the Property Diagram is used to identify all the relationships that exist in each of the ontology. The Properties are grouped according to its type: Object Properties, Datatype Properties or Annotation Properties. Figure 3 and 4 presents an example of Class and Property Diagram for Traditional Medicine Ontology.

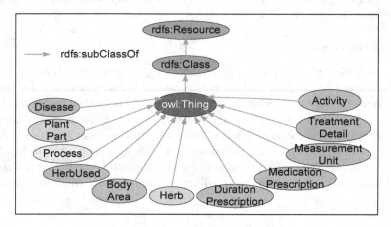

**Fig. 3.** Class Diagram for Traditional Medicine Ontology

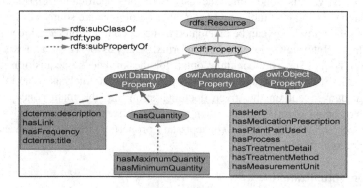

**Fig. 4.** Property Diagram for Traditional Medicine Ontology

Gomez-Perez et al, [9] introduced Data Dictionary to describe Concepts, Instances and Attributes in the form of table. Herbal Medicine Knowledge Repository outlines classes in the Class Table. The Class Table specifies the Class Name, Class Label (*rdfs: label*), Type (*rdf:type*) and subclass (*rdfs: subClassOf*) (see Table 2). Fernandez et. al. [5] proposed the use of Verbs Dictionary to express the meaning of verbs in a declarative way and it is represented in the form of table. Herbal Medicine Knowledge Repository describes properties in the Property table. The Property Table describes the Property Name, Property Label (*rdfs: label*), Domain (*rdfs: domain*), Range (*rdfs: range*) and Type (*rdf:type*) (see Table 3). In addition to that, table of formulas and table of rules will also being used to describe formulas and rules. Table 2 and 3 depicts the examples of Class and Property Table.

**Table 2.** An Example of Class Table for Traditional Medicine Ontology

| Traditional Medicine Class | | | |
|---|---|---|---|
| **Class** | **rdfs:label** | **rdf:type** | **rdfs:subClassOf** |
| Activity | Activity | owl:Class | owl:Thing |
| BodyArea | Body area | owl:Class | owl:Thing |
| Disease | Disease | owl:Class | owl:Thing |
| EthnobotanyUse | Ethnobotany use | owl:Class | owl:Thing |
| HerbUsed | Herb used | owl:Class | owl:Thing |
| MedicationPrescription | Medication | owl:Class | owl:Thing |
| TreatmentDetail | Treatment detail | owl:Class | owl:Thing |

**Table 3.** An Example of Property Table for Traditional Medicine Ontology

| Traditional Medicine Property | | | | |
|---|---|---|---|---|
| **Property name** | **rdfs:label** | **rdfs:domain** | **rdfs:range** | **rdf:type** |
| deriveFromHerb | derive from herb | EthnobotanyUse | HerbUsed | owl:ObjectProperty |
| hasHerb | has herb | HerbUsed | Herb | owl:ObjectProperty |
| hasMedication Prescription | has medication prescription | TreatmentDetail | Medication Prescription | owl:ObjectProperty |
| hasDuration | has duration | MedicationPrescription | xsd:string | owl:DatatypeProperty |
| hasFrequency | has frequency | MedicationPrescription | xsd:string | owl:DatatypeProperty |

The Herbal Medicine Knowledge Repository comprises four main ontologies; Herb, Traditional Medicine, Process and Pharmacology. Herb has 27 concepts, 42 properties and 1663 instances. While Traditional Medicine contains 14 concepts, 27 properties and 719 instances. On the other hand, Process contains 2 concepts, 5 properties and 283 instances. Last but not least, Pharmacology consists of 14 concepts, 18 properties and 239 instances.

## 4.4     Integration

There are number of prominent vocabularies publicly available, such as SKOS [13], FOAF [14], Dublin Core Metadata Intiative (DCMI) [15] and Countries [17]. Throughout the development of the four ontologies for the Herbal Medicine Knowledge Repository, we identified few Concepts (or Classes) that can be reused from other vocabularies; such as *dcterms:Location*, *dcterms:FileFormat*, *foaf:Image*, and *countries:Country*. Meanwhile, the Properties that are reused from other ontologies include *skos:altLabel*, *skos:prefLabel*, *dcterms:source*, *dcterms:format*, and *foaf:depiction*. The Herb knowledge base make reference to various plant taxonomies such as GRIN Taxonomy for Plant, IPNI (The International Plants Names Index), Flora of China, NCBI (National Center for Biotechnology Information) and TROPICOS. We refer to the id number of resources that exists in these taxonomies. Since these taxonomies are not available in RDF format, we populated our knowledge base with the respective id number instead of referencing these resources (using URI). In future, once the taxonomies are available in the form of RDF, we will take the initiative to refer the resource in herb knowledge base with the resource found in these taxonomies.

## 4.5     Implementation

TopBraid Composer [16] is used as the ontology engineering tool. We utilized a standard template for instances population. The standard templates allow us to import the instances in batches so that the population process can be fastened.

## 4.6     Evaluation

In order to verify the ontology, series of review session with team member and domain expert are arranged to guarantee the correctness of the ontologies. List of SPARQL [18] query also generated to ensure the instances was populated correctly. Other than that, we also use the three visualization tools that were developed in-house which able to view, browse and query the ontologies. The tool is Multimodal Semantic Browser (MMSB) [19] which enables Knowledge Engineers to search, view, and browse the ontology in text or graphical mode. Semantic Query Interface (SQI) [20] enables Knowledge Engineers to query the ontology using natural language while Visual Semantic Query (VSQ) [21] enable Knowledge Engineers to query the ontology using friendly graphical way.

## 4.7     Documentation

There are no consensuated guidelines on how to document ontologies. But, documentation is continuous activity that should be done during the whole ontology development process [5]. All information about the requirement specification, conceptual model and other related information to the Herbal Medicine Knowledge Repository will be documented in one formal document.

We have produced several documents to assist in the development of Herbal Medicine Knowledge Repository which are Graphical Model Document and Requirement Specification Document. Graphical Model Document helps Knowledge Engineers to model the ontology as well as to identify and propose classes and properties involved in Herbal Medicine Knowledge Repository. Knowledge Engineers use this document to communicate with the Domain Expert to verify and validate whether the knowledge are modeled correctly and to ensure that Knowledge Engineers and Domain Experts have the same understanding on the knowledge base that will be developed. At the same time, we also prepare the Requirement Specification Document of Herbal Medicine Knowledge Repository to describe the details about the requirement specification (eg. purpose, scope, detail description of class and properties).

## 5    Conclusion

Ontologies have been adopted in various application areas as a key component of Semantic Web. Building domain ontology is a challenging task especially when the Knowledge Engineers have insufficient background knowledge of the domain as well as on the knowledge engineering methodologies.   In this paper, we adopt METHONTOLOGY to model the Herbal Medicine Knowledge Repository.

As the future work, the Herbal Medicine Knowledge Repository will exploit rules, constraints and restrictions which will further enrich the repository.  In addition to that, we are also looking at identifying the limitation of this methodology and considering the possibilities of integrating METHONTOLOGY with other mature methodologies to create a highly systematic methodology for ontology modeling and engineering.

**Acknowledgements.** This research and development is supported by the Ministry of Science, Technology and Innovation (MOSTI), Malaysia.

## References

1. Traditional Medicine, http://www.wpro.who.int/internet/ resources.ashx/RCM/RC52-07.pdf (last visited : June 2011)
2. The KACTUS Booklet Version 1.0., http://hcs.science.uva.nl/projects/ NewKACTUS/Reports.html (last visited : June 2011)
3. Corcho, O., Fernandez-Lopez, M., Gomez-Perez, A. : Methodologies, Tools and Languages for Building Ontologies. Where is their meeting point? Elsevier Science B.V (2003)
4. Fernandez-Lopez, M., Gomez-Perez, A., Pazos-Sierra, J.: Building A Chemical Ontology Using Methontology and the Ontology Design Environment. IEEE Intelligent System 14 (1999)
5. Fernandez, M., Gomez-Perez, A., Juristo, N.: METHONTOLOGY: From Ontological Art Towards Ontological Engineering. AAAI Technical Report, Madrid, Spain (1997)

6. Brusa, G., Caliusco, M.L., Chiotti, O.: A Process for Building a Domain Ontology: An Experience in Developing a Government Budgetary Ontology. In: Australasian Ontology Workshop, Australia (2006)
7. Fernandez Lopez, M.: Overview Of Methodologies For Building Ontologies. In: IJCAI, August 2 (1999)
8. Aguado, G., Bateman, J., Bernardos, S., Fernandez, M., Gomez-Perez, A., et al.: Ontogeneration: Reusing Domain and Linguistic Ontologies for Spanish Text Generation. In: Workshop on Applications of Ontologies and Problem Solving Methods, pp. 1–10. European Coordinating Committee for Artificial Intelligence (1998)
9. Gomez-Perez, A., Fernandez, M., De Vicente, A.: Towards A Method to Conceptualize Domain Ontologies. In: Working Notes of the Wokshop Ontological Engineering (1996)
10. Nantiruj, T., Maneerat, N., Varakulsiripunth, R., Izumi, S., Shiratori, N., Kato, T., Kato, Y., Takahashi, K.: An E-Health Advice System With Thai Herb and an Ontology. In: The 3rd International Symposium on Niomedical Engineering, pp. 315–319 (2008)
11. Gruninger, M., Fox, M.S.: Methodology for the Design and Evaluation of Ontologies. In: IJCAI Workshop on Basic Ontological in Knowledge Sharing, Montreal, Canada (1995)
12. Uschold, M., King, M.: Towards a Methodology. In: IJCAI 1995 Workshop on Basic Ontological Issues in Knowledge Sharing (1995)
13. SKOS, SKOS Simple Knowledge Organization System Reference (2009), http://www.w3.org/TR/skos-reference/ (last visited : June 2011)
14. FOAF, Vocabulary Specification 0.98 Namespace Document (2000), http://xmlns.com/foaf/spec/ (last visited June 2011)
15. DCMI (2010). DCMI Metadata Terms, October 11 (2010), http://dublincore.org/documents/dcmi-terms/ (last visited : June 2011)
16. TopQuadrant (2007). TopBraid Composer: Getting Started Guide (Version 2.0), July 21, (2007), http://www.topquadrant.com/docs/marcom/ TBC-Getting-Started-Guide.pdf (last visited : June 2011)
17. Country ontology, http://topbraid.org/examples/ (last visited : June 2011)
18. SPARQL New Features and Ratioinale, W3C Working Draft (July 2009), http://www.w3.org/TR/sparql-features/ (last visited : June 2011)
19. Sadanandan, A.A., Onn, K.W., Zadeh, M.R.B., Lukose, D.: P120092121
20. Onn, K.W., Lukose, D.: Visualizing and Navigating Large Multimodal Semantic Webs. In: Proceedings of the I-KNOW, pp. 175–185 (2010)
21. Sadanandan, A.A., Onn, K.W., Lukose, D.: Ontology Based Graphical Query Language Supporting Recursion. In: Setchi, R., Jordanov, I., Howlett, R.J., Jain, L.C. (eds.) KES 2010, Part I. LNCS, vol. 6276, pp. 627–638. Springer, Heidelberg (2010)
22. Zaharudin, I., Shahrul Azman, M.N., Mahanem, M.-N.: Knowledge Acquisition from Textual Documents for the Construction of Medical Herbal Domain Knowledge. Journal of Applied Science, 794–798 (2009)
23. Park, J., Sung, K., Moon, S.: Using the METHONTOLOGY Approach to a Graduation Screen Ontology Development: An Experiential Investigation of the METHONTOLOGY Framework. Asia Pacific Journal of Information Systems (2010)

# A Data Structure between Trie and List
# for Auto Completion

Vooi Keong Boo and Patricia Anthony

UMS-MIMOS Center of Excellence in Semantic Agent, School of Engineering and Information
Technology, University Malaysia Sabah, Malaysia
`nogias1685@yahoo.co.uk, panthony@ums.edu.my`

**Abstract.** Auto completion is one of the useful features available as a web
service. This technique can be implemented in many applications, ranging from
a small scale of service like auto complete for items sold in a shop to a large
scale of service like Google suggestions which involve suggesting a huge
dataset. One of the challenges in implementing auto complete service with a
large dataset and a limited computer power is on how to achieve a fast lookup
without consuming a lot of memory. This paper presents a data structure that
can implement auto complete service that contains up to millions of concepts in
a standard computer. The proposed data structure increases the search
complexity in return to saving a large amount of memory. A service similar to
DBpedia lookup service which contains 9 million words as completion
candidates is developed to test the performance of this data structure. The
testing shows that such data structure requires less memory than ternary search
tree and more importantly a lookup can be performed within milliseconds.

**Keywords:** trie, search tree, auto completion.

## 1 Introduction

Auto completion is a technique that enables someone to find something with less
typing efforts. It has been applied in various types of applications since very early
time such as auto complete in UNIX shell where a user can use a tab key to get a list
of available suggestion [1]. A modern version of auto completion involves application
such as text editor or window explorer where suggestions given to the user are based
on some historical records. Since the emergence of Ajax technology, auto complete
has become one of the useful features available in web services since a simple
communication can be done on the server without reloading the whole page. This
makes communication between the client and the server much easier. Auto complete
is applied in various web services, from a small scale of service like auto complete for
items sold in a shop to a large scale of service like Google suggestions which involve
a huge dataset. In general, to apply an auto completion that targets on a small database
is simple. If the size of the database is in thousands only, then a brute force search on
the dataset is enough to retrieve a list of suggestions that match a query in a short
time. However, one of the challenges for implementing auto complete service with a

D. Lukose, A.R. Ahmad, and A. Suliman (Eds.): KTW 2011, CCIS 295, pp. 303–312, 2012.

large dataset with limited computer power is on how to achieve a fast lookup without consuming a lot of memory.

This paper proposes a data structure that is able to provide auto complete service in a fast manner while consuming less memory compared to some existing data structures. The remainder of this paper is organized as follows. Section 2 reviews some of the technologies that are related to auto completion while Section 3 describes the proposed data structure that is able to provide the auto complete service. The evaluation of the proposed data structure is described in Section 4 and Section 5 concludes with some possible enhancements in the future.

## 2    Related Technology in Auto Complete

In general, an auto complete service can be applied using either a well defined search tree such as trie or ternary search tree [2] or an inverted indexing system such as Lucene. The first solution involves building a search tree which contains all the words within a dataset such that a fast search can be performed to retrieve all possible suggestions for the dataset. Trie is used because of its low complexity in searching process and the building of the tree is fast.

Lucene is another common choice of technology that can apply auto completion. This can be achieved by making a prefix indexing on all the words in a dataset so that each prefix contains a set of built in completion. Figure 1 shows an example that provides an auto complete service with Lucene[1]. The left frame in the web page provides a search bar where user can type the word to be searched. Auto completion of the word is provided to the user based on the names of existing Wikipedia pages.

For example, since "Albert 2" is an available article in Wikipedia, it is displayed as one of the suggestions. When a user clicks on one of the suggestions provided, the frame on the right hand side will point to that particular matching page in Wikipedia. Another web service that uses Lucene to provide a prefix search web service is the DBpedia lookup service [3] where the service enables client to search for any resources in DBpedia by providing a partial complete query only. This service is also used to provide some extra services such as categorization of concepts that are entered by the client [4][5].

Various researches have used these techniques to increase the efficiency in searching for data. For example, inverted indexing is applied in [1], [6] to develop a new indexing data structure called HYB which claimed to be able to process a given query faster but yet consumed less space compared to the standard inverted indexing structure. Besides, the structure of trie is enhanced in [7] by applying an adaptive algorithm to build a cache efficient trie search. This author claimed that different parts in trie should have different structures on the node because the node that is closest to the root will tend to have more sub nodes than nodes that are far away from the root. Also, the data structure is tuned such that it can fit well in the computer memory to enable a fast search. [8] also proposed a trie compaction algorithm for a large set of

---

[1] http://sematext.com/demo/ac/index.html (retrieved 28 April 2011).

keys which will speed up the retrieval of information while [9] applied a trie partitioning to reduce the memory required while loading and searching. This paper is not focused on optimizing a search tree but proposes an alternative data structure that draws some features from trie.

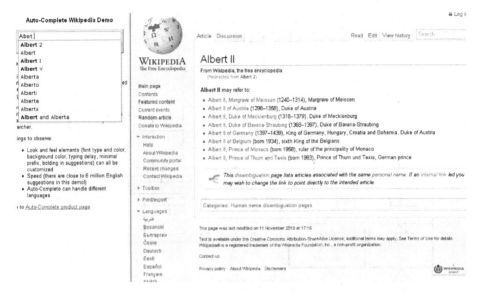

**Fig. 1.** A demo from sematext.com that provides an auto complete service

## 3    The Data Structure Concept

In a standard search tree like trie or ternary search tree, to implement a fast auto complete, the basic method is to insert the possible completion in each of the nodes such that if a query is given, the searching mechanism will find the node corresponding to the query to get a list of completions. However, this method will require each node to have a set of completion, so if there are millions of words in a dataset, then up to millions of completions need to be built in the search tree.

Consider a set of word-rank pair X as follows:

X = { (Actor, 0.6) , (Actress, 0.65) , (Academic, 0.9) , (Account, 0.5) , (Acceleration, 0.55) , (Animal,0.8) , (Anime, 0.2) , (Ancient, 0.5) , (Anchor, 0.45) , (Art, 0.75) , (Arm, 0.3) , (Arun, 0.4) , (Asia, 0.7)  }

This dataset can be loaded into a trie as shown in Figure 2. For this type of search tree, each node will contain a list of completion. If a query 'Ac' is submitted, then the searching will go to node 'Ac' to retrieve the completion while a query likes 'Aca' or 'Acad' will get one hit only. However, the nature of this type of data structure  is that the amount of nodes required for building the tree is generally greater than the amount of words in the dataset. As such, this structure is not scalable for a huge dataset especially if the structure is implemented using a programming language like Java which is not designed for performance. For example, a ternary search tree[2] that is written in Java is not able to load more than 7 million distinct words with 20 suggestions even with an allocation of 1.4 GB of memory.

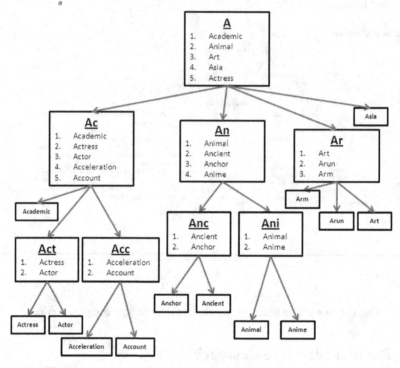

**Fig. 2.** An example of trie data structure that applies auto complete

To apply auto complete in a less powerful computer, an alternative data structure needs to be designed. The main concept behind the data structure is to sacrifice a little of the search complexity in return to saving a large amount of memory. This is based on two arguments as follow:

- Searches that take 1 millisecond is not so different with searches that take 10 milliseconds since 1 and 10 milliseconds are insignificant to humans.
- There is a difference between loading an application that consumes 1GB memory and one that consumes 2GB memory.

The logic behind these arguments is that the search time required for auto complete does not need to be too fast. In general, it is not easy for human to differentiate a search that can be done in one millisecond or one nanosecond and so there is no need to build a perfect data structure to implement the auto complete. A data structure for auto complete that requires a reasonable search time but consumes less memory than

---

[2] http://suggesttree.sourceforge.net/ (version in 20 April 2011).

the standard search tree is sufficient. One approach to achieve the proposed concept is to build a search tree in which each node contains only a partial complete list of suggestions such that even a brute force search is enough to complete the suggestion list in a short time. This node is illustrated in Figure 3. This node holds the same dataset shown in Figure 2. The special feature of the data structure is that each node not only has a list of suggestions but it also has fewer lists of partial complete suggestions. So if a query of 'A' is asked, then the suggestion list of the node A will be returned. However, if a query 'Anc' is asked, then a dynamic searching will be performed on the partial list of 'An' in node A to find all the suggestions that match the query. In this case, 'Ancient' and 'Anchor' will be returned. With an average computer power, searching from a thousand of words can be achieved in just milliseconds. In this setting, holding a partial complete list in a node is not a problem as long as the list is maintained according to a certain size. Moreover, by doing so, we can reduce a considerable amount of nodes required to build a search tree.

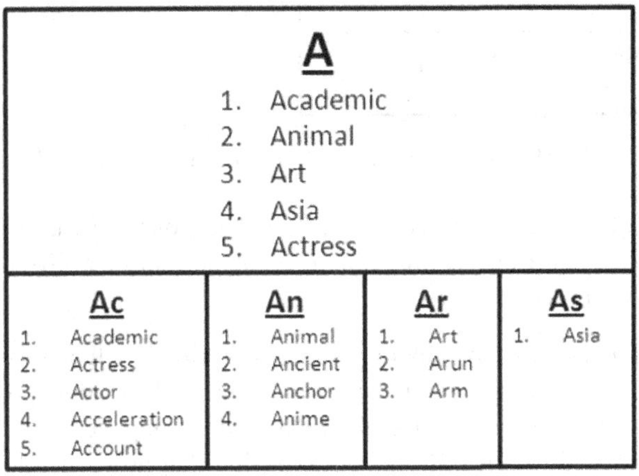

| **A** | | | |
|---|---|---|---|
| 1. Academic<br>2. Animal<br>3. Art<br>4. Asia<br>5. Actress | | | |
| **Ac** | **An** | **Ar** | **As** |
| 1. Academic<br>2. Actress<br>3. Actor<br>4. Acceleration<br>5. Account | 1. Animal<br>2. Ancient<br>3. Anchor<br>4. Anime | 1. Art<br>2. Arun<br>3. Arm | 1. Asia |

**Fig. 3.** A single node in a search tree

When the partial complete list in a node is too big for a dynamic search to be executed in a short time, then that list will be rebuild into another node in the tree. For example, let's say the threshold for a partial complete list is 5, then the list for 'Ac' in node 'A' will need to be rebuilt as shown in Figure 4. Using this concept, the node as well as the suggestion lists in a node can be reduced. The node reduced in the example given here may not be significant. However, when we consider a dataset that contains millions of words that require millions of nodes to be build in a standard trie or ternary search tree, applying the data structure here with a threshold of 3000 words for partial complete list can reduce the memory required. A dynamic generation of suggestion lists from 3000 words can be performed on a computer with an average processor.

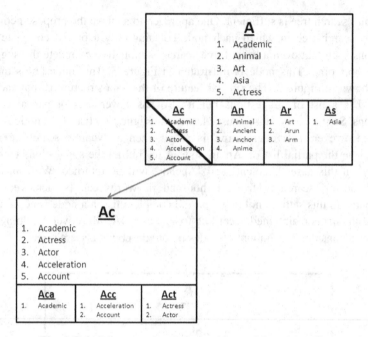

**Fig. 4.** Rebuilding a new node from a big partial complete list

The building of the whole search tree with $x$ suggestions and a threshold value of $y$ for partial complete list from a source file that contains pair of word-rank can be done using the following process:

1. Load all pairs of word-rank to a single node which will be the root node first. $x$ words with the highest rank are picked during the loading process.
2. After all the pairs are loaded, for each partial complete list, if the number of words exceeds $y$, then all the words within that group is reloaded to a new node named with the symbol for that partial complete list.
3. Step 1 and 2 are repeated when new nodes are created.

The building process only involves two main actions, loading the list of suggestions for each node and verifying the size of a partial complete list. This process will be repeated until there is no more partial complete list from the nodes that exceed y. On the other hand, the searching process can be described as follow:

1. Start searching from the root node.
2. If the query matches with the label for the node, returns a list of suggestion corresponding to that node.
3. If the query is longer than the label, then search the next node with a prefix that matches the query. Repeat Step 2. If there are no more nodes that are available for further search, then go to Step 4.

4. A dynamic search is performed on the partial complete lists that share the same prefix with the query to retrieve the suggestion lists. If there is no partial complete list that shares the same prefix is found, then the search is deemed to have failed (this means that not a single word in the tree matches with the query).

The searching process also involves two main actions, firstly to check the node label and secondly to search the next candidate node. Step 4 is only executed if there is no more candidate node found. The search time will depend on whether Step 4 is executed. For example, based on Figure 5, if the search query is "Ac", then the search will start from root node to node "A" and node "Ac". When node "Ac" is reached, the suggestions from the node is returned since its label matches with the search query. In this case, the searching speed will be the same as trie.

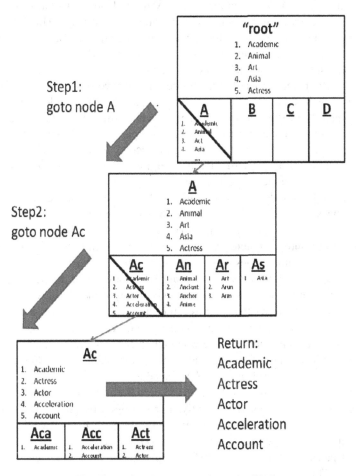

**Fig. 5.** A searching process for query "Ac"

## 4     Evaluation

We conducted a set of experiments to evaluate the performance of the proposed data structure. We used the PageRank dataset that was computed from a DBpedia wikilink dataset consisting of more than 9 millions distinct pair of concepts and its ranking. We used Java as the programming language. By using a standard ternary search tree like SuggestTree, it is found that the loading of all the concepts is not possible if each node requires at most 20 suggestions of words in a 32-bit machine even when the heap size is maximized. The loading still failed even when the dataset size is reduced to 7 million words.

However, when we applied the proposed data structure with threshold limit of 3000, it was able to load the whole dataset within 1.4GB of memory and a building time of around 10 minutes even without optimizing the memory structure of the nodes used. By reducing the heap space limit to 800MB, the loading was still successful but it took approximately 37 minutes to build the whole data structure (In JVM setting, if the memory required is very close to the maximum memory allocated, an extra time is consumed for memory handling process to build a new node). When the threshold of the partial complete list in each node is set to 3000, only 26 thousand nodes are required to be built for the whole search tree. In other word, the node size is reduced by a factor of more than hundred times. By using a higher threshold, the node can be further decreased in exchange for a longer searching time. Even though each partial complete list in a node can consist of up to 3000 pair of word-ranks, the retrieval of 20 suggestions can be achieved under 1 millisecond in an Intel Core Quad CPU. When we run it on a standard laptop with Intel Core 2 Duo, it only took 10 milliseconds to retrieve the suggestion list.

We developed a simple web service to test the auto complete service as shown in Figure 6. The service is similar to one shown on Figure 1 and it is hosted on a standard 32-bit processor desktop. This service involves a text field which will invoke a request for the auto complete at each key stroke through Ajax. A list of names representing Wikipedia articles which are sorted by the PageRank values will be returned so that the user can choose which article to view based on the suggestion. No delay is set between key strokes to ensure that each key stroke will invoke a request to the server. The result shows that even when each key stroke invokes a request to the server, it was able to provide the auto completion. As shown in Figure 5, "Albert Einstein" is ranked as a second choice when the query "Albert" is entered. This is an acceptable result because the PageRank value of "Alberta" is higher than the PageRank value for "Albert Einstein" in Wikipedia.

A further increase of the threshold of the partial complete list in each node will reduce the number of nodes significantly. For example, by setting the threshold to 5000, a total of 15 thousand nodes are required to be built for the whole search tree. On the other hand, setting the threshold to 10000 will reduce the number of node to 7000 and it can be build within 800MB heap limit of Java language in around 19 minutes only. This is better than the case with threshold of 3000.

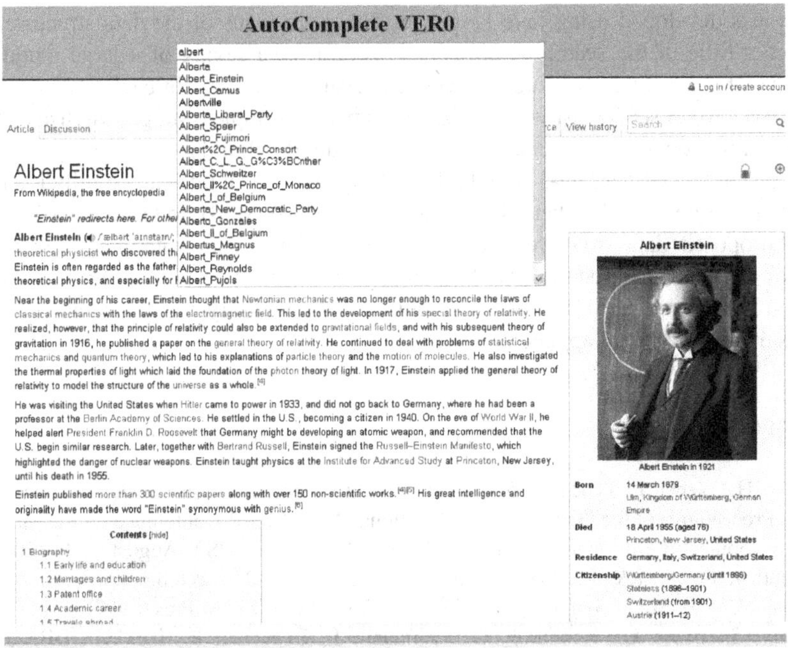

**Fig. 6.** A web service deployed to test the auto complete with the proposed data structure

The data structure has several features:

1. The time complexity required for building the whole search tree is O(n). Building the tree involves loading all the words to a root node first, followed by the creation of a new node when the partial complete list exceeds a threshold.
2. The time complexity for searching a record is at most proportional to the length of the query since the searching mechanism is similar to trie.
3. The proposed data structure does not require a sorted dataset to be built as the whole dataset is loaded to the root node first before the tree is expanded.
4. The threshold value for the partial complete list determines how much computation time can be sacrificed in exchange for lower memory consumption. If the threshold is infinite, then the data structure will be similar to a pure link list, which will cause auto complete to be an impossible task (searching for 20 words from millions of words is not an easy task). On the contrary, if the threshold is set to 1, the search tree will become a trie based data structure.

## 5    Conclusion

A data structure that implements auto complete service is discussed in this paper. This data structure can easily process up to millions of concepts in a standard computer even though it is developed using Java language. The main focus of the data structure is to sacrifice a little of the search complexity in return for a saving of a large amount of memory. This is a data structure between a trie and a list since in each node of a trie, there exist several sets of list which represent a partial complete suggestion list with the purpose of reducing the total amount of nodes required to build a search tree. The result shows that a dataset that failed to be build with ternary search tree can be fully loaded to the data structure described here. For future work we would like to test the performance of the proposed data structure with an extremely large data with many concurrent requests. The data structure can be further optimized by considering "a structure between Patricia trie and a list" or "a structure between ternary search tree and a list" as long as the main concept applied here remains.

# References

1. Bast, H., Weber, I.: Type less, find more: fast autocompletion search with a succinct index. In: Proceedings of the 29th Annual International ACM SIGIR Conference on Research and Development in Information Retrieval, Seattle, Washington, USA, August 06-11 (2006)
2. Bentley, Sedgewick, R.: Fast algorithms for sorting and searching strings. In: Eighth Annual ACM-SIAM Symposium on Discrete Algorithms. SIAM Pres (1997)
3. Auer, S., Bizer, C., Kobilarov, G., Lehmann, J., Cyganiak, R., Ives, Z.: DBpedia: A Nucleus for a Web of Open Data. In: Aberer, K., Choi, K.-S., Noy, N., Allemang, D., Lee, K.-I., Nixon, L.J.B., Golbeck, J., Mika, P., Maynard, D., Mizoguchi, R., Schreiber, G., Cudré-Mauroux, P. (eds.) ASWC 2007 and ISWC 2007. LNCS, vol. 4825, pp. 722–735. Springer, Heidelberg (2007)
4. Mirizzi, R., Ragone, A., Di Noia, T., Di Sciascio, E.: Ranking the Linked Data: The Case of DBpedia. In: Benatallah, B., Casati, F., Kappel, G., Rossi, G. (eds.) ICWE 2010. LNCS, vol. 6189, pp. 337–354. Springer, Heidelberg (2010)
5. Softic, S., Taraghi, B., Halb, W.: Weaving Social E-Learning Platforms Into the Web of Linked Data. In: Proceedings of I-SEMANTICS 2009 - International Conference on Semantic Systems, pp. 559–567 (2009)
6. Bast, H., Weber, I.: When you 're lost for words: Faceted search with autocompletion. In: SIGIR 2006 Workshop on Faceted Search, Seattle, Washington, USA (August 2006)
7. Acharya, A., Zhu, H., Shen, K.: Adaptive Algorithms for Cache-Efficient Trie Search. Selected Papers from the International Workshop on Algorithm Engineering and Experimentation, January 15-16, pp. 296–311 (1999)
8. Aoe, J.-I., Morimoto, K., Shishibori, M., Park, K.-H.: A Trie Compaction Algorithm for a Large Set of Keys. IEEE Transactions on Knowledge and Data Engineering 8(3), 476–491 (1996)
9. Kastrinakis, D., Tzitzikas, Y.: Advancing Search Query Autocompletion Services with More and Better Suggestions. In: Benatallah, B., Casati, F., Kappel, G., Rossi, G. (eds.) ICWE 2010. LNCS, vol. 6189, pp. 35–49. Springer, Heidelberg (2010)

# Agent for Mining of Significant Concepts in DBpedia

Vooi Keong Boo and Patricia Anthony

UMS-MIMOS Center of Excellence in Semantic Agent,
School of Engineering and Information Technology, University Malaysia Sabah, Malaysia
nogias1685@yahoo.co.uk, panthony@ums.edu.my

**Abstract.** DBpedia.org is a community effort that tries to extract structured information from Wikipedia such that the extracted information can be queried just like a database. This information is opened to public in the form of RDF triple which is compatible with the semantic web standard. Various applications are developed for the purpose of utilizing the structured data in DBpedia. This paper makes an attempt to apply PageRank analysis on the link structure of DBpedia using a mining agent to mine significant concepts in DBpedia. Based on the result, popular concepts have the tendency to be ranked higher than the less popular ones. This paper also proposes an alternative view on how PageRank analysis can be applied to DBpedia link structure based on special characteristics of Wikipedia. The result shows that even concepts with a low PageRank value can be used as a valuable resource for recommending pages in Wikipedia.

**Keywords:** DBpedia, PageRank, concept ranking.

## 1    Introduction

DBpedia.org is a community effort that tries to extract structured information from Wikipedia to a form that can be queried just like a database. Wikipedia has a large source of information available on the World Wide Web. Its data describes a large number of concepts. This description of concept is a useful source of knowledge that can be further analyzed and explored. However, the content of Wikipedia is designed for human's consumption only and it is not intended for the machine. For this purpose, the knowledge in Wikipedia has been extracted by the DBpedia community into a form of RDF triple defined by W3C. Semantic technology can then be used to explore and extract the information from the RDF triple dataset. Every page in Wikipedia has a standard page structure. It has a title, a short description of the title, an index table, and related images (found on the right hand of the page). The bottom left hand side of a Wikipedia page shows the categorization of the topic. Furthermore, the topic in various languages is captured on the left hand side of the page. Some Wikipedia pages also contain an *infobox* on the right of its page to consistently present a summary of some unifying aspects that the articles share and to improve navigation to other interrelated articles. Since information in Wikipedia is semi-structured, it is possible to extract such information from Wikipedia pages to generate a huge graph [1].

D. Lukose, A.R. Ahmad, and A. Suliman (Eds.): KTW 2011, CCIS 295, pp. 313–322, 2012.
© Springer-Verlag Berlin Heidelberg 2012

The DBpedia dataset includes an estimated 1.95 million concepts [2] and this number is still growing. As of March 2010, the DBpedia dataset describes about 3.4 million of "things" and over 1 billion of "facts". Since the information extracted in DBpedia follows the RDF triple standard, one can just retrieve any information provided in DBpedia by querying the database. The implementation of DBpedia dataset enables one to write SPARQL queries like: "Give me all the sitcoms that are set in NYC" or "Give me all the tennis players from Moscow" [3].

DBpedia knowledge base can be accessed through the web using four methods [4]. The first method is through the SPARQL endpoint[1] provided for querying DBpedia knowledge base, where client application can send SPARQL query to the endpoint which supports several extensions of the query language such as full text search, aggregate functions and COUNT(). The second method is through the Lookup Index[2], which is a service that enables a client to find a DBpedia resource URI through an input string. The string matching in DBpedia resource utilizes Lucene indexing and a relevance ranking to find the most likely matching resource. The third method to access DBpedia knowledge base is through the RDF dumps provided. The DBpedia community has divided the DBpedia knowledge base into a few parts of N-triple dataset which can be downloaded from the DBpedia website. The separation of the knowledge base is based on the type of information extracted from Wikipedia. For example, all information about the category of an article is grouped into one single N-triple dataset while all information about an *infobox* from an article is grouped into another N-triple dataset. The final method to access DBpedia knowledge base is through the linked data[3] which is a method of publishing RDF data on the web based on HTTP URI where retrieval of a resource can be done with the HTTP protocol. The DBpedia resource is published under the namespace of *http://dbpedia.org/resource* such that if a semantic web agent wants to retrieve information about the concept "Berlin" from the knowledge base, the URI of *http://dbpedia.org/resource/Berlin* will be sent to the DBpedia server, where an RDF description will be returned. This allows the knowledge base from DBpedia to be linked with similar concepts defined in other domains.

However, the data available from these methods represents the raw data extracted from Wikipedia only. As described by DBpedia community, these data can be further expanded with extra analysis such as data mining, PageRank, or similar algorithms for the Wikipedia link structure. This paper attempts to apply PageRank analysis on the DBpedia link structure to see how well it performs in ranking related concepts in DBpedia. The remainder of this paper is organized as follows: Section 2 reviews some works related to link structure analysis and its implementation to DBpedia dataset. The implementation of a mining agent using PageRank analysis to mine the DBpedia dataset is discussed in Section 3. Section 4 proposes an alternative approach to rank concepts in DBpedia and Section 5 concludes with future work.

---

[1] http://dbpedia.org/sparql
[2] http://lookup.dbpedia.org/api/search.asmx
[3] http://www.w3.org/DesignIssues/LinkedData.html

# 2      Related Work

This section discusses works related to ranking the importance of concepts with link structure analysis and also on the application of ranking algorithms on the structured data in DBpedia. The first part discusses some of the possible ways to determine the importance of each node quantitatively, given a web graph that contains a link between each node. In this case, the web graph can be the WWW or the link structure in the DBpedia dataset. The second part describes the works on how the structured data extracted from DBpedia is analyzed based on its link structure and applied as web services.

## 2.1      Link Structure Analysis

Link structure in a graph contains important information for analysis because each link between nodes shows how the nodes are interrelated. Due to this, various researches have been conducted to find this hidden information behind the link structure of a graph. The simplest way to perform a link analysis on a graph is to count the number of inlinks and outlinks in each of the nodes. This is based on the logic that if a node is important, then most probably that node will be linked by many other nodes. However, this logic is not always true especially in the case of the WWW where a web page can be linked to any other pages without any limitation. Hence, a better link analysis model should always consider this factor.

PageRank [5] is among the most successful examples of link structure analysis. It is one of the important system features that led to the success of Google. The concept of PageRank is similar to what happens in an academic citation where the importance of a paper is proportional to the citation that a paper has received from other papers and also the quality of the paper that gives the citation. In this case, since PageRank not only considers the inlink quantity (citation number), but also considers the inlink quality (paper quality), this gives a better approximation on the importance of a node in the graph. The advantage of PageRank is that link structure is the only information that is required for analysis, which makes it generic enough to be applied on any similar dataset.

However, Haveliwala [6] commented that the score computed by PageRank is general over that whole web, so the accuracy may be reduced if a search is to be done on the particular domain of a topic only. Hence, a concept of "Topic-Sensitive PageRank" is proposed, which introduces a biased PageRank to improve the ranking. In this approach, rather than computing a vector of PageRanks for all web pages, they computed a set of PageRank's vector corresponding to the set of possible topics, where each will contain certain biases on their topics. For example, the topic of "artificial intelligence" can have a unique set of PageRanks for all web pages such that web pages which talk about "artificial intelligence" have a higher PageRank while topic "physics" will have another set of PageRanks for all web pages.

The work presented in [7] is focused on the link structure analysis. The concept proposed in this paper is to rank the linked data, particularly in the case of DBpedia dataset, based on some unique properties contained in the link structure. For example, we can find a node called "category" and "page". In this case a "page" node is defined as pages that are grouped together under a "category" in DBpedia. These special properties are considered in the ranking process to generate a domain specific ranking. One of the claims made about the ranking concept is that it yields a relative ranking system, where each node ranking will vary based on the query given, by considering how a node and the query are overlapped in relation to the category (logically if two concepts are under the same category, then there is a higher possibility that both concepts are related). The ranking of a concept can be further merged with the search results from a search engine such as Google or Yahoo! and a social bookmarking system (Delicious) to yield a hybrid ranking approach on the DBpedia resource.

## 2.2     Ranking Algorithm Implementation

The DBpedia dataset has been applied in various works such as analyzing the link structure analysis of DBpedia dataset to figure out the importance of concepts inside DBpedia [2]. The lookup service provided in DBpedia applied some ranking algorithms within the DBpedia dataset. This algorithm can be used to search for any DBpedia resources by providing a related keyword and the result will be ranked based on the number of inlinks pointing to that concept. A more advance search for a concept can be done by specifying the category of the concept to be searched. Although the ranking concept used in the DBpedia lookup service only considers the number of inlinks, it was still able to yield an acceptable result where a popular topic will be ranked higher.

Another research work that applies ranking on the DBpedia dataset is described in [7]. As stated in Section 2.1, the ranking concept involved in this work is a hybrid ranking approach that considers factors such as search engine, social bookmarking system and also some unique properties that are provided in DBpedia. This work is based on the requirement of smart tools such as an annotating tool that can detect similar concepts from a given query. A website was developed to test the performance of the ranking system. The testing was performed by 50 volunteers who are considered as experts in the IT domain. During the testing, the participants submitted queries to the system, which then returned a list of keywords that are related to the query to be rated. For example, a query of "Drupal"[4] returned the result of "Ubercart", "PHP", and "MySQL", etc, which the author claimed to be relevant. "Ubercart" is returned as a result because it is a popular e-commerce module for "Drupal", while "PHP" is the programming language used in "Drupal". The term "MySQL" is returned because it is often used together with "Drupal". These experimental results validate their ranking approach.

---

[4] http://www.drupal.org/

# 3    PageRank Analysis on DBpedia

This section discusses the ranking of concepts on DBpedia dataset with PageRank analysis. The link structure in DBpedia (which is derived from Wikipedia) is slightly different from the link structure in the WWW. So even though PageRank may be able to evaluate the importance of web pages, it may not work well on Wikipedia. Hence, an experiment was conducted to evaluate the performance of PageRank in ranking related concepts in DBpedia

For this purpose, we created a simulation to mimic the PageRank surfer's model for DBpedia dataset. The simulation is described in [8], in which the PageRank for a given page can be computed by simulating the behavior of a random surfer's model. We simulate the random surfing by providing random link structures that were drawn from the links provided in the web pages that the surfer visited. We developed a mining agent to determine the likelihood that the surfer will stay on a page after a certain number of steps. The role of the mining agent is to compute the PageRank of a page. This mining agent analyzes the page_link dataset which describes all the links between Wikipedia articles captured by DBpedia. Simulation of the surfer is performed based on the damping factor of 0.85 and 50 clicks. The mining agent uses this information to compute the PageRank for a given DBpedia concept and rank the related concepts.

We created another control agent that uses DBpedia lookup service based on a direct inlink count to rank the related concepts of a given DBpedia concept. To evaluate the performance of our mining agent, its performance is compared with the control agent. For each DBpedia concept, the agents will calculate and rank the related concepts associated to the DBpedia concept. For example, a DBpedia concept "Steve" is related to other concepts such as "Steven Spielberg", "Steve Martin", "Steve Vai" and etc. The results are shown in Table 1 and Table 2.

It can be seen from the result that there is no significant difference between the ranking generated by the mining agent and the one generated by the control agent. Table 1 and 2 show the ranking of the related concepts based on "Steve" and "Albert"[5]. The first column shows the ranking obtained by the control agent while the second column shows the ranking generated by the mining agent. The parenthesized figures are the number of inlinks for each related concept. These two tables also show that the top result is still dominated by the concept that has the highest inlink count. However, "Steve Vai" in Table 1 is ranked third by the control agent whereas the miner agent ranked it in the fifth place. This is possible because there may be many concepts with high PageRank that are linked to "Steve Vai". It can also be seen that "Steve Jobs" and "Steve Albini" concepts are ranked highly by the mining agent even though the number of inlinks for both concepts are lower than the number of inlinks for the "Steve Vai" concepts. In this situation, having a high number of inlinks does not necessary means that the concepts are highly relevant. In contrast, the control agent only ranks the related concepts based on the number of inlinks, such that the higher the inlink count is, the more relevant it is. In Table 2, the result shows that the ranking generated by the mining agent and the control agent are exactly the same.

---

[5] Statistic based on 25 April 2011.

**Table 1.** Comparison between inlink and PageRank for the word "Steve"

| Control Agent (inlink) | Mining Agent (PageRank) |
|---|---|
| Steven Spielberg (1847) | Steven Spielberg (1847) |
| Steve Martin (702) | Steve Martin (702) |
| Steve Vai (563) | Steve Jobs (468) |
| Steve Winwood (547) | Steve Albini (417) |
| Steve Earle (512) | Steve Vai (563) |

**Table 2.** Comparison between inlink and PageRank for word "Albert"

| Control Agent (inlink) | Mining Agent (PageRank) |
|---|---|
| Alberta (9702) | Alberta (9702) |
| Albert Einstein (2105) | Albert Einstein (2105) |
| Albert Camus (530) | Albert Camus (530) |
| Albertville (404) | Albertville (404) |
| Alberta Liberal Party (398) | Alberta Liberal Party (398) |

Although a plain analysis using the mining agent did not show a much better ranking compared to a traditional inlink calculation, it can be observed that PageRank can be used to evaluate general and popular concepts since such topics will have a higher tendency to receive links from other pages. The next section proposes an alternative view on applying PageRank on DBpedia based on the special feature of the dataset.

## 4    Inverse PageRank on DBpedia

The traditional approach of evaluating PageRank value to concepts in Wikipedia may not perform well because the structure of Wikipedia is different from the World Wide Web based on the following reasons:

1) There is no such concept of a useful page or a junk page in Wikipedia. A page with a high PageRank usually refers to general topic while a page with a low PageRank usually refers to a specific topic.
2) If a link exists between article A and article B, then somehow there should be a way to relate concept A to concept B and vice versa.

In the context of the WWW, there exists a concept of useful (important) web page and junk web page since there is more freedom to put information to the web. For example, anyone can build a website that describes "cat's behavior". However, out of all these web pages, there will be some that provide high quality information about "cat's behavior" (considered as useful page), while some may just give irrelevant information about "cat's behavior" (considered as junk page). Hence, PageRank is a way to identify the quality of a page for the user (where quality page will be referred by many other pages and as a result yields a higher PageRank).

However, in Wikipedia, the concept of junk pages or useful pages is non-existent as there is no way to judge the importance between two concepts. Every page in Wikipedia represents a distinct concept, either a general concept or a specialized concept. Hence, classifying a given page with a low PageRank value as a low quality page does not make sense. A page with low ranking value does not mean that the page is not worth exploring. In contrast, if a domain specific search is conducted, these pages can be recommended as a special topic.

Cat coat genetics can produce a variety of coat patterns. Some of the most common are:

Bicolor, Tuxedo and Van

    This pattern varies between the tuxedo cat which is mostly black with a white chest, and possibly markings on the
    the way to the Van pattern (so named after the Lake Van area in Turkey, which gave rise to the Turkish Van breed
    colored parts of the cat are the tail (usually including the base of the tail proper), and the top of the head (often inc
    are several other terms for amounts of white between these two extremes, such as Harlequin or jellicle cat. Bicolo
    primary (non-white) color black, red, any dilution thereof, and tortoiseshell (see below for definition).

Tabby

    Striped, with a variety of patterns. The classic blotched tabby (or marbled) pattern is the most common and consi
    bullseyes. The mackerel or striped tabby is a series of vertical stripes down the cat's side (resembling the fish). Th
    spots is referred to as a spotted tabby. Finally, the tabby markings may look like a series of ticks on the fur, thus
    is almost exclusively associated with the Abyssinian breed of cats. The worldwide evolution of the cat means that
    are associated with certain countries; for instance, blotched tabbies are quite rare outside NW Europe, where they
    type.

Tortoiseshell and calico

    This cat is also known as a calimanco cat or clouded tiger cat, and by the abbreviation 'tortie'. In the cat fancy, a t
    patched over with red (or its dilute form, cream) and black (or its dilute blue) mottled throughout the coat. Addition
    white spots in its fur, which make it a 'tortoiseshell and white' cat; if there is a significant amount of white in the fu
    colors form a patchwork rather than a mottled aspect, in North America the cat will be called a calico. All calicos
    carry both black and red), but not all tortoiseshells are calicos (which requires a significant amount of white in the
    than mottling of the colors). The calico is also sometimes called a tricolor cat. The Japanese refer to this pattern a
    fur"), while the Dutch call these cats lapjeskat (meaning "patches cat"). A true tricolor must consist of three color
    or light; white; and one other color, typically a brown, black, or blue.[106] Both tortoiseshell and calico cats are typ
    the coat pattern is the result of differential X chromosome inactivation in females (which, as with all normal female
    chromosomes). Conversely, cats where the overall color is ginger (orange) are commonly male (roughly in a 3:1 ra
    ginger tom, the females will be tortoiseshell or ginger. Male tortoiseshells can occur as a result of chromosomal a
    linked to sterility) or by a phenomenon known as mosaicism, where two early stage embryos are merged into a si

**Fig. 1.** Hyperlinks provided in Wikipedia page can be referred as relationship between two concepts. ("Tabby" will be redirected to "Tabby cat").

The second point is more obvious and straightforward. In a Wikipedia page, if a certain paragraph talks about another concept which is also available in Wikipedia, then usually there will be a link provided for the user to access that concept. For example, as shown in Figure 1, there will be a link from "cat" to "tabby cat" because a certain paragraph in the "cat" page talks about "tabby cat". The same reasoning can be used for hyperlink like "tuxedo cat" and "Turkish van". Hence, it is possible to find some related topics from a concept by just analyzing the link structure of that concept even though some relationships may not make sense like "Freyja" and "Ancient Egypt". "Cat" is linked to "Freyja" because "Freyja" is depicted as riding a chariot drawn by cats[6] while "cat" is linked to "Ancient Egypt" because cat was a sacred animal at that time.

By taking into account both points, this paper proposes an alternative view to apply PageRank on DBpedia link structure by retrieving topic sensitive keywords from a specific concept based on an inverse PageRank measurement of DBpedia. Inverse here refers to ranking related concepts based on the PageRank value in ascending order (low to high). Consider a given task, "find me some specific concepts that are relevant to cat". In this setting, the first step is to search for all concepts that are linked to a "cat" page in DBpedia. During the search for the concept, a list of concepts that is related to "cat" is then displayed. Table 3 shows the comparison of some related concepts that are linked by "cat" based on the PageRank of a concept. The left column shows concepts with high PageRank values. It can be seen that the concepts with high PageRank value are general topics which are too general to be suggested to the user. On other hand, the right column shows concepts with low PageRank values, which refer to specific topics. For example, "cat communication" is a specific topic in communication for "cat". Hence, these concepts are better candidates for page recommendation in Wikipedia.

Table 3. List of concepts related to "cat" sorted by PageRank

| *High PageRank concept* | *Low PageRank concept* |
| --- | --- |
| United States | Cat communication |
| Canada | Melanin |
| Gene | Bicolor cat |
| Animal | Cat heath |
| Europe | Cat allergy |
| German language | Onychectomy |
| The New York Times | Hygiene |
| Chordate | Cat genetics |
| Bird | High-rise syndrome |
| Species | White tiger |
| Hawaii | X-inactivation |
| Dutch language | Taurine |
| Cyprus | Hiss |
| Family (biology) | Polygene |
| Fish | Jellicle |

The limitation of this approach is that there may be some concepts that do not have a strong relationship with "cat" but will still be suggested. Hence, this approach cannot be applied to a field that requires a high precision in suggestion quality. However, one example in which this approach can be used is in the analysis of clusters of text to determine their relevancy. This type of analysis only requires

---

[6] http://en.wikipedia.org/wiki/Cat (last access time: 19 May 2011).

limited text for analysis such that mapping can be performed on the text clusters that has low PageRank values. The logic behind this, is that the concept with low PageRank in DBpedia normally represents some specific topics which may not appear so often in a text especially if the concept has no relationship with the topic. So if those concepts did appear in a certain text cluster, then somehow that cluster may be a relevant cluster. For example, consider two titles, "A site for cat health" and "A site for cat, dog and bird". If a mapping is done from the title to the concept with a low PageRank, a matching will occur for the first title with concept "cat health" while no matching is found for the second title if only 15 concepts with lowest PageRank is considered in the mapping. Hence, a system can state that the first title provides more information about "cat" since the keyword "cat health" is used in the title.

To make use of this concept in a wider area of analysis, not only should PageRank be considered in the ranking, but some analysis such as simrank can be applied in the ranking. For example, consider the low PageRank concept in Table 3. This list of concepts can be further filtered because the similarity ranking between "cat" with a concept like "melanin" and "white tiger" is low due to their long distance from the hierarchy tree in Wikipedia. Hence, these concepts can be the potential suggestions for the "cat" concept.

## 5    Conclusion/Future Work

In this paper, we simulated the environment of the PageRank surfer's model for DBpedia dataset. We also described a mining agent to mine significant concepts in DBpedia. The general result shows that ranking using PageRank is similar to the ranking using inlink property of a page only. This is due to the nature of Wikipedia, in which every page is a distinct concept, and hence there is no point in comparing the importance between concepts. However, PageRank is still a suitable approach to evaluate general or specific concepts in DBpedia. This paper also proposes an alternative way to apply PageRank on DBpedia dataset, based on the unique feature contained in Wikipedia. In this approach, a page with a low PageRank value can be recommended to a user when a topic specific search is performed. This approach may be applied in ontology agent to search for relationship between concepts in linked data. The future work will include a large scale of evaluation based on the concept of inverse PageRank proposed here as well as the evaluation on whether applying similarity analysis will help in mining relevant concepts from DBpedia link structure.

## References

1. Jentzsch, A.: PDF DocumentDBpedia – Extracting structured data from Wikipedia. Presentation at Semantic Web In Bibliotheken (SWIB 2009), Cologne, Germany (November 2009)
2. Auer, S., Bizer, C., Kobilarov, G., Lehmann, J., Cyganiak, R., Ives, Z.: DBpedia: A Nucleus for a Web of Open Data. In: Aberer, K., Choi, K.-S., Noy, N., Allemang, D., Lee, K.-I., Nixon, L.J.B., Golbeck, J., Mika, P., Maynard, D., Mizoguchi, R., Schreiber, G., Cudré-Mauroux, P. (eds.) ASWC 2007 and ISWC 2007. LNCS, vol. 4825, pp. 722–735. Springer, Heidelberg (2007)

3. Bizer, C., Auer, S., Kobilarov, G., Lehmann, J., Cyganiak, R.: DBpedia.org - Querying Wikipedia like a Database. In: Developers Track at 16th International World Wide Web Conference (WWW 2007), Banff, Canada (May 2007)
4. Bizer, C., Lehmann, J., Kobilarov, G., Auer, S., Becker, C., Cyganiak, R., Hellmann, S.: DBpedia - A crystallization point for the Web of Data. Journal of Web Semantics: Science, Services and Agents on the World Wide Web (7), 154–165 (2009)
5. Page, L., Brin, S., Motwani, R., Winograd, T.: The PageRank citation ranking: Bringing order to the web. Technical report, Stanford University, Stanford, CA (1998)
6. Haveliwala, T.H.: Topic-sensitive PageRank. IEEE Transactions on Knowledge and Data Engineering 15, 784–796 (2003)
7. Mirizzi, R., Ragone, A., Di Noia, T., Di Sciascio, E.: Ranking the Linked Data: The Case of DBpedia. In: Benatallah, B., Casati, F., Kappel, G., Rossi, G. (eds.) ICWE 2010. LNCS, vol. 6189, pp. 337–354. Springer, Heidelberg (2010)
8. Keong, B.V., Anthony, P.: PageRank: A Modified Random Surfer Model. In: 7th International Conference on IT in Asia, Kuching, Malaysia (2011) (in press)

# Prediction of Protein Residue Contact Using Support Vector Machine

Weng Howe Chan and Mohd Saberi Mohamad

Artificial Intelligence and Bioinformatics Research Group,
Faculty of Computer Science and Information Systems,
Universiti Teknologi Malaysia, 81310 Skudai, Johor, Malaysia
stephenchanwh@gmail.com, saberi@utm.my

**Abstract.** Prediction of protein residue contact is one of the important two-dimensional prediction tasks in protein structure prediction. The residue contact map of protein contains information which represents three-dimensional conformation of protein. However the accuracy of the prediction is dependent on the type of protein information used to distinguish between contacts or non-contacts. According to CASP (Critical Assessment of Techniques of Protein Structure Prediction) the accuracy of protein contact map prediction is still low due to the behaviour of the predictors developed where the predictors only effective against specific type of protein structure. In order to further improve the performance of the predictor, effective features must be identified and used. Therefore, this research is conducted to determine the effectiveness of the existing features used in protein contact map prediction.

**Keywords:** protein residue contact map, support vector machine, protein structure prediction.

## 1    Introduction

Bioinformatics is defined as a field of science that involve the application of statistics and computer science in the field of biology. It is an emerging field undergoing rapid growth in the past few decades. Bioinformatics at first is applied in the creation and maintenance of database of biological information and currently also applied in tasks like interpretation and analysis of biological data includes deoxyribonucleic acid (DNA) sequences, ribonucleic acid (RNA) sequences, protein structures, protein sequences and protein domains which referred as computational biology. The branch of bioinformatics that consists of the analysis and prediction of three dimensional structures of biological macromolecules such as DNA, RNA and proteins referred as structural bioinformatics.    In structural bioinformatics, one of the challenges is the prediction of protein structure.

Protein is one of the most important compounds in human body. Function of a protein is defined by its structure. Protein structures are divided into few categories such as primary structure, secondary structure, tertiary structure, and quaternary structure.    A protein consists of more than one linear chain of amino acids that

D. Lukose, A.R. Ahmad, and A. Suliman (Eds.): KTW 2011, CCIS 295, pp. 323–332, 2012.

further fold into polypeptides of different structures and features. Protein structure prediction has played an important role in protein design which is essential is several fields for instance medicine. In protein structure prediction, different information of protein has been used and one of them is protein contact map which is used in this research. Protein contact map is a compact representation of three-dimensional conformations of a protein. A contact map is a two-dimensional Boolean matrix representation of protein structure, each of the dimensions is represented by residue number, while the value is true when the corresponding residues are spatial neighbours and false otherwise [2]. Protein contact map is binary symmetric matrices where non-zero values represent the residue in contact [3] that is shown in Fig. 1.

According to Fig. 1, Secondary structures are highlighted along the both axis. Both α-helices and β-strands represented by black and grey respectively. While on the left side, the structural protein features are shown: (a) Anti-parallel sheet contacts; (b) parallel sheet contacts; (c) contacts between helical regions. Generally, a residue pair considers as a contact when the distance between the residues within a pair is below a defined distance threshold. The distance threshold is calculated in angstrom (Å) which measure as 0.1 nanometre or 1 x 10-10 metres. In CASP (Critical Assessment of Techniques for Protein Structure Prediction), default distance threshold used for assessment is 8Å which is same as the threshold used in this research.

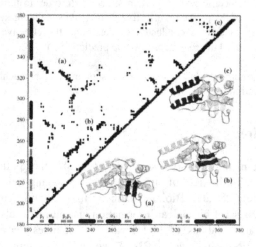

**Fig. 1.** Example of contact map of HSP-60 protein fragment

Based on the researches done in the past decade, many techniques and algorithms have been developed to predict contact map of a protein. Among those methods, machine learning algorithms have been widely used such as support vector machine (SVM). Machine learning is an artificial intelligent technique where it is a scientific discipline that is concern about design and development of algorithms that allows computer to learn and evolve behaviours based on the empirical data. It learns to recognize complex patterns thus make decision to gain useful output. In this research,

SVM has been used and implemented. SVM is a supervised learning method that analyses and recognizes pattern for classification and regression. SVM constructs a hyperplane or a set of hyperplanes in high or infinite dimensional space for classification and regression.

Since the prediction of protein contact map is significant in contributing the three-dimensional structure prediction of protein, refer to the state of art of protein contact map prediction, one of the main concerns is the performance issues of several predictors for protein contact map. According to the CASP (Critical Assessment of Techniques of Protein Structure Prediction), the accuracy and the coverage of the prediction of protein contact map are still low and performance varies with the type of structure of the tested protein. In fact, many of the predictors that had been developed tend to predict different correct contacts with implementation of different types of information obtained from protein such as protein profile, predicted secondary structure, solvent accessibility and so forth. Therefore consensus combination of predictors may lead to a better accuracy in protein contact map prediction. To date, researchers are still working on protein contact map prediction in order to enhance the predictor to obtain better and more accurate prediction. Besides, challenges also faced during the prediction of long sequence with many non-local contacts, non-local contacts which had appeared to be a problem because the global topology of the proteins is defined by non-local contacts (also known as long range-contacts) but the methods developed so far are more accurate on local contacts only [2].

This research concentrate in the performance related problem faced in the protein contact map prediction and thus with the use of support vector machine (SVM) method plus different combination of features, studies and experiments have been done in order to identify and determine the effectiveness of the features used in the prediction.

In this paper, details regarding the dataset and the features used as well as the methods are presented in section 2 while results and analysis are discussed and presented in section 3. Conclusion of this research is outlined in section 4.

# 2     Materials and Methods

## 2.1     Dataset

The dataset used in this research consist of 424 proteins and 48 proteins for both training and testing set respectively. This dataset had been used in previous research [1] which consist of information regard to the particular protein such as predicted secondary structure and predicted solvent accessibility generated from SSpro [4], protein sequence, beta partners as well as three-dimensional coordinates of alpha carbon for each residue in the protein. The dataset is redundancy reduced where the pairwise sequence identity of two sequences is less than 25%. Fig. 2 shows the example format of one of the data entry in the dataset.

```
1EJGA
46
T  T  A  B  P  S  I  V  A  R  S  N  F  N  V  C  R  L  P  G  T  P  E  A
C  C  C  C  C  C  E  E  E  C  C  C  E  E  E  C  C  C  C  C  H  H  H  H  H  C  C  C  C  E  E  E  C  C  C  C  C  C  C  C  C  C
0   34 33 0   0   0   0   0   0   0   0   0   0   0   0   0   0   0   0   0   0   0   0   0
0   0   0   0   0   0   0   0   0   0   0   0   0   0   0   0   0   0   0   0   0   0   0   0
50 50 10 10 50 50 50 50 10 50 50 10 10 10 10 10 10 50 50 50 50 50 10 10 10 50 10 10 50 10
16.9 12.8 4.2    13.9 11.5 5.9    13.6 10.7 9.6    10.7 8.9 11.3    9.5 9.1 14.9    8.8 5.3 15.2
```

**Fig. 2.** Example of the dataset format

## 2.2    Input Features

This research is to construct different prediction models that consist of different combination of features [1] in order to analyze and compare the effectiveness of the features implemented in the prediction model.   There are total of five different kinds of features applied in this research which are local window feature, pairwise information feature, central segment window feature, segment average information feature, and protein information feature.    These features consist of different information extracted from the dataset proteins including predicted secondary structure and solvent accessibility as well as the amino acid composition of the corresponding protein.

Local window feature is a 9-residue window feature which centered at each residue in each potential residue pair at which the distance of the residue in the pair is not less than the separation value set.   In this features, each position within the window, there are 27 inputs which include 21 inputs for amino acids plus a gap, 3 inputs for predicted secondary structure (helix, coil, and sheet), 2 for the predicted solvent accessibility (exposed and buried) and 1 for the entropy.   So this feature will have a size of 486.

In pairwise information feature, for each pair of position (i,j) in a multiple sequence alignment, 7 inputs are calculated, one input corresponds to the mutual information of the profiles of the two positions $\sum_{kl} p_{kl} \log(p_{kl}/(p_k p_l))$, where $p_{kl}$ is the empirical probability of residues (or gap) k and l appearing at the two positions i and j simultaneously.   While $p_k$ and $p_l$ refer to the probability of appearance of residues k and l respectively.   Another two inputs are computed using cosine and correlation and one input for the amino acid type.   Finally the last three inputs are regard to the pairwise potential values from three different pairwise potentials which are Levitt's pairwise potential [5], Jernigan's pairwise potential [6] and Braun's pairwise potential [7] for the residue pairs in the target sequence.   This feature has a size of 16.

Third is the central segment window feature where this feature has a window size of 5 which locate at the position of (i+j)/2 which is the center of the potential residue pair.   For each position of the window, 27 inputs are used same as in the local window features which are 21 for amino acids plus a gap, 3 for predicted secondary structure, 2 for predicted solvent accessibility, 1 for the entropy.   Therefore, central segment window feature has a size of 135.   Another similar feature which is segment average information feature also using the information extracted from the segment between the residue pair.   This feature has a size of 42 and it consist the information about the predicted secondary structure, solvent accessibility and the segment length information. Lastly is the protein information feature. This feature has a size of 30. In this feature, the global amino acid composition, secondary structure, and relative solvent accessibility of the target sequence are calculated.

## 2.3    Construction of Prediction Models

This research combines different features into several combinations and used to construct several prediction models. In order to compare and analyse the effectiveness of the features and with the availability of high performance computers from Centre of Information and Communication Technology (CICT) UTM, a total of ten prediction models with different combination pairs of features are constructed. Table 1 shows the ten prediction models that constructed in this research.

**Table 1.** Prediction models with corresponding features and size

| Model | Features | Size |
|---|---|---|
| 1 | Pairwise Information + Local Window | 502 |
| 2 | Pairwise Information + Central Segment Window | 151 |
| 3 | Pairwise Information + Segment Average Information | 58 |
| 4 | Pairwise Information + Protein Information | 46 |
| 5 | Local Window + Central Segment Window | 621 |
| 6 | Local Window + Segment Average Information | 528 |
| 7 | Local Window + Protein Information | 516 |
| 8 | Central Segment Window + Segment Average Information | 177 |
| 9 | Central Segment Window + Protein Information | 165 |
| 10 | Segment Average Information + Protein Information | 72 |

The process of the construction of the prediction models consists of two major steps:

*Step 1.* Generation of the SVM input for all the combination of features for the training set proteins. In this steps, corresponding information that needed are generated according to the feature involved and result in generation of an input data with SVM compatible format. This process continues executed on the protein sequence in the dataset and append to a single output file. Fig. 3 shows the process to generate SVM input features file.

*Step 2.* The SVM compatible input files generated is then used in the SVM learning process using SVM Light and generates the prediction models.

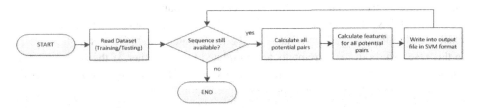

**Fig. 3.** Process of generating SVM input features

## 2.4    Learning Using Support Vector Machine

Support vector machines (SVMs) are used to predict an input feature vector which associated with a pair of residues to see if the two residues are in contact (positive) or not (negative). SVM provides several of classification and regression method that uses linear to non-linear way to solve corresponding problem which is controlled by the kernel methods. Kernel methods or kernel functions can re-map the data points into a higher dimensionality feature space solving the problem that are not solvable using linear method.

One of the key property of kernel method is the embedding does not need to be given in explicit form. Given a set of training data points, $S = S^+ \cup S^-$ where $S^+$ represent the positive samples and $S^-$ represent the negative samples, using the theory of risk minimization, support vector machines learn a classification function $f(x)$ as follow where $a_i$ are non-negative weights and b is the bias. $K(x, x_i)$ is the kernel method used, $x_i$ is the training data points and x is the target data point that is predicted to be positive or negative by taking the sign of $f(x)$.

$$f(x) = \sum_{x_i \in S^+} a_i K(x, x_i) - \sum_{x_i \in S^-} a_i K(x, x_i) + b \tag{1}$$

In this research, radial basis function (RBF) kernel is used in this research to train the prediction models. The RBF kernel is used with the gamma parameter $(\gamma)$ set to 0.025 which is same as the previous setting [3]. This is because this will ease the comparison of the results later. The RBF kernel can be presented by following

$$K(x, y) = e^{-\gamma \|x - y\|^2} \tag{2}$$

## 2.5    Performance Measurement

In order to justify the results obtained and the performance of the prediction models, data are compared among the prediction models in terms of prediction performance. In this research, the performance is measured by accuracy and coverage where accuracy is the number of correct predictions per total number of predictions; its value shows the ability of the prediction model to get correct prediction out of the total number of prediction. Higher accuracy implies that the models are able to get more correct prediction. Meanwhile for coverage is the number of correct predictions per total number of true contacts. This parameter is similar to the sensitivity, where it shows the ability of the prediction model to identify true contacts. Higher value of sensitivity implies that the percentage of the true contacts identified is high as well. In this research, both measurements are correlated to each other, if the accuracy of a model is high; the coverage also shows high value. This implies that the model is efficient in predicting true contact out of the prediction.

# 3    Results and Analysis

The performance of the constructed prediction models are evaluated by comparing the prediction to the true contacts information. Measurement is done in terms of accuracy and coverage where accuracy is defined as the total number of correct prediction per total number of predictions while coverage is defined as the total number of correct prediction per total number of true contacts in the protein. The overall accuracy results for all prediction models are shown in Table 2 while for the coverage results are shown in Table 3.

**Table 2.** Results from different prediction models (Accuracy)

| Protein | length | TC | Type | ACCURACY | | | | | | | | | |
|---------|--------|----|------|---------|---------|---------|---------|---------|---------|---------|---------|---------|----------|
| | | | | Model 1 | Model 2 | Model 3 | Model 4 | Model 5 | Model 6 | Model 7 | Model 8 | Model 9 | Model 10 |
| 1CTJA | 89 | 95 | alpha | 0.023 | 0.011 | 0.180 | 0.023 | 0.023 | 0.124 | 0.023 | 0.135 | 0.000 | 0.214 |
| 1C75A | 71 | 95 | alpha | 0.170 | 0.141 | **0.338** | **0.085** | 0.239 | 0.352 | 0.141 | 0.296 | 0.056 | **0.268** |
| 1CQYA | 99 | 225 | beta | 0.172 | **0.283** | 0.131 | 0.051 | 0.192 | 0.131 | 0.182 | 0.253 | **0.212** | 0.121 |
| 1BMGA | 98 | 220 | beta | 0.184 | 0.163 | 0.102 | 0.071 | 0.255 | 0.265 | 0.255 | 0.163 | 0.071 | 0.071 |
| 1MWPA | 96 | 197 | a+b | 0.135 | 0.010 | 0.052 | 0.063 | 0.146 | 0.104 | 0.125 | 0.135 | 0.052 | 0.042 |
| 1G2RA | 94 | 126 | a+b | **0.394** | 0.106 | 0.170 | 0.053 | 0.223 | 0.362 | **0.426** | 0.138 | 0.011 | 0.043 |
| 1CXQA | 143 | 211 | a/b | 0.287 | 0.014 | 0.070 | 0.021 | 0.280 | 0.357 | 0.280 | 0.035 | 0.021 | 0.021 |
| 1F4PA | 147 | 293 | a/b | 0.320 | 0.088 | 0.054 | 0.061 | **0.327** | **0.374** | 0.265 | 0.136 | 0.027 | 0.048 |
| 1AIHA | 85 | 85 | small | 0.118 | 0.012 | 0.188 | 0.000 | 0.235 | 0.294 | 0.059 | **0.318** | 0.012 | 0.247 |
| 1EJGA | 46 | 59 | small | 0.261 | 0.044 | 0.065 | 0.044 | 0.152 | 0.239 | 0.217 | 0.065 | 0.000 | 0.065 |
| 1AA0A | 113 | 63 | coil-coil | 0.115 | 0.089 | 0.177 | 0.009 | 0.168 | 0.257 | 0.115 | 0.230 | 0.044 | 0.177 |

**Table 3.** Results from different prediction models (Coverage)

| Protein | Length | TC | Type | COVERAGE | | | | | | | | | |
|---------|--------|----|------|---------|---------|---------|---------|---------|---------|---------|---------|---------|----------|
| | | | | Model 1 | Model 2 | Model 3 | Model 4 | Model 5 | Model 6 | Model 7 | Model 8 | Model 9 | Model 10 |
| 1CTJA | 89 | 95 | alpha | 0.0211 | 0.011 | 0.168 | 0.021 | 0.021 | 0.116 | 0.021 | 0.126 | 0.000 | 0.200 |
| 1C75A | 71 | 95 | alpha | 0.126 | 0.105 | **0.253** | **0.063** | 0.179 | 0.263 | 0.105 | 0.221 | 0.042 | 0.200 |
| 1CQYA | 99 | 225 | beta | 0.076 | 0.124 | 0.058 | 0.022 | 0.084 | 0.058 | 0.080 | 0.111 | **0.093** | 0.053 |
| 1BMGA | 98 | 220 | beta | 0.082 | 0.073 | 0.046 | 0.032 | 0.114 | 0.118 | 0.114 | 0.073 | 0.032 | 0.032 |
| 1MWPA | 96 | 197 | a+b | 0.066 | 0.005 | 0.025 | 0.031 | 0.071 | 0.051 | 0.061 | 0.066 | 0.025 | 0.020 |
| 1G2RA | 94 | 126 | a+b | **0.294** | 0.079 | 0.127 | 0.040 | 0.167 | 0.270 | **0.318** | 0.103 | 0.0080 | 0.032 |
| 1CXQA | 143 | 211 | a/b | 0.194 | 0.010 | 0.047 | 0.014 | 0.190 | 0.242 | 0.190 | 0.024 | 0.014 | 0.014 |
| 1F4PA | 147 | 293 | a/b | 0.160 | 0.044 | 0.027 | 0.031 | 0.164 | 0.188 | 0.133 | 0.068 | 0.014 | 0.024 |
| 1AIHA | 85 | 85 | small | 0.118 | 0.012 | 0.188 | 0.000 | 0.235 | 0.294 | 0.059 | 0.318 | 0.012 | 0.247 |
| 1EJGA | 46 | 59 | small | 0.203 | 0.034 | 0.051 | 0.034 | 0.119 | 0.186 | 0.170 | 0.051 | 0.000 | 0.051 |
| 1AA0A | 113 | 63 | coil-coil | 0.206 | **0.159** | 0.159 | 0.016 | **0.302** | **0.460** | 0.206 | **0.413** | 0.079 | **0.318** |

Based on the results reviewed, and also based on the prediction performance data shown in Table 2 and Table 3, accuracy of the contact map prediction is directly correlated to the information or features integrated into the prediction model. This can be seen in this research, the prediction results of model 4, model 9 and model 10 which integrating protein information features as one of the information to predict contact map. However, the results obtained is very low in accuracy and performance is not balance and consistent on all types of proteins. While for model 1, model 5, model 6 and model 7, these models obtained good results among others. This can be clearly seen by observing the average accuracy and coverage obtained as shown in Fig. 4 and Fig. 5. These four models yielded overall consistent results throughout this research, four of this model have similarity where each of the models also implemented local window information as one of the feature. This further implies the effectiveness of the information of local window feature in distinguishing residue contact from protein sequence. Based on the findings from previous researches, performance of the prediction is affected by the reliability of the information used such as multiple sequence alignment, predicted secondary structure, predicted solvent accessibility and so forth.

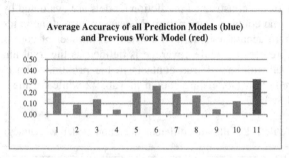

**Fig. 4.** Average prediction accuracy of all models (blue) and previous work model (red)

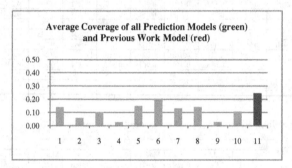

**Fig. 5.** Average coverage of all models (green) and previous work model (red)

Based on Fig. 4, prediction model 6 manages to get accuracy near the accuracy obtained by the previous work model done by Cheng and Baldi [3]. This shows the significant of the features within the model especially the local window feature which shows significance on model 1, 5, and 7. Besides, based on Fig. 5, the coverage of model 6 is very near to the coverage obtained by previous work model. This shows that significance of the features used in model 6 has a high recall rate on the true contacts of the proteins.

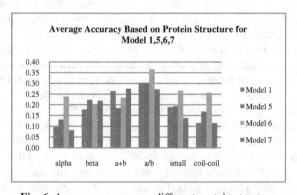

**Fig. 6.** Average accuracy on different protein structure

Meanwhile, the performance affection in terms of effectiveness of the features used to build the prediction models, the performance also affected by the type of the proteins that used for testing.  Fig. 6 shows the average accuracy of model 1, 5, 6, 7 based on different types of protein structure of the tested proteins.

According to Fig. 6, clearly shown that the types of structure such as beta, a+b, and a/b tend to be predicted with higher accuracy.  Refer to the research done previously, the contacts that within beta-sheets are predicted with higher accuracy than contacts that between alpha helix and a beta strands or between alpha helix [4, 8].  This is probably because of the strong restraints between beta-strands such as hydrogen bond gives the increased accuracy.  This are shown more clearly in Fig. 7 by average the accuracies obtained for all models based on different type of structure.

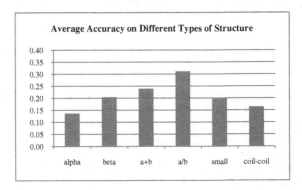

**Fig. 7.** Average accuracy on different types of structure

# 4    Conclusion

Different with previous research, this research concentrate on the determination and analysis of the effectiveness of the features used in protein contact map prediction to contribute and improve protein contact map prediction which the main advantage of this research that is not so concentrated in previous research. Even though the accuracy achieved by the constructed prediction models is lower than previous research, however, based on the results that obtained by combination of two features, highest average results achieved is 81% of the average accuracy from previous research (5 features). This implies that the feature information (2 features) used in model 6 is efficient in predicting protein contact map, and indirectly implies the existence of unnecessary or inefficient features. This also shortens the execution time of the process with more experiments can be conducted. However, due to the time constraint, this research is done using combination of two features, more variety of combination can be made with implement of more features can be done in future. Therefore, in this research, with the construction of the multiple prediction models with different combination of features, effectiveness of the features that affect the performance of the prediction are identified, and further improve the knowledge regard to the effective information to be used in protein residue contact prediction.

We believe that, in order to further increase the accuracy of the predictions for all kind of proteins, a more informative feature of proteins is needed even combination of informative features that able to distinct the contacts among residues. This research had shown that the use of local window feature in the prediction model yield decent results among others, while on the other hand, this research also shows that combination of local window feature and segment average information (model 6) produce balance results among all structures. By the identification of these information, through combining others effective features with the one shown in this research, it is believed that this can help to improve the accuracy of the prediction.

**Acknowledgments.** We also would like to thank Universiti Teknologi Malaysia for supporting this research by UTM GUP research grant (Vot number: Q.J130000.7107.01H29).

# References

1. Cheng, J., Baldi, P.: Improved Residue Contact Prediction Using Support Vector Machines and A Large Feature Set. BMC Bioinformatics 8, 113 (2007)
2. Yuan, X., Bystroff, C.: Protein Contact Map Prediction. In: Xu, Y., Xu, D., Liang, J. (eds.) Computational Methods for Protein Structure Prediction and Modeling, pp. 255–277. Springer, Heidelberg (2007)
3. Bartoli, L., Capriotti, E., Fariselli, P., Martelli, P.L., Casadio, R.: The Pros and Cons of Predicting Protein Contact Maps. In: Zaki, M., Bystroff, C. (eds.) Protein Structure Prediction Second Edition, pp. 199–217. Humana Press, Totawa (2008)
4. Cheng, J., Randall, A., Sweredoski, M., Baldi, P.: SCRATCH: a protein structure and structural feature prediction server. Nucleic Acids Research 72–76 (2005)
5. Huang, E., Subbiah, S., Tsai, J., Levitt, M.: Using a Hydrophobic Contact Potential to Evaluate Native and Near-Native Folds Generated by Molecular Dynamics Simulations. J. Mol. Biol. 257, 716–725 (1996)
6. Miyazawa, S., Jernigan, R.: An empirical energy potential with a reference state for protein fold and sequence recognition. Proteins 36, 357–369 (1999)
7. Zhu, H., Braun, W.: Sequence specificity, statistical potentials, and three-dimensional structure prediction with self-correcting. Protein Sci. 8, 326–342 (1999)
8. MacCallum: Striped Sheets and Protein Contact Prediction. Oxford Bioinformatics 20, 224–231 (2004)

# Optimized Local Protein Structure with Support Vector Machine to Predict Protein Secondary Structure

Yin Fai Chin, Rohayanti Hassan, and Mohd Saberi Mohamad

Artificial Intelligence and Bioinformatics Research Group,
Faculty of Computer Science and Information Systems,
Universiti Teknologi Malaysia,
81310 Skudai, Johor, Malaysia
{chinyinfai,rohayanti}@gmail.com, saberi@utm.my

**Abstract.** Protein includes many substances, such as enzymes, hormones and antibodies that are necessary for the organisms. Living cells are controlled by proteins and genes that interact through complex molecular pathways to achieve a specific function. These proteins have different shapes and structures which distinct them from each other. By having unique structures, only proteins able to carried out their function efficiently. Therefore, determination of protein structure is fundamental for the understanding of the cell's functions. The function of a protein is also largely determined by its structure. The importance of understanding protein structure has fueled the development of protein structure databases and prediction tools. Computational methods which were able to predict protein structure for the determination of protein function efficiently and accurately are in high demand. In this study, local protein structure with Support Vector Machine is proposed to predict protein secondary structure.

**Keywords:** Local Protein Structure, Support Vector Machine, Protein Secondary Structure Prediction.

## 1 Introduction

In recent years, human genome project has successfully generated tremendous amount of newly protein sequences in the biological database. Ironically, most of them are completely unknown in function and structure and cause complete genome sequencing gives much less understanding on the organism than initially hoped for [1]. Proteins control and mediate many of the biological activities of cells. Hence, to gain an understanding of cellular function, the structure of every protein must be understood [2]. This has shown that the study the sequence of a single protein or small complexes is no longer sufficient in helping the current genome development.

Protein structure predictions represent a key step in studying and understanding protein functions. The fact that protein function do not only depends on protein sequence but also the shape and structure induces the important goal of the proteomic studies which is identification of protein structure. Given a protein sequence, the

D. Lukose, A.R. Ahmad, and A. Suliman (Eds.): KTW 2011, CCIS 295, pp. 333–342, 2012.

secondary structure prediction problem is to predict whether each amino acid is in a helix, strand or neither. H, E and C represent helix, strand and non-routine structure, respectively [3]. The simple definition of secondary structures hides various limitations. The complexity of fundamentals for secondary structure assignments induce the creation of numerous assignment methods based on different criteria or characteristics. Due to certain limitation in secondary structure, a more precise assignment for secondary structure is presented which is local protein structure. Local protein structure is defined as the description of complete set of small prototype or protein structures. Analysis of local protein structures represents an evaluation of every parts of protein backbone. Hence, focusing on local protein structure might develop a new milestone in the future of protein secondary structure prediction.

The aim of this research is to predict protein secondary structure using machine learning algorithms based on RS126 as the dataset. RS126 is important as the core dataset to be trained and tested using machine learning algorithm because the dataset contains 126 non-redundant proteins where the number pairs of proteins in the set have more than 25% similarity over a length of 80 residues. Given the small similarity of the dataset sequences, this represents a situation that is rather close to real-world settings and it can be considered as the ideal environment for protein secondary structure prediction. The machine learning algorithm, implemented in this study is Support Vector Machine (SVM). The reason SVM is being used is because they are known to be a powerful algorithm for making binary decisions. The results will be able to show the higher accuracy of computational prediction system based on SVM for protein secondary structure prediction.

## 2     Materials and Methods

### 2.1     Materials

Materials briefly explain the dataset used and also the source of data such as the background of the dataset and how to obtain it. Details of dataset preparation and usage will be explained in the following section.

### 2.1.1     RS126 Dataset

The dataset used in this study is RS126. The initiation of the research is to obtain the protein sequence datasets in order to predict protein secondary structure.

RS126 is one of the oldest dataset with the longest history to evaluate for protein secondary structure prediction. The scheme is created by Rost and Sander [4]. RS126 being the most commonly used datasets to predict protein structure are applied in most of the study including this research. It contains 23,347 residues with an average protein sequence length of 185. 32% of RS126 are alpha helix, 21% as beta strand and 47% as coil.

RS126 dataset can be collected from various supplementary data files in previous research or study. Besides that, it can also be obtained from online database such as Protein Data Bank (PDB). Fig. 1 shows the list of RS126 dataset used in protein secondary structure prediction.

| PDB ID | Chain | PDB ID | Chain | PDB ID | Chain | PDB ID | Chain | PDB ID | Chain |
|--------|-------|--------|-------|--------|-------|--------|-------|--------|-------|
| 1BMV | 1 | 2UTG | A | 4SGB | I | 1PYP | - | 3CD4 | - |
| 4RHV | 1 | 3GAP | A | 1MCP | L | 1RBP | - | 3CLA | - |
| 1BMV | 2 | 3HMG | A | 2OR1 | L | 1RHD | - | 3CLN | - |
| 1R09 | 2 | 3TIM | A | 1GD1 | O | 1SO1 | - | 3EBX | - |
| 1LMB | 3 | 4SDH | A | 2TMV | P | 1SH1 | - | 3ICB | - |
| 4RHV | 3 | 4TS1 | A | 2WRP | R | 1UBQ | - | 3PGM | - |
| 2MEV | 4 | 4XIA | A | 5CYT | R | 2AAT | - | 3RNT | - |
| 4RHV | 4 | 5HVP | A | 1ACX | - | 2ALP | - | 4BP2 | - |
| 1BBP | A | 7CAT | A | 1AZU | - | 2CAB | - | 4CMS | - |
| 1CDT | A | 9API | A | 1BDS | - | 2CYP | - | 4CPV | - |
| 1FXI | A | 9WGA | A | 1CBH | - | 2FOX | - | 4GR1 | - |
| 1GP1 | A | 1WSY | B | 1CC5 | - | 2FXB | - | 4PFK | - |
| 1IL8 | A | 2LTN | B | 1CRN | - | 2GBP | - | 4RXN | - |
| 1OVO | A | 2SOD | B | 1ECA | - | 2GCR | - | 5LDH | - |
| 1TNF | A | 3HMG | B | 1ETU | - | 2GN5 | - | 5LYZ | - |
| 1WSY | A | 9API | B | 1FDX | - | 2I1B | - | 6ACN | - |
| 256B | A | 9INS | B | 1FKF | - | 2LHB | - | 6CPA | - |
| 2AK3 | A | 1FC2 | C | 1FND | - | 2MHU | - | 6CPP | - |
| 2CCY | A | 5ER2 | E | 1GDJ | - | 2PCY | - | 6CTS | - |
| 2GLS | A | 6TMN | E | 1HIP | - | 2PHH | - | 6DFR | - |
| 2HMZ | A | 1FDL | H | 1L58 | - | 2SNS | - | 6HIR | - |
| 2LTN | A | 1CSE | I | 1LAP | - | 2STV | - | 7ICD | - |
| 2PAB | A | 1TGS | I | 1MRT | - | 3AIT | - | 7RSA | - |
| 2RSP | A | 2TGP | I | 1PAZ | - | 3B5C | - | 8ABP | - |
| 2TSC | A | 4CPA | I | 1PPT | - | 3BLM | - | 8ADH | -| |
| 9PAP | - | | | | | | | | |

**Fig. 1.** List of RS126 dataset used to predict protein secondary structure

### 2.1.2    Dihedral Angle (DA)

Generally, dihedral angle is defined as the angle between two planes. In terms of proteomics, the backbone dihedral angles of proteins are called phi ($\varphi$), psi ($\psi$) and omega ($\omega$). Every different angle has its own functions. Dihedral angle is used as feature vector in this research due to its nature form of representation, which is the numerical or integer form. Besides that, dihedral angles play a key role in defining or 'tightening' the secondary structure of protein structures during the structure refinement process. The importance of dihedral angle information tends to increase with the size of the protein being studied as the quality and quantity of other restraints.

In this study, all the dihedral angles are obtained through ramachandran function in Matlab. Ramachandran function generates the dihedral angle for the protein specified by the PDB database identifier PDBid. PDBid is a string specifying a unique identifier for a protein structure record in the PDB database. Each structure in the PDB database is represented by a four-character alphanumeric identifier. The PDBid is similar to the identifier of protein in RS126. For example, 4hhb is the identifier for hemoglobin. The results will return the dihedral angles for each protein in RS126 as 3 columns which include phi angle, psi angle and omega angle.

### 2.1.3    DSSP

The DSSP program was designed by Wolfgang Kabsch and Chris Sander as the standard method for assigning secondary structure to the amino acids of a protein, given the atomic-resolution coordinates of the protein. DSSP is a database of secondary structure assignments for all protein entries in the Protein Data Bank (PDB). DSSP is also the program that calculates DSSP entries from PDB entries.

DSSP has eight types of protein secondary structure, depending on the pattern of hydrogen bond. The list bellows shows the different types of protein secondary structure in DSSP:

i)      H = alpha helix
ii)     B = residue in isolated beta-bridge
iii)    E = extended strand, participates in beta ladder
iv)     G = 3-helix (3/10 helix)
v)      I = 5 helix (pi helix)
vi)     T = hydrogen bonded turn
vii)    S = bend
viii)   L = others

These eight types are usually assigned into three larger groups: helix (G, H and I), strand (E and B) and loop (all others). In this research, DSSP used as feature class are from the three classes, which is helix (H), strand (E) and coil (C). DSSP dataset can be obtained from the RS126 sequence data which contain secondary structures and will be implemented as the feature class to fit into SVM for prediction.

## 2.2    Methods

The study of protein secondary structure prediction will focus on its feature representation which is the local protein structure. Using the conventional methods of machine learning algorithm, which is applying only Support Vector Machine is not effective in protein structure prediction. This is due to the nature behavior where biological features are known to be dynamic rather than being taken as static data in pattern recognition problem solving. With this issue in mind, a preprocessing step is taken into consideration as an extra biological feature in order to enhance the performance of the system and accurately predict protein secondary structure from local protein structure. It is to be believed that considering biological features such as local protein structure, protein sequences information in feature selection is crucial in machine learning approaches. The reason why local protein structure is used as the additional feature in the study is because local protein structure able to analyze small sets of protein and approximate every part of protein backbone.

With DSSP and dihedral angle available in the workspace, secondary structure and DA can be segmented into different local protein structure with different segment lengths. Every local protein structure will have their own DA and DSSP after segmentation and by implementing them as feature vector and feature class, the data can now fit into SVM for classification to predict protein secondary structure.

Support vector machines (SVM) are a group of supervised learning methods that can be applied to classification or regression. The Support Vector Machine (SVM) is a binary classification algorithm and with this attribute, it is suitable for the task of predicting protein secondary structure. SVM has shown that it is able to classify data precisely in the field of protein secondary structure prediction, functional classification of proteins, protein fold recognition, and prediction of subcellular location. SVM has previously been used in the prediction of protein secondary structure [5][6][7][8]. 10 fold cross validation is implemented in support vector machine to classify and predict protein secondary structure.

By using 10 fold cross validation, the datasets are partitioned into 10 samples. From the 10 samples, 1 of them is assigned as testing model to validate the data and the rest are used as testing model. The process of cross validation is repeated 10 times, where each of the 10 samples is used once as the validation model. All of the results can be used to produce estimations for prediction. Kernel implemented is the RBF kernel. By using non-linear kernel, the margin hyperplanes can be optimized. The algorithm still works similarly with a linear algorithm, just that a RBF kernel is applied to every dot product.

The performance of the system is tested and output of the system will be analyzed right after it is released. The performance and accuracy of protein structure prediction is measured and evaluated by how well the system can predict protein secondary structure with higher accuracy and less false positive rate. To enhance the measurement system, widely used evaluation measurement for classification problem such as accuracy, true positive rate (sensitivity) and false positive rate will be applied.

Accuracy measures the probability of true results (true positives and true negatives) in the whole population (true positives, false positives, false negatives, true negatives). Accuracy can be calculated as follow:

$$Accuracy = \frac{TP + TN}{TP + FP + FN + TN} \tag{1}$$

True positive rate which is also known as sensitivity or recall defines the proportion of actual positives which are correctly identified as such. It measures the probability of the true positive value among true positives and false negatives. The formula of sensitivity is shown as below:

$$Sensitivity = Recall = \frac{TP}{TP + FN} \tag{2}$$

False positive rate measures the probability of the positive prediction result when the proteins are non-secondary structure. It can be calculated as follow:

$$FPR = \frac{FP}{TN + FP} = 1 - Specificity \tag{3}$$

Besides applying the evaluation method mentioned above, a statistical method, t-test is implemented for validation of the results obtained. A t-test is any statistical hypothesis test in which the test statistic follows a Student's t distribution, if the null hypothesis is supported. In the research, t-test is applied on two samples of result which represents different local protein structures

## 3    Results and Discussions

Initially, to understand the importance of optimizing local protein structure, the prediction is conducted using machine learning algorithm SVM without any feature representations. The native RS126 dataset is used as the dataset to fit into SVM for training and testing followed by evaluation. The native RS126 is the original sequence and protein structure obtained from the dataset without any pre-processing step being applied. The output is recorded and tabulated in Table 1.

**Table 1.** Evaluation results of the prediction of native RS126 dataset

|  | Helix | Strand | Coil | Overall |
|---|---|---|---|---|
| Accuracy | 0.54 | 0.42 | 0.07 | 0.43 |
| TPR | 0.39 | 0.32 | 1 | 0.45 |
| FPR | 0.09 | 0.11 | 0.49 | 0.34 |

Chen proposed that by selecting numerous lengths for local protein structure, it will assist in improving the accuracy of protein secondary structure prediction [9]. The initial result shows that the accuracy of the prediction without using any feature selection or representations is very low even compare to the other existing methods. Hereby, this research proposed an optimization using local protein structure to predict protein secondary structure.

This study is carried out using 3 different segment lengths, length 13, 15 and 17. The definition of applying different segment length is taking in to account 13, 15 and 17 continuous residues or amino acids in the protein sequence. For each protein in RS126, local protein structure with 3 different segment lengths will be applied. The optimal length for local protein structure will be determined using the best overall accuracy from the results of evaluation. With t-test validation, the significance of the optimal local protein structure compare to the initial method can be observed.

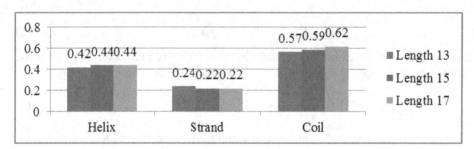

**Fig. 2.** Accuracy of each local protein structure based on secondary structural state

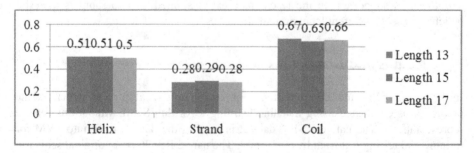

**Fig. 3.** True positive rate of each local protein structure based on secondary structural state

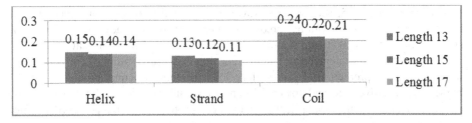

**Fig. 4.** False positive rate of each local protein structure based on secondary structural state

Most of the prediction results are evaluated by accuracy as depicted in Fig. 2. According to Fig. 2, for local protein structure with segment length 13, the highest accuracy is achieved by coil followed by helix and then strand. Similar results are collected from other local protein structures where coil having the highest accuracy among all secondary structural states. In terms of secondary structure, for helix, segment length 15 and 17 record the highest accuracy compare to others. Meanwhile, strand structure with length of 13 has the highest accuracy in compare to length 15 and 17. As for coil, length 17 records the highest accuracy among all.

In this research, other than accuracy, to provide a more reliable result, true positive rate and false positive rate are also used to analyze the prediction result. The results for true positive rate and false positive rate are illustrated in Fig. 3 and 4. For true positive rate, length 17 has the highest score for helix, length 15 for strand and length 13 for coil. As for false positive rate, length 17 has the lowest score for all secondary structural state.

From the tables and figures illustrated, it is obvious that generally, segment length 17 has the better accuracy compare to other local protein structures with the score of 0.44, 0.22 and 0.62. Most of the accuracies achieved is either the highest or is merely behind the highest score. Similar in true positive rate, most of the score that length 17 achieved is in the top range while in false positive rate, length 17 has the lowest rate among all local protein structures. It can be concluded that segment length 17 is the best local protein structure in this research.

A comparison of the prediction with optimal local protein structure with the prediction using native protein dataset is being conducted and analyzed. The proposed method with optimized local protein structure is expected to have better performance compare to the conventional prediction method in terms of accuracy, true positive rate and false positive rate. The comparison of the performance of both methods is illustrated in Fig. 5.

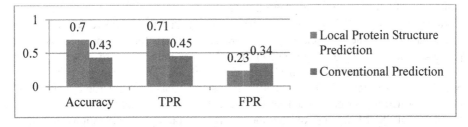

**Fig. 5.** Line graph of local protein structure prediction versus conventional prediction

According to Fig. 5, accuracy of local protein structure prediction is very much higher compared to conventional prediction. The score of accuracy for local protein structure prediction is 0.70 and it almost doubles the score of conventional prediction. This shows that by implementing feature selection or representation, there will be an improvement in prediction. Besides that, local protein structure prediction gives higher true positive rate and lower false positive rate. All the evaluation methods above indicate that the implementation of local protein structure achieved drastic improvement compare to the prediction method without any pre-processing or optimization.

Further validation of the results has been proposed to ensure the reliability of the prediction. A statistical validation, t-test, is conducted to test the significance of the results returned by the prediction. Table 2 shows the results of t-test for accuracy of the prediction system between optimal local protein structure and native structure. Only 11 samples are tabulated due to the large amount of protein sequence in RS126 dataset. It is noted that most of the t-test results returned h value as 1. This proves that the difference of accuracy predicted from the secondary structure prediction between optimal local protein structure and native structure is significant. The improvement of the accuracy, true positive rate and false positive rate is convincing and reliable.

**Table 2.** Sample of t-test results for accuracy between optimal local protein structure and native structure

|  | Significance | Lower Bound | Upper Bound |
|---|---|---|---|
| 1azu | Yes | 0.47 | 0.57 |
| 1bbpa | Yes | -0.39 | -0.33 |
| 2aat | Yes | 0.30 | 0.40 |
| 3ait | Yes | 0.07 | 0.26 |
| 4bp2 | Yes | -0.29 | -0.17 |
| 5cytr | Yes | 0.27 | 0.41 |
| 6acn | Yes | 0.30 | 0.41 |
| 7cata | Yes | 0.30 | 0.43 |
| 8abp | Yes | 0.20 | 0.29 |
| 9apia | Yes | 0.22 | 0.34 |
| 256ba | Yes | 0.12 | 0.21 |

Finally, a comparison of accuracy between proposed method (optimal local protein structure), initial research (native structure) and other prediction methods is conducted. This is to observe the level of optimization of the proposed method compare to the conventional or other methods.

According to Table 3, it can be clearly observed that the initial research has the lowest accuracy due to lack of feature representations for the predictions. The proposed method which implement optimal local protein structure has the higher accuracy even compared to other prediction methods. This might be because by breaking down a native protein structure into small local protein structure segment, more information can be learned by the algorithm and will yield better predictions. Besides that, SVM is one of the most efficient binary classification algorithm compare to the algorithm used by other methods such as N-grams and others.

**Table 3.** Comparison of accuracy between different methods of protein secondary structure prediction

| Methods | Reference | Accuracy |
|---|---|---|
| Extreme learning machine, Improved propensity score in binary scheme. Fixed window size. | Wang *et al.* [10] | 69.0 |
| Context sensitivity vocabulary, N-grams. | Yan *et al.* [11] | 69.8 |
| Initial Study: Native RS126 dataset, SVM | - | 43.0 |
| Proposed Method: Optimal Local Protein Structure, DSSP as Feature Class, DA as Feature Vector, SVM | - | 70.0 |

# 4    Conclusion

Optimized local protein structure with SVM has been proposed to predict protein secondary structure. There were several interesting outcome faced during the study. The importance of protein secondary structure prediction, comparison of the study with previous work, influence of local protein structure to predict protein secondary structure, application of statistical method to enhance the reliability of evaluation methods have been conducted extensively and make great contributions to the research of protein secondary structure. Some future works are suggested to enhance the current prediction of protein secondary structure prediction such as use different datasets other than RS126, develop more feature representations and use various parameters in the classification process such as different cross validation and kernel. It is important to study more details about protein secondary structure because it will help us to understand more about their functions. With the knowledge of proteomics, contribution can be made to various fields such as development of cure in medicine sector.

**Acknowledgements.** We would like to thank Universiti Teknologi Malaysia for supporting this research by UTM GUP research grant (Vot number: Q.J130000.7123.00H67).

# References

1. Walhout, A.J., Vidal, M.: Protein interaction maps for model organisms. Nat. Rev. Mol. Cell Biol. 2(1), 55–62 (2001)
2. Legrain, P., Wojcik, J., Gauthier, J.: Protein-protein interaction maps: a lead towards cellular functions. Trends in Genet. 17, 346–352 (2001)
3. Jing, N., Xia, B., Zhou, C.G., Wang, Y.: Protein Secondary Structure Prediction Methods based on RBF Neural Networks. Computational Methods 10, 1037–1043 (2006)
4. Rost, B., Sander, C.: Combining evolutionary information and neural networks to predict protein secondary structure. Proteins: Struct., Funct., Genet. 19, 55–72 (1994)

5. Hua, S., Sun, Z.: A novel method of protein secondary structure prediction with high segment overlap measure: Svm approach. J. Mol. Biol. 308, 397–407 (2001)

6. Guo, J., Chen, H., Sun, Z.R., Lin, Y.L.: A novel method for protein secondary structure prediction using dual-layer SVM and profiles. PROTEINS: Structure, Function, and Bioinformatics 54, 738–743 (2004)

7. Hu, H.J., Pan, Y., Harrison, R., Phang, C.T.: Improved protein secondary structure prediction using support vector machine with a new encoding scheme and an advanced tertiary classifier. Nanobioscience 3, 265–271 (2004)

8. Yang, B.R., Hou, W.Z., Zhuna, Quan, H.: KAAPRO: An approach of protein secondary structure prediction based on KDD in the compound pyramid prediction model. Expert Systems with Applications 36(5), 9000–9006 (2010)

9. Chen, C.T., Lin, H.N., Sung, T.Y., Hsu, W.L.: Hyplosp: A Knowledge-Based Approach to Protein Local Structure Prediction. Journal of Bioinformatics and Computational Biology 4(6), 1287–1308 (2006)

10. Wang, G., Zhao, Y., Wang, D.: A protein secondary structure prediction framework based on the extreme learning machine. Neuorocomputing 72, 262–268 (2008)

11. Yan, L., Carbonell, J., Seetharaman, J.K., Gopalakrishnan, V.: Context sensitive vocabulary and its application in protein secondary structure prediction. ACM 1(58113), 881 (2004)

# Ensuring Security and Availability of Cloud Data Storage Using Multi Agent System Architecture

Amir Mohamed Talib, Rodziah Atan, Rusli Abdullah,
and Masrah Azrifah Azmi Murad

Faculty of Computer Science & IT, Information System Department,
University Putra Malaysia, 43400 UPM,
Serdang, Selangor, Malaysia
ganawa53@yahoo.com, {rodziah,rusli,masrah}@fsktm.upm.edu.my

**Abstract.** We believe that data storage security in cloud computing, an area full of challenges and of paramount importance, is still in its infancy now, and many research problems are yet to be identified. Cloud Computing is the utilization of many servers/ data centers or cloud data storages (CDSs) hosted in many different locations. To ensure the availability of users' data in the cloud we propose architecture of Multi Agent System (MAS). This architecture consists of two main agents: Cloud Service Provider Agent (CSPA) and Cloud Data Availability Agent (CDAA). CDAA has two types of agents: Cloud Data Distribution Preparation Agent (CDDPA) and Cloud Data Retrieval Agent (CDRA). This paper is also illustrated with an example of how our MAS architecture detects distributed cloud server colluding attacks.

**Keywords:** Cloud Computing, Cloud Data Storage, Cloud Service Provider, Cloud Data Security Availability and Multi-Agent System.

## 1    Introduction

Cloud Computing refers to both the applications delivered as services over the internet and the hardware and systems software in the datacenters that provide those services [1,2,3].

Cloud Data Storage (CDS) is a model of networked computer data storage where data is stored on multiple virtual servers, generally hosted by third parties, rather than being hosted on dedicated servers [3, 4].

Cloud data availability is to ensure that the cloud data processing resources are not made unavailable by malicious action. The problem of verifying availability of CDS becomes even more challenging [5,6,7]. Therefore, this paper proposed MAS architecture to ensure the availability of the cloud user's data.

Multi Agent Systems (MASs) are techniques in the artificial intelligence area focusing on the system where several agents communicate with each other. In [6, 7], MAS is defined as "a loosely coupled network of problem-solver entities that work together to find answers to problems that are beyond the individual capabilities or knowledge of each entity".

D. Lukose, A.R. Ahmad, and A. Suliman (Eds.): KTW 2011, CCIS 295, pp. 343–347, 2012.

The contributions of the paper are two-fold: First, our MAS architecture is able to tolerate multiple failures in cloud distributed storage systems. Second, as well as enable the cloud user to reconstruct the original data by downloading the data vectors from the cloud servers.

## 2     Related Work

The computing power in a cloud computing environments is supplied by a collection of datacenters, which are typically installed with hundreds to thousands of servers [8]. The authors built architecture of a typical cloud based datacenter consist of four layers. At the lowest layers there exist massive physical resources (storage servers and application servers) that power the datacenters [8]. In a distributed servers work [9] built on this model and constructed a random linear function based homomorphic authenticator which enables unlimited number of queries and requires less communication overhead. In [11], the authors proposed an improved framework for POR protocols that generalizes [10] work. Later in their subsequent work [11], extended POR model to distributed systems. However, all these schemes are focusing on static data. The effectiveness of their schemes rests primarily on the pre-processing steps that the user conducts before outsourcing the cloud data file. In a data possession work [12], defined the "provable data possession" (PDP) model for ensuring possession of file on untrusted storages. Their scheme utilized public key based homomorphic tags for auditing the data file, thus providing public verifiability. However, their scheme requires sufficient computation overhead that can be expensive for an entire file. In their subsequent work [12] described a PDP scheme that uses only symmetric key cryptography. This method has lower-overhead than their previous scheme and allows for block updates, deletions and appends to the stored file, which has also been supported in our work. However, their scheme focuses on single server scenario and does not address small data corruptions, leaving both the distributed scenario and data error recovery issue unexplored. In another data possession work [13], aimed to ensure data possession of multiple replicas across the distributed storage system. They extended the PDP scheme to cover multiple replicas without encoding each replica separately, providing guarantees that multiple copies of data are actually maintained.

## 3     Cloud Attack Detection in Our MAS Architecture

Before describing how the intrusion detection is performed by our MAS architecture, we explain the notions of global and local cloud attack blueprints. To detect intrusions, the CDAA receives a set of goals representing the global cloud attack blueprints. To recognize this global cloud attack blueprint, it must be decomposed in local cloud sub-blueprints used locally by the different agents distributed in the CDS. In general agents can detect only local cloud attacks because they have a restricted view of the CDS. So, we make a distinction between a global cloud attack blueprint and local cloud sub-blueprints. A global cloud blueprint is an attack blueprint, derived from the security policies specified at a high level by the CSPs, that the MAS must detect and the detection of this blueprint will be notified only to CDAA. A local cloud blueprint is a blueprint derived from the global cloud blueprint but that must be detected by local agents. For a CDAA over-viewing the global cloud attack blueprint the probability of an attack is equal to 1, while for the local agent it is below 1. For this, we describe briefly an example for detecting cloud server colluding attacks. To detect this attack, we propose the following scenario (see Fig. 1).

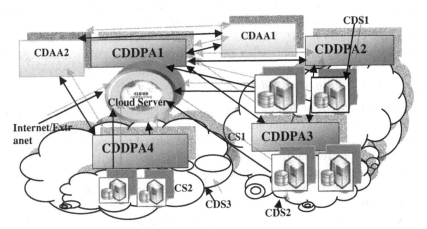

**Fig. 1.** Cloud server colluding attacks scenario

We have a distributed cloud servers of two CDSs (CDS1 and CDS2) and a cloud hosting (cloud server), hosted in the CDS1. Concerning the MAS:

In CDS1, we have three agents (CDDPAs): CDDPA1, CDDPA2 and CDDPA3. CDDPA1 has to monitor the cloud server. CDDPA2 responsible of monitoring specific cloud servers, respectively CS1 and CS2, constituted of a set of cloud hosts. CDDPA and CDAA are managed by CDAA1, who is responsible of detecting attacks occurring in CDS1.

In CDS2, we have CDAA2, who detects attacks in CDS2 and CDDPA4, who monitor the CS3. The two CDAAs are controlled by one CDAA, who communicate with the security policy of CDAA.

In cloud server colluding attacks, the attacker sends a high rate of cloud server colluding attacks packets to calculate the value of the errors to flood the cloud server or a particular cloud host. The CDAA minimizing the effect brought by the cloud data errors or cloud server failures. The goal of each CDDPA is to detect the cloud server colluding attacks, which means, to detect a certain threshold of cloud server colluding attacks echo request packets to calculate the value of the errors. This threshold is the same for CDDPA2, CDDPA3, CDDPA4 but it is different for CDDPA1, who play a preponderant role, because of the sensibility of the monitored cloud host. The CDDPAs monitor all cloud server colluding attacks requests coming to their CS and for the CDDPA1 it concerns the cloud server. These requests could come from different sources of the CDS: 1) local cloud hosts of the same CDS, 2) local cloud host of other CDSs belonging to the same intranet, 3) external cloud hosts i.e.: from the outside. The CDDPAs don't receive the same detection goal. The CDDPA1 receives a goal G1 and the other CDDPAs receive a goal G3, because the threshold of cloud server colluding attacks requests is different. Thus, the detection goals that the agents will receive contain the following cloud attack blueprints:

*Goal G1 < blueprint AMIR <name = cloud server colluding attacks, secure source = cloud server log file, even type = CDS, even name = cloud server colluding attacks echo requests, even source = anywhere, even destination = cloud server, result = none, value = a, calculate the value of errors = e1>, action>*

*Goal G3 < blueprint AMIR <name = cloud server colluding attacks, secure source = cloud server log file, even type = CDS, even name = cloud server colluding attacks echo requests, even source = anywhere, even destination = cloud server, result = none, value = i, calculate the value of errors = e1>, action>*

The values (a) and (i) of the cloud server colluding attacks requests threshold depend on the usual cloud server colluding attacks errors in the cloud server. In fact, the CSP will have to specify these values according to the normal threshold of the cloud server colluding attacks errors in the cloud server.

An internal cloud attack could come from the CDS2, CS1 or CS2. If the CDDPA1 observes a certain threshold of cloud server colluding attacks echo request i.e.: threshold equal to n, coming from CS1, CS2, CS3, it informs all the CDDPAs of the same CDS1 and the CDAA1, which informs the CDAA2 and in its turn inform the CDDPA4. The goal of each local agents, will be then modified in terms of cloud server colluding attacks echo request packets threshold and monitoring period of time, which will be lower than those specified in the initial goal. So that CDDPA1 will have to reach a new goal G2:

*Goal G2 < blueprint AMIR <name = cloud server colluding attacks, secure source = cloud server log file, even type = CDS, even name = cloud server colluding attacks echo requests, even source = anywhere, even destination = cloud server, result = none, value = m, calculate the value of errors = e1>, action>*

The sense of monitoring will also change, for all the agents except for CDDPA1, which means that they must observe cloud server colluding attacks packets directed to the cloud server. So, the agents CDDPA2, CDDPA3, CDDPA4, will have to reach a new goal G4:

*Goal G4 < blueprint AMIR <name = cloud server colluding attacks, secure source = cloud server log file, even type = CDS, even name = cloud server colluding attacks echo requests, even source = anywhere, even destination = cloud server, result = none, value = r, calculate the value of errors = e1>, action>*

If one agent reaches its goal an alarm will be sent to the Cloud Service Provider Agent (CSPA) to translate the attack in terms of goal and delete it.

## 4     Conclusions

In this paper, we investigated the problem of cloud data availability in cloud computing, to ensure the availability of users' data in CDS; we proposed MAS architecture. This architecture consists of two main agents: Cloud Data Distribution

Preparation Agent (CDDPA) and Cloud Data Retrieval Agent (CDRA). This paper is also illustrated with an example how our MAS architecture detects distributed cloud server colluding attacks.

# References

1. Chappell, D.: A Short Introduction to Cloud Platforms. An Enterprise-Oriented View, Principal of Chappell & Associates in San Francisco, California. pp. 1–13 (2008), http://www.davidchappell.com
2. Talib, A.M., Atan, R., Abdullah, R., Murad, M.A.A.: A Framework of Multi-Agent System to Facilitate Security of Cloud Data Storage. In: Annual International Conference on Cloud Computing and Virtualization, CCV 2010, Singapore, p. 241 (2010)
3. Talib, A.M., Atan, R., Abdullah, R., Murad, M.A.A.: Multi Agent System Architecture Based on Data Life Cycle Management to Ensure Cloud Data Storage Security. In: The Second APSEC 2009 Workshop & Tutorial and Software Engineering Postgraduates Workshop (SEPoW 2010), Penang, Malaysia, December 22, pp. 41–45 (2010)
4. Rittinghouse, J., Ransome, J.F.: Cloud Computing: Implementation, Management, and Security, ch.6, p. 153. CRC Press (2009)
5. Wang, C., Wang, Q., Ren, K., Lou, W.: Ensuring Data Storage Security in Cloud Computing, pp. 1–9 (2009)
6. Talib, A.M., Atan, R., Abdullah, R., Murad, M.A.A.: Security Framework of Cloud Data Storage Based on Multi Agent System Architecture: Semantic Literature Review. CCSE 3, 175 (2010)
7. Talib, A.M., Atan, R., Abdullah, R., Murad, M.A.A.: Formulating a Security Layer of Cloud Data Storage Framework Based on Multi Agent System Architecture. GSTF International Journal on Computing 1(1), 120–124 (2010) ISSN: 2010-2283
8. Buyya, R., Murshed, M.: Gridsim: A toolkit for the Modeling and Simulation of Distributed Resource Management and Scheduling for Grid Computing. In: Concurrency and Computation: Practice and Experience, pp. 1175–1220 (2002)
9. Smith, J.E., Nair, R.: Virtual Machines: Versatile Platforms for Systems and Processes. Morgan Kaufmann Pub. (2005)
10. Shacham, H., Waters, B.: Compact Proofs of Retrievability. In: Pieprzyk, J. (ed.) ASIACRYPT 2008. LNCS, vol. 5350, pp. 90–107. Springer, Heidelberg (2008)
11. Juels, A., Kaliski Jr., B.S.: PORs: Proofs of Retrievability for Large Files, pp. 10–1145. ACM (2007), doi: http://doi.acm.org/10.1145/1315245.1315317
12. Ateniese, G., Di Pietro, R., Mancini, L.V., Tsudik, G.: Scalable and Efficient Provable Data Possession, pp. 1–10. ACM (2008), doi: http://doi.acm.org/10.1145/1460877.1460889
13. Curtmola, R., Khan, O., Burns, R., Ateniese, G.: MR-PDP: Multiple-Replica Provable Data Possession, pp. 411–420. IEEE (2008)

# Multiple Object Classification
# Using Hybrid Saliency Based Descriptors

Ali Jalilvand and Nasrollah Moghadam Charkari

IPPP Lab., Electrical and Computer Engineering Department,
Tarbiat Modares University, Tehran, Iran
ali.jalilvand@ieee.org, Moghadam@modares.ac.ir

**Abstract.** We propose an Automatic approach for multi-object classification, which employs support vector machine (SVM) to create a discriminative object classification technique using view and illumination independent feature descriptors. Support vector machines are suffer from a lack of robustness with respect to noise and require fully labeled training data. So we propose a technique that can cope with these problems and decrease the influence of viewpoint changing or illumination changing of a scene (noise in data) named the saliency-based approach. We will combine the saliency-based descriptors and construct a Feature vector with low noise. The Proposed Automatic method is evaluated on the PASCAL VOC 2007 dataset.

**Keywords:** Image classification, Image categorization, Object recognition, Support Vector Machine, saliency based descriptors, pattern recognition.

## 1    Introduction

The visual object categorization is considered as open fields of research in areas machine vision, machine learning. In this paper we take all the phase used to make an automatic method which is accurate enough to objects categorization, in to consideration. The proposed automatic method is shown in fig. 1. There are two important phase in our method, which will be described in continue. Feature extraction phase: One of the most popular methods for this phase is appearance-based recognition approaches, which can be categorized into two levels (a) global [1], and (b) local [2][3]. the most popular method with respect to robustness and stability to local regions is the Scale Invariant Feature Transform (SIFT), proposed in [2], Another popular method is, that is several times faster than SIFT and first presented in [3].The classifier methods phase: the classification is another important phase in object category recognition approaches, One of the most popular and dominant machine learning algorithms from all the computer vision researches in object categorization is the support vector machine (SVM).

Recently, there have been efforts that leverage manually segmented foreground objects from the cluttered background to improve categorization. We used of [4], which proposed a novel framework for generating and ranking plausible objects hypotheses in an image using bottom-up processes and mid-level cues.

D. Lukose, A.R. Ahmad, and A. Suliman (Eds.): KTW 2011, CCIS 295, pp. 348–351, 2012.
© Springer-Verlag Berlin Heidelberg 2012

Rest of the paper is organized as follows: Section 2 describes the steps of feature extraction procedures used in this paper. The SVM classifier method is introduced in section 3. In section 4, our experiment results are presented and conclusion is presented in section 5.

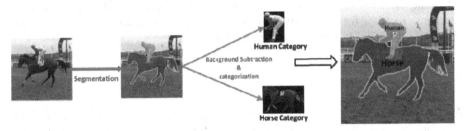

**Fig. 1.** Illustration of multi objects categorization obtained by our automatic approach without user interaction

## 2    Feature Extraction

Feature extraction is an important step in the object recognition systems. In the object recognition tasks, the local feature should be invariant to viewpoint transformation, scale changes and changes in illumination and noise. There are techniques that can cope with these problems named the saliency-based approach, such as SIFT [2] and SURF [3]. The used Feature Extraction method in this paper is the proposed method in [5] along with a new implementation for flexibility and some other little changes, including the color descriptors and SURF. The feature extraction has three phases: **A. Region Detector:** In feature extraction, local feature is a property of an object located on a given region or single point. One of the most popular interest point detector or Region detector is the HARRIS corner based detector.

In [2] their evaluation shows that the laplacian function selects the highest percentage of correct characteristic scales, and as a result they introduce the scale invariant Harris-Laplace (H-Lap) detector. So for this phase, the Harris-Laplace point detector is used. **B. Descriptors' Construction:** The object categorization systems need effective image descriptors to obtain suitable performance. In this section we use opponentSIFT, C-SIFT, rgSIFT and SUFT, because they show a good performance in the object categorization [12][3].**C.feature vector adjustment**: First, we assign visual descriptors to discrete codewords predefined in a codebook. Then, we use the frequent distribution of the codewords as a feature vector representing an image. A comparison of codeword assignment is presented according to by [6].

## 3    Classifier

For classification, we use Support Vector Machines (SVM) with RBF-kernel, which widely used and well known classifier. The decision function of a support vector machine classifier for a test sample with feature vector x has the following form:

$$g(X) = \sum_i y_i \alpha_i K(X, s_i) + \beta \qquad (4)$$

To classify the test sample X where K is a kernel function, the $s_i$ are the support vectors and the $\alpha_i$ are the learned weight of the training sample corresponding, $\beta$ is a bias term. Here, an image X is represented by a variable number of above local features. To classify an image, we classify each of these local features x individually determine the class of the whole image by combining the individual classification decisions. To allow for effectively combining, not only the resulting class but also the distance to the decision hyperplane is considered.

# 4    Experiments

As previously mentioned, we are interested in the change of relative performance in the object categorization's accuracy. Then we present experimental results for two different dataset: PASCAL 2007. Next we compare the applied method with one state of the arts methods in this scope [7][8] which are amongst the most difficult datasets currently in use to benchmark object categorization systems.

To evaluate each of 20 object categories' performance (2007), we measure the average precision (AP) score, just obtained from the test images that actually contain the current object class. Table 2 summarizes the resulting scores in numeric form. Since the model is appearance based, it works well for categories which are characterized by such properties (e.g. the vehicle classes and some animal classes like horse). Performance is less appropriate for some wiry objects (e.g. potted plant) or objects more defined by their function and context than by their appearance (e.g. chair, dining table).

**Table 1.** Recognition accuracy rate of our method and other methodes. the y-axis represents the categories and The x-axis represents the recognition accuracy (%).

| | plane | bike | Bird | Boat | bottle | bus | car | Cat | chair | Cow |
|---|---|---|---|---|---|---|---|---|---|---|
| OUR METHOD | 0.27 | 0.23 | 0.12 | 0.15 | 0.08 | 0.37 | 0.39 | 0.3 | 0.07 | 0.23 |
| [7] | 0.18 | 0.41 | 0.09 | 0.09 | 0.24 | 0.34 | 0.39 | 0.11 | 0.15 | 0.16 |
| [8] | 0.11 | 0.12 | 0.09 | 0.06 | 0 | 0.25 | 0.14 | 0.36 | 0.09 | 0.14 |

| | dtable | dog | Horse | mbike | person | pplant | sheep | sofa | train | Tv |
|---|---|---|---|---|---|---|---|---|---|---|
| OUR METHOD | 0.23 | 0.22 | 0.41 | 0.3 | 0.1 | 0.12 | 0.24 | 0.29 | 0.33 | 0.41 |
| [7] | 0.11 | 0.06 | 0.3 | 0.33 | 0.26 | 0.14 | 0.14 | 0.15 | 0.2 | 0.33 |
| [8] | 0.24 | 0.32 | 0.27 | 0.34 | 0.03 | 0.02 | 0.09 | 0.3 | 0.3 | 0.08 |

## 5     Conclusion

In this paper we used techniques that can cope the influence of viewpoint changing or illumination changing of a scene (noise in data) named the saliency-based approach, such as CSIFT and SURF. Furthermore we combine the descriptors in order to construct a feature vector with low noise and increase accuracy rate in different categories. According to the experiments conducted on the standard test datasets PASCAL, Since our proposed future work, We plan to make use of more context information in improves the systems performance well.

**Acknowledgements.** This project was sponsored by Telecommunication Research Center (TRC).

## References

1. Wang, J.Z., Li, J., Wiederhold, G.: SIMPLIcity: Semantics-sensitive integrated matching for picture Libraries. IEEE Transactions on Pattern Analysis and Machine Intelligence 23(9), 947–963 (2001)
2. Lowe, D.G.: Distinctive image features from scale-invariant keypoints. International Journal of Computer Vision 60(2), 91–110 (2004)
3. Bay, H., Tuytelaars, T., Van Gool, L.: SURF: Speeded Up Robust Features. In: Leonardis, A., Bischof, H., Pinz, A. (eds.) ECCV 2006. LNCS, vol. 3951, pp. 404–417. Springer, Heidelberg (2006)
4. Carreira, Sminchisescu, C.: Constrained parametric min-cuts for automatic object segmentation. In: CVPR (2010)
5. van de Sande, K.E.A., Gevers, T., Snoek, C.G.M.: Evaluating Color Descriptors for Object and Scene Recognition. IEEE Transactions on Pattern Analysis and Machine Intelligence 32(9) (2010)
6. van Gemert, J.C., Veenman, C.J., Smeulders, A.W.M., Geusebroek, J.M.: Visual word ambiguity. IEEE Transactions on Pattern Analysis and Machine Intelligence (2010)
7. Felzenszwalb, McAllester, D.A., Ramanan, D.: A discriminatively trained, multiscale, deformable part model. In: CVPR (2008)
8. Blaschko, M., Lampert, C.H.: Object localization with global and local context kernels. In: BMVC (2009)

# Parameter Estimation for Simulation of Glycolysis Pathway by Using an Improved Differential Evolution

Chuii Khim Chong, Mohd Saberi Mohamad, Safaai Deris,
Yee Wen Choon, and Lian En Chai

Artificial Intelligence and Bioinformatics Research Group, Faculty of Computer Science
and Information Systems, Universiti Teknologi Malaysia, 81310 Skudai, Johor, Malaysia
{ckchong2,ywchoon2}@live.utm.my, {saberi,safaai}@utm.my,
enrius@gmail.com

**Abstract.** An improved differential evolution (DE) algorithm is proposed in
this paper to optimize its performance in estimating the germane parameters for
metabolic pathway data to simulate glycolysis pathway for *Saccharomyces
cerevisiae*. This study presents an improved algorithm of parameter sensitivity
test into the process of DE algorithm. The result of the improved algorithm is
testifying to be supreme to the others estimation algorithms. The outcomes from
this study promote estimating optimal kinetic parameters, shorter computation
time and ameliorating the precision of simulated kinetic model for the
experimental data.

**Keywords:** Parameter Estimation, Differential Evolution, Evolutionary
Algorithm, Optimization, Metabolic Engineering.

## 1    Introduction

Current studies basically have concentrated on the modification of the computer
readable data from the biological activity which allows the mean of analysis. Thus,
studying in the metabolic pathway permits scientists to simulate the process inside
the cell by a mathematical modeling. To develop a valid pathway model that
functions as biological functions simulator is the goal for the study of system
biology. Parameter estimation is one of the key steps in mathematical model.
Regrettably, it has encountered a few problems such as increasing number unknown
parameters and equations in the model which contributes to high complexity of the
model [1] and low accuracy due to the existence of noise data [2]. Therefore, the aim
of this study is to propose an intelligent algorithm of incorporate DE and parameter
sensitivity test to resolve the rising unknown parameters which lead to the
complexity of the model. The advantages of DE are effectiveness, speed, simplicity
and ease of use as it consists of only few control parameters [3]. Moreover,
parameter sensitivity test also plays an important role in generating a model with less
irrelevant parameters which can minimize computational burden that leads to less
computational effort and time.

D. Lukose, A.R. Ahmad, and A. Suliman (Eds.): KTW 2011, CCIS 295, pp. 352–355, 2012.
© Springer-Verlag Berlin Heidelberg 2012

## 2    Method

Parameters that are comprised of the glycolysis pathway model for *Saccharomyces cerevisiae* will first undergo parameter sensitivity test to identify relevant parameters. Next, these identifiable parameters will undergo DE to estimate its optimal value.

### 2.1    Parameter Sensitivity Test [4]

I.    Model checking and retrieve necessary information (parameters, states and reactions).
II.   Determine steady-state for nominal model.
III.  Determine steady-state for perturbed model.
IV.   Output steady-state sensitivity in graph.

### 2.2    Differential Evolution Algorithm [5]

I.    Generate random population of m x n solutions for the problem within the higher and lower bound where m = number of identifiable parameters and n= number of generation. Solutions are presented in the form of floating points.
II.   Evaluate the fitness function f(x) of each individual for n solution where each individual represents a candidate parameter value.
III.  Create new population by repeating the following steps until new population is complete:

o    Randomly select three parent individuals from a population for each n by the following formula where *i* is the parent index.

$$individual\_i = floor(rand() * population\ size) + 1 .\qquad (1)$$

o    New generation mutate by the following formula for each population size.

$$temp\_population(i) = Pop(individual3) + F*(Pop(individual1) - Pop(individual2)) .\qquad (2)$$

*Where F is differentiation constant and Pop is the original population matrix.*

o    If randb(j)<CR or j=rnbr(i)
         Crossover occurs and generated new population.
  Else
         No crossover and original population remain.

*Where Randb(j)= jth random evaluation of a uniform random number generator [0,1] and Rnbr = random chosen index {1,2,...D}.*

IV.   If end condition is satisfied stop and return optimal parameter in current population else go to step II.

# 3 Result and Discussion

Execution times with 87 unknown parameters shown in Table 1 are greater than the execution time shown in Table 2 with 4 unknown parameters. It is shown that the lesser the execution time needed when the smaller the number of parameters for the

**Table 1.** Execution time (without parameter sensitivity test with 87 unknown parameters)

|                | Nelder-Mead | SA | GA | DE |
|----------------|-------------|---------|---------|---------|
| Execution time | 3:42:57 | 5:44:43 | 0:08:53 | 0:07:52 |

**Table 2.** Execution time (with parameter sensitivity test with 4 unknown parameters)

|                | Nelder-Mead | SA | GA | DE |
|----------------|-------------|---------|---------|---------|
| Execution time | 0:07:13 | 0:09:48 | 0:08:52 | 0:07:17 |

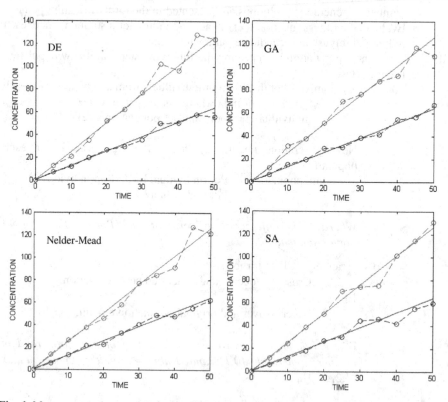

**Fig. 1.** Measurement data versus simulated data (-o simulated result, -- measurement result; red line- species 4, blue line – species 10)

algorithm. In Table 2 Nelder-Mead and DE are the two algorithms that required less execution time. Unfortunately for Nelder-Mead as the number of unknown parameters increase, the execution time increases dramatically whereas this is not occurs in DE. DE is shown need a minimum execution time even with bigger population size. Even though the execution time of DE is short consistently, the result that generated by DE are almost similar with the other estimation algorithms respectively which shown in Figure 1. Thus, the proposed method of the incorporation of parameter sensitivity test and DE solves the parameter estimation problems which reduces the number of unknown parameters and the computational time and raises the accuracy of the simulated model with the actual model.

## 4     Conclusion and Future Work

In conclusion, DE is shown outperform than other stochastic optimization algorithms in this study but regrettably it may easily be captured in local minima. Basically, the concentration will be on global minima rather than local minima. To evade being captured in local minima, DE is required for large population. Wang and Chiou showed that DE needs larger population and more computational time in order to produce the optimal result [6]. State observer which acts as initiator for the starting point such as Kalman Filter can be implemented into the DE algorithm to improve its performance in the future. DE seems to be very sensitive to control parameters: population size, crossover constant and differentiation constant. Therefore self adapting approach to these control parameters can be used to optimize the performance of the traditional as well as the proposed DE algorithm.

**Acknowledgments.** The work is financed by Institutional Scholarship MyPhd provided by the Ministry of Higher Education of Malaysia. We also would like to thank Universiti Teknologi Malaysia for supporting this research by UTM GUP research grant (Vot number: Q.J130000.7123.00H67).

## References

1. Chou, I.C., Voit, E.O.: Recent developments in parameter estimation and structure identification of biochemical and genomic systems 219, 57–83 (2009)
2. Lillacci, G., Khammash, M.: Parameter Estimation and Model Selection in Computational Biology. PLos 6(2), 1–17 (2010)
3. Wang, F.S., Chiou, J.P.: Differential Evolution for Dynamic Optimization of Differential-Algebraic Systems, 531–536 (1997)
4. Schmidt, H., Jirstrand, M.: Systems Biology Toolbox for MATLAB: a computational platform for research in systems biology. Bioinformatics 22(4), 514–515 (2006)
5. Storn, R., Price, K.: Differential Evolution – A Simple and Efficient Heuristic for Global Optimization over Continuous Spaces. Journal of Global Optimization 11, 341–359 (1997)
6. Wang, F.S., Chiou, J.P.: Estimation of Monod model parameters by hybrid differential evolution. Bioprocess and Biosystems Engineering 24, 109–113 (2001)

# A Multi-objective Genetic Algorithm for Optimization Time-Cost Trade-off Scheduling

Hadi Aghassi[1], Sedigheh Nader Abadi[2], and Emad Roghanian[3]

[1] Iran University of Science and Technology, Tehran, Iran
hadi.aghassi@gmail.com
[2] Islamic Azad University of Arak, Arak, Iran
g.naderabadi@gmail.com
[3] K.N. Toosi University of Technology, Tehran, Iran
e_roghanian@kntu.ac.ir

**Abstract.** In this paper, we present a new genetic algorithm for Project Time-Cost Trade-off (TCTO) Scheduling problem. In the proposed GA, the selection of genes for mutation is adopted to be based on chromosome value, as solution convergence rate is high. This paper also offers a new multi attribute fitness function for the problem. This function can vary by DM preferences (time or cost). The algorithm is described and evaluated systematically. The computational outcomes validate the effectiveness of the suggested approach.

**Keywords:** Time-Cost Trade-off, Genetic Algorithm, Project Scheduling, Multi Attribute fitness function.

## 1    Introduction

Having various methods and techniques in business, managers are able to reduce necessary time of performing projects. Apparently, these shifts in time will cause some changes in required cost of the projects. Generally, trying to decrease time length of a given project will increase direct costs, and will decrease indirect costs of activities of the project.

If maximum and minimum times of performing a project were prepared, we can select a time between them as sum of direct and indirect costs be minimum. Fig. 1 illustrates trends of direct and indirect costs and sum of these costs. Obviously, the best economical time length of executing projects is when sum costs curve be minimum with respect to time. Direct costs are necessary costs and properties used in the project directly [9]. Another type of tasks would cost during the project indirectly, e.g. management, accounting, etc. In the TCTO also there is another variable called cost slope. Cost slope is the amount of extra direct costs, which are going to be spent on activity execution while duration of the activity is decreased one time per unit [8]:

$$C = |\frac{C_f - C_n}{D_f - D_n}|. \tag{1}$$

While $C$ is cost slope, $C_f$ is forced cost of the activity, $C_n$ is normal cost of the activity, $D_f$ is forced time of the activity, and $D_n$ is normal time of the activity.

D. Lukose, A.R. Ahmad, and A. Suliman (Eds.): KTW 2011, CCIS 295, pp. 356–359, 2012.
© Springer-Verlag Berlin Heidelberg 2012

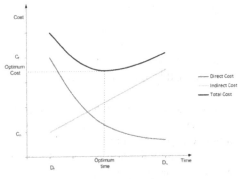

**Fig. 1.** Direct, indirect and total costs

**Table 1.** Properties of activities [10] including prerequisite activities, normal/forced time, normal/forced cost (1000$).

| Activity | Prerequisites | $D_n$ | $D_f$ | $C_n$ | $C_f$ |
|----------|---------------|-------|-------|-------|-------|
| A | - | 24 | 14 | 12 | 23 |
| B | A | 25 | 15 | 1 | 3 |
| C | A | 33 | 15 | 4.5 | 3.2 |
| D | A | 20 | 12 | 30 | 45 |
| E | B, C | 30 | 22 | 10 | 20 |
| F | D | 24 | 14 | 18 | 40 |
| G | E, F | 18 | 9 | 30 | 22 |

Genetic algorithm offered by John Holland for first time [1]. GA is not only an issue of human existence, but also it simulates nature evolution [2]. This algorithm is a suitable approach for encountering optimization problems [3]. There are a large variety of mutation and crossover operators and also various methods of using them, including selection, initial population generating, and selection parent mechanisms for mating. This large variety caused researchers to be interested in GA issues [4, 5].

One of the well-known techniques for project scheduling is CPM method [6]. In this method, no change is possible on project schedules; this denotes that all of the activities of the project have to be executed during project time limit. All activities are assumed to be continuous during the project. Consequently, when an activity's execution starts, it works without interruption until the end.

## 2    Proposed Model of Time-Cost Trade-off

In this paper, we use discrete points for relating cost and time of activities to our GA model be more similar to real conditions. The duration of each activity can vary between a lower bound, that is the forced duration of the activity $D_f$, and an upper bound, that is the normal duration of the activity, $D_n$. For instance, two chromosomes $C_1=(15,18,18,15,23,20,9)$, $C_2=(24,23,16,12,28,14,17)$ represent two different feasible solutions for the project [10] including the activities shown in Table 1.

### 2.1    Multi Attribute Optimized Fitness Function

The objective of fitness is to highlight the best optimum point of the project:

$$f(x) = [w_t \, \Delta T + w_c(\Delta Direct - \Delta Indirect)]/(w_t + w_c). \qquad (2)$$

Where $\Delta T$ is the reduced duration of the project in regard to normal duration, and $\Delta$Indirect Cost and $\Delta$Direct Cost are respectively the reduction in indirect cost and the addition of direct cost. By changing preferences for cost and time identities ($w_c$, $w_t$), DM can make the results ideal.

# 3    Implementation and GA Model

Indirect cost of the project is often a constant value. We have used 150 cost units per day in our research for indirect cost. We also used values of 2 and 1 for $w_c$, $w_t$, i.e. the time of the project is more important than cost. In this section, our proposed GA algorithm implementation is described in details.

First, initial population size, iterations, crossover and mutation rates are identified. Notice that increasing size of initial population, iterations and crossover rate leads to convergence of algorithm. Moreover, mutation operator is an implicit operator, and it is better to assign a low mutation rate to prevent a purely random process. We used 0.1 for mutation rate and 0.75 for crossover rate in our research. Also selecting the size of population 2.5 times larger than chromosome length gives a good answer in short time [8]. We sort activities according to their ES (observe prerequisite relations). Gene value of initial strings is assigned by a random number between forced and normal time of activity. In the next stage, we use two-point crossover operator for mating process [7], and Elitism method for choosing parents.

Moreover, mutation operator makes random variations in the population. These changes are not accidental totally, and based on what is the problem structure, they lead to solution as targeted [8]. We define a changing criterion for genes as important genes according to this criterion have better chance for changing. We called gene value decrease coefficient $P_1$, and its increase coefficient $P_2$:

$$P_1 = \alpha * (1 - \frac{C_i - C_{Min}}{C_{Max} - C_{Min}}). \tag{3}$$

$$P_2 = \beta * (\frac{C_i - C_{Min} + C_{Avr}}{C_{max} - C_{Min}} - 0.5). \tag{4}$$

Where $C_{min}$ is the smallest cost slope among activities, $C_{max}$ is maximum cost slope of activities, $C_i$ is cost slope of activity $i$, $\alpha$ and $\beta$ are coefficients identified by user. Default value of them is 1. Genes have the chance of $P_1$ to be decreased in mutation.

# 4    Evaluation

In this section, our proposed algorithm is compared to CPM method. We evaluated this research using 10, 25 and 50 activities. It is implemented in C Sharp .Net 2008 under windows 7 with Pentium IV, 2.13 MHz processor and 2 gigabyte RAM. CPM algorithm is executed one time, but because of random behavior of GA, each project is executed 15 times, and average execution time is compared to CPM execution time. The results are illustrated in Table 2. In this table a simple GA (without the mutation criterions and with earlier fitness), the proposed GA and CPM methods are compared.

Having these results, we can say that CPM method is not a reliable method for projects with a large number of activities. We also compared the results of the proposed method by considering different values for algorithm parameters that describe the importance of time or cost for project.

**Table 2.** Results of comparing Proposed Genetic Algorithm, basic GA and CPM

| Chromosome Length | 10 Activities | | 25 Activities | | 50 Activities | |
|---|---|---|---|---|---|---|
| | Avg CPU Time | % deviation from the optimum | Avg CPU Time | % deviation from the optimum | Avg CPU Time | % deviation from the optimum |
| CPM method | 0.102 | 0.0 | 0.356 | 0.0 | 0.978 | 0.0 |
| Basic Genetic Algorithm | 0.198 | 3.1 | 0.633 | 4.1 | 1.114 | 4.1 |
| Proposed GA ($w_t = w_c$) | 0.169 | 2.2 | 0.536 | 2.3 | 1.10 | 1.2 |
| Proposed GA ($w_t = 2w_c$) | 0.155 | 1.5 | 0.512 | 1.9 | 1.05 | 0.7 |
| Proposed GA ($2w_t = w_c$) | 0.171 | 0.05 | 0.551 | 0.1 | 1.13 | 0.4 |

# 5    Conclusion

In this paper, we proposed a new genetic algorithm for Project TCTO scheduling problem. In this GA, we introduced a new selection of genes criterion for mutation based on gene values to solution convergence rate was high. In the mutation operator, we used two control variables to manage the search of problem space.

We also evaluated our proposed algorithm by changing DM preferences $w_t$ and $w_c$, and by comparing with a famous algorithm called CPM. Evaluations showed that for small sized examples, there is not a great difference between two algorithms, but for complicated examples, our proposed method was more trustworthy. This paper also presented a new multi attribute fitness function for the problem. The future research issue that we are interested in is to study other evolutionary heuristic algorithms.

# References

1. Holland, J.H.: Adaptation in Natural and Artificial Systems. University of Michigan Press, Ann Arbor (1975); re-issued by MIT Press (1992)
2. Gen, M., Cheng, R.: Genetic Algorithms and Engineering Design. John Wiley &Sons (1997)
3. Goldberg, D.: Genetic Algorithms in Search, Optimization, and Machine Learning. Addison-Wesley (1989)
4. Kea, H., Maa, W., Ni, Y.: Optimization models and a GA-based algorithm for stochastic time-cost trade-off problem. Applied Math. and Computat. 215, 308–313 (2009)
5. Wuliang, P., Chengen, W.: A multi-mode resource-constrained discrete time-cost tradeoff problem and its genetic algorithm based solution. International Journal of Project Management 27(6), 60–609 (2009)
6. Siemens, N.: A Simple CPM Time-Cost Tradeoff Algorithm. Management Science, Application Series 17(16), 354–363 (1971)
7. Reeves, C.R.: Modern heuristic techniques for combinatorial problems. John Wiley & Sons, Inc., New York (1993)
8. Hooshyar, B., Tahmani, A., Shenasa, M.: A Genetic Algorithm to Time-Cost Trade off in project scheduling. In: IEEE Congress on Evolutionary Comput., CEC, pp. 3081–3086 (2008)
9. Chen, P.H., Weng, H.J.: A Two-Phase GA Model for Resource-Constrained Project Scheduling. Automation in Construction 18(4), 485–498 (2009)
10. Zheng, D.X.M., Ng, S.T., Kumaraswamy, M.M.: Applying a Genetic Algorithm-Based Multi-objective Approach for Time-Cost Optimization. Journal of Construction Eng. and Management, ASCE 130(2), 168–176 (2004)

# Finding Web Document Associations Using Frequent Pairs of Adjacent Words

Jason Yong-Jin Tee, Lay-Ki Soon, and Bali Ranaivo-Malançon

Faculty of Information Technology
Multimedia University Cyberjaya, Selangor, Malaysia
{yjtee,lksoon,ranaivo}@mmu.edu.my

**Abstract.** This paper presents an approach to find associations between Web documents using collocated word pairs. Given two Web documents which are connected via a hyperlink, we attempt to find the contextual association of these two Web pages by using collocations of word pairs from a statistical point of view. Our preliminary experimental results show that our approach is able to extract fairly coherent word pairs to derive associations between hyperlinked Web documents.

**Keywords:** document association, frequent word pairs, collocation.

## 1  Introduction

According to Oxford Dictionaries[1] association is defined as the state of co-occurrence. Therefore, Web document association refers to the relation of multiple Web documents containing textual overlap. The textual overlap that become candidates for association should be at least fairly meaningful based on the content of the two Web documents.

Note that for the purpose of this paper, we have used a smaller set of Web documents in the experiment, which serves as our preliminary effort before embarking on a larger set of Web documents.

This paper will first elaborate on related work before describing the methodology and experiment to find associations between two hyperlinked Web documents. Subsequent sections will focus on results and evaluation. Finally, the paper is concluded with discovered limitations and possible direction for future work.

## 2  Related Work

Ahonen-Myka [1] had investigated on frequent word sequences in text and have favourable results. In her work, the existence of word pairs were recorded once per document, no matter how many times the word pairs occured in a single document. This paper's approach will take into account all co-occurrances in each Web documents.

---

[1] http://www.oxforddictionaries.com/

D. Lukose, A.R. Ahmad, and A. Suliman (Eds.): KTW 2011, CCIS 295, pp. 360–363, 2012.
© Springer-Verlag Berlin Heidelberg 2012

# 3    Methodology and Experimental Setup

## 3.1    Corpus

To verifiy the feasibility of our proposed method, the corpus used in the experiment presented in this paper is based on a Web crawl of the Multimedia University[2] Web site. The crawl started from the main page as the only seed page. A total of 6,597 Webpages were found. The crawler, which was developed by Qureshi et al. [2], naturally terminated after approximately 8 hours of crawling due to a stopping point where there is no more hyperlinks for the crawler to follow.

Then we proceeded to extract Web documents that are maximally hyperlinked to each other using *sccmap*, a component of Graphviz[3] which contains many other tools for visualization and processing. A total of 12 collections were found.

## 3.2    Extracting and Selecting Word Pairs

The Web documents that are used are Web documents that are maximally hyperlinked together. Therefore, collections of such Web documents are extracted from the entire corpus. The algorithm to extract word pairs is be expressed in Figure 1.

```
for each d in D
        s[] <- parse(d) //parse document d into tokens
        for each s in s[] until 2nd last element
        //extracts current and next token to form a word pair
            pair <- extract(s[i],s[i+1])
            store(pair) //stores the word pair into database
            //if word pair already exists, increment count by 1
        end for
    end for
```

**Fig. 1.** Algorithm used to extract word pairs

Word pairs with the count less than the number of Web documents in the collection are removed. Also, word pairs that begin with a non-alphabet, such as symbols and digits, are removed. These steps help to prune a significant number of word pairs. The total number of word pairs before pruning is 4463. After pruning, the total was reduced by 93.

# 4    Results

We have manually checked on these results, and found that the results seem to be fairly coherent with the contents of the Web documents. Some of the word

---

[2] http://www.mmu.edu.my/

[3] http://www.graphviz.org/

pairs are just part of a longer phrase. This approach successfully finds fairly meaningful Web document associations such as "faculty of" and "information technology" for Web documents of the faculty's main page.

We noticed that */sig/nlp/* and */sig/nlp* have no associations. We found that the content of one of the URLs is actually a redirection. Due to the severe lack of content in one Web document, we were unable to derive any association of textual content from these two Web documents.

**Table 1.** Web Document Association of Collection number 10 (Web documents from http://fit.mmu.edu.my/)

| WebDoc1 | WebDoc2 | Association |
|---|---|---|
| /research/centres.php | /main/ | faculty of<br>of information<br>information technology<br>multimedia university |
| /sig/nlp/ | /sig/nlp | |
| /main/ | /sig/nlp/ | faculty of<br>of information<br>information technology<br>multimedia university |
| /research/centres.php | /sig/nlp/ | natural language<br>language processing<br>faculty of<br>of information<br>information technology<br>multimedia university |

At the moment, due to the size of the dataset, manual evaluation was done on each Web documents that were associated. The associated terms were judged to be relevant or not relevant to the overall theme or topic of the Web document. Based on the results shown in Table 1, the associated terms do relate to the overall topic of the Web documents. For example, the Web document "/research/centres.php" contains an listing of research centres in the University. Coincidently, the Web document "/sig/nlp/" relates to the research group that speciailizes on Natural Language Processing. Therefore the terms "natural language" and "language processing" are deemed relevant to the two Web documents. As the for remaining terms, we discover that both Web documents mention "faculty of information technology" and "multimedia university". Therefore a textual overlap of frequent terms occur and these terms qualify as associations between the two Web documents as well.

## 5   Discussion and Future Work

Evaluation methods should be automated while working on a larger dataset. We plan to work on larger datasets in the near future, and manual evaluation methods such as human judgement is inefficient and wastes time. For our subsequent work, we plan to find mechanisms to facilitate evaluations automatically. This gives us not only the chance to compare the performance of our work with others, but also proper channels to prove our hypotheses.

There is a concern where word pairs that were extracted may be a form of advertising content. During pre-processing, only text that were enclosed by tags such as <p> and </p> were collected. If the advertising content is in the form of plain text, we can only assume that the advertisement is related to the overall theme of the document. If the advertising content does not pertain to the overall theme of the document, then the word pairs should not be frequent enough to affect the results. However, the future work shall address this in more detail. A bigger dataset containing Web documents with advertisements should be used.

Adjacent word pairs were used instead of other n-grams, because these pairs can be connected to form longer phrases. The use of n-grams will require the value of $n$, that dictates the length of the phrase. Also, the words have to be adjacent. If we allow a skip in any of the pairs, then the pairs might not join up to be longer phrases. In future work, a comparison can be done between the results produced by the different methods for better understanding.

The system does not check for typographical errors on associated terms. However, an improvement can be done to heuristically check for spelling errors, which can be earmarked as a possible improvement to the system in the near future.

# References

1. Ahonen-Myka, H.: Discovery of Frequent Word Sequences in Text. In: Hand, D.J., Adams, N.M., Bolton, R.J. (eds.) Pattern Detection and Discovery. LNCS (LNAI), vol. 2447, pp. 180–328. Springer, Heidelberg (2002)
2. Qureshi, M., Younus, A., Rojas, F.: Analyzing Web Crawler as Feed Forward Engine for Efficient Solution to Search Problem in the Minimum Amount of Time through a Distributed Framework. In: Proceedings of 1st International Conference on Information Science and Applications, Seoul, Korea (2010)

# Representing E-Mail Semantically
# for Automated Ontology Building

Majdi Beseiso, Abdul Rahim Ahmad, and Roslan Ismail

Universiti Tenaga Nasional (UNITEN),
Km 7, Jalan Kajang-Puchong, 43009 Kajang, Selangor, Malaysia
majdibsaiso@yahoo.com, {abdrahim,roslan}@uniten.edu.my

**Abstract.** The Semantic Web represents enormous data available on the World
Wide Web in a machine readable format for effective extraction of ontologies
providing for effective and relevant information extraction. Research shows that
an enormous amount of time is spent on E-Mails for communication and
information exchange. Adding semantics to the existing e-mail systems could
not only provide for efficient usage of time and resources but also refresh the
meaning of e-mail communication. This paper gives future scenario in applying
semantic web for E-Mail applications by proposing a new framework intended
to add a semantic web layer to the mail server in order to improve the process
of linking, integrating, and searching of E-Mails. Also, a new method for
semantic E-Mails representation is proposed that will improve the process of
ontology extraction from semantic E-Mail.

**Keywords:** Semantic Web, Semantic E-Mails, Ontologies.

# 1 Introduction

Dynamicity makes any application attractive. The effectiveness of any information
and technology system has much to do with the interactive nature in the functions it
offers. Semantic web goes beyond the two way interactions because it goes further to
facilitate understanding, arranging and coordinating the content it handles. The
situation is more complicated in handling E-Mails as the scenario is more complicated
due to the need for continuous updating and the steady efficacy in the expected
performance.

McDowell et al (2004) investigated on how the idea of semantic web can be
carried over to E-Mails. Kassoff et al (2006) also explored usage of semantic
technology in E-Mails by introducing the notion of Semantic E-Mail Addressing
(SEA). SEA facilitates the delivery of E-Mails to a determined semantic recipient of a
recipient group, which may be changed dynamically over time. Riss el al. (2009)
introduced semantic technology in E-Mail management. Zhang et al (2010) tackles on
semantic element analysis on E-Mails. This comes from the fact that E-Mail contents
are semantic body formed by large semantic number elements.

D. Lukose, A.R. Ahmad, and A. Suliman (Eds.): KTW 2011, CCIS 295, pp. 364–367, 2012.
© Springer-Verlag Berlin Heidelberg 2012

Our research is targeting on developing a representation of E-Mails semantically which will be used for automatic ontology generation. The challenge is to eventually build a semantic mail application that can understand E-Mail context without human intervention. This paper is organized as follows. Section 2 proposes a framework for semantic E-Mails, followed by the method for representing E-Mails semantically in section 3. A conclusion to the feasibility is made in the final section.

## 2     Semantic E-Mail Architecture

Semantic E-Mail is achieved through semantic E-Mail framework which identifies the fundamental components and their connections in semantic E-Mail servers. The proposed framework is intended to add a semantic web layer to the current mail server in order to improve the process of linking, integrating, and searching for E-Mails. The semantic E-Mail framework is as shown in Figure 1. The proposed framework is divided into two components:

1. Ontology Learning: the module that builds knowledge for applications which make use of domain-specific knowledge .
2. Semantic Query Engine: The semantic query engine will look for keywords provided by user based on the related ontology. In case of multiple results, the document will be kept for document ranking. Finally, semantic query identifies objects based on similarity measurement between concept of ontology and the extracted documents.

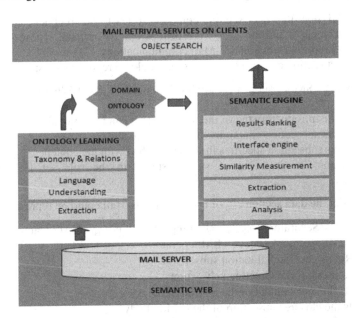

**Fig. 1.** Semantic E-Mail Architecture

# 3     Semantic E-Mail Representation

This section describes in detail the proposed method for representing the semantic E-Mails. Let us consider a set of E-Mails represented as follows:

$E = \{e_1, e_2, e_3, \ldots\ldots e_m\}$ The semantic E-Mail em is represented as:

$e_m = \{$ <tag>$e_{fm}$<tag>$e_{tm}$<tag>$e_{rm}$<tag>$e_{sm}$<tag>$e_{bm}$<tag>$e_{am}\}$, Where:

<tag>:  Represent the XML tags for identification of the sections

$e_{fm}$:  Represents the sender as the from section of the E-Mail.

$e_{tm}$:  Represents the time the mail was sent; basically the time stamp.

$e_{rm}$:  Represent the recipients.

$e_{sm}$:  Represents the subject section of the E-Mail.

$e_{bm}$:  Contains the body of the E-Mail.

$e_{am}$:  Contains the attachment enclosed with the E-Mail

For efficient ontology extraction from E-Mail, further subsections and their definitions are created. Subsections are marked using XML tags. The subject subsection and the body subsection are identified during the representation process.

The $e_{rm}$ section of the semantic E-Mail $e_m$ may have the subsections represented as

$e_{rm} = \{$ <tag> $e_{rm\_to}$ <tag> $e_{rm\_cc}$ <tag> $e_{rm\_bcc}$ <tag>$\}$, Where:

<tag>:  Represent the XML tags for identification of the sections

  $e_{rm\_to}$ : Represents the To subsection

  $e_{rm\_cc}$ : Represents the cc subsection

  $e_{rm\_bcc}$: Represents the bcc subsection

It is assumed that the $\mathcal{E}_{rm}$ section contains at least one recipient in the $e_{rm\_to}$ subsection. The $e_{rm\_cc}$ and the $e_{rm\_bcc}$ subsections can be optional. These subsections are regularly used in professional E-Mail communications.

The Subject section $e_{sm}$ with its subsections could be represented as:

$e_{sm} = \{$ $e_{sm\_tag}$ <SMSeprator1> $e_{sm\_txt}$<SMSeprator2> $e_{sm\_dt}$ $\}$ Where:
Where:

$e_{sm\_tag}$ : Is the introductory tag of the Subject. This can be optional.

$e_{sm\_txt}$ : Is the subsection containing the subject matter

$e_{sm\_dt}$ : This subsection specifies the time attached with the content

<SMSeprator1,2>: is a list of strings stored in the rule set for subsection identification.

The content or the data section $e_{bm}$ of the semantic E-Mail could be represented as

$e_{bm} = \{$ <BMSeparator1> $e_{bm\_sal}$ <BMSeperator2> $e_{bm\_grt}$ <BMSeparator3> $e_{bm\_dat}$
        <BMSeperator4> $e_{bm\_cls}$ <BMSeparator5> $e_{bm\_sig}$ $\}$, Where:

$e_{bm\_sal}$: Represents the Salutation subsection

$e_{bm\_grt}$: Represents the Greeting subsection

$e_{bm\_dat}$: Is the data content of the E-Mail in the body

$e_{bm\_cls}$: Is the closure subsection of the E-Mail body

$e_{bm\_sig}$: Is the sign off subsection of the E-Mail.

<BMSeparator1 To N>: Set of key strings or grammars stored in the Rule Set.

For effective building of semantics, E-Mails need to be preprocessed. An E-Mail that is not grammatically correct can have some misspelled words which need to be preprocessed for efficient ontology extraction. To facilitate spelling and grammar checking, support of external word processing editors like Microsoft Word, Open Office etc. can be used.    The E-Mail representation proposed in this paper incorporates the benefits of semantic tagging and use of dictionaries for effective ontology extraction. For ease of ontology annotation the named entity recognition (NER) method which proposed in [5] was used to identify name entities from the informal text.

## 4    Conclusion

This paper describes part of an ongoing research to build a semantic E-Mail server. A semantic E-Mail framework was introduced. It was intended to add a semantic web layer to the E-Mail server in order to improve the process of linking, integrating, and searching of E-Mails. A new method for semantic E-Mails representation was also proposed in order to improve the process of ontology extraction from E-Mails.

## References

1.  McDowell, L., Etzioni, O., Halevy, A.: Semantic E-Mail: Theory and Applications. Journal of Web Semantics 2(2), 153–183 (2004)
2.  Kassoff, M., Petrie, C., Zen, L.M., Genesereth, M.: Semantic E-Mail addressing: sending E-Mail to people, not strings. In: Proc. of AAAI 2006 (2006)
3.  Riss, U., Jurisch, M., Kaufman, V.: E-Mail in Semantic Task Management. In: Proceedings of 2009 IEEE Conference on Commerce and Enterprise Computing, Vienna, Austria, pp. 468–475. IEEE Comp. Society (2009)
4.  Zhang, Q., Yang, H., Yuan, Z., Sun, Z.: Studies on the Semantic Body-Based Spam Filtering. In: ISME 2010, pp. 233 - 236 (2010)
5.  Chang, Y., Sung, Y.: Applying Name Entity Recognition to Informal Text, Ling 237 Final Projects (2005)

# A Nodal Approach to Agent-Mediation in Personal Intelligence

Shahrinaz Ismail[*] and Mohd Sharifuddin Ahmad

College of Graduate Studies, Universiti Tenaga Nasional, Malaysia
sha905@gmail.com
sharif@uniten.edu.my

**Abstract.** A nodal approach to personal knowledge management is proposed in which a knowledge worker works cooperatively with intelligent software agents in a virtual workspace called a Node. A node is a cooperative architecture of a knowledge worker and one or more software agents, also known as role agents, to perform some roles for the knowledge worker. The nodal approach analyzes personal knowledge management processes by characterizing these nodes as intelligent entities. In this paper, we propose the characterization of a node and show how such node could be replicated and animated to perform a knowledge management function.

**Keywords:** Personal Intelligence, Nodal Approach, Software Agent Technology.

## 1 Introduction

The trend in today's working environment, which brings out the reality of how people manage their own knowledge, has brought the opportunities for research in people-oriented management strategies where individual knowledge workers become the focus of artificial intelligence application and agent-based mediation. Knowledge workers are the most fundamental base of organisational processes, where the real work of creating, using, exchanging, and sharing knowledge with other knowledge workers are being done to achieve some common goal. Each individual knowledge worker performs structured, semi-structured and unstructured tasks, while interacting with other knowledge workers. The diverse ways of performing these tasks constitutes their personal intelligence in managing their knowledge, in which a software agent could assist in mediating them.

A knowledge worker and the mediating agents can be abstracted as a node, which consists of the knowledge worker, the agents and their functions. The agents perform their functions to assist the knowledge worker in performing the human functions, forming a symbiotic relationship between human and agents that can be construed as personal intelligence.

D. Lukose, A.R. Ahmad, and A. Suliman (Eds.): KTW 2011, CCIS 295, pp. 368–374, 2012.

## 2    Related Works

### 2.1    Intelligent Software Agents

Software agents are entities that function continuously and autonomously in a particular environment that is often inhabited by other agents and processes [1]. Agents are expected to have the ability to learn from their experiences, communicate and cooperate with people and other agents, including the "ability to move around some network" [2], such as within private networks and Semantic Web. There are many definitions of software agents, with similarities being that the autonomous behaviour of the agents is for interaction with its environment or surrounding. Among the most significant and latest definitions are:

i)    an encapsulated computer system that is situated in some environment and is capable of flexible action in that environment in order to meet its design objectives [3];

ii)    autonomous agents are computational systems that inhabit some complex dynamic environment; sense and act autonomously in this environment and by doing so realise a set of goals or tasks for which they are designed [4].

For the purpose of applying a nodal approach to agent-mediation in personal intelligence, this paper is based on the agent being autonomous, reactive, proactive, able to communicate, adaptive, goal-oriented, capable to cooperate, reason, and flexible, as adapted from the features listed by Paprzycki and Abraham [5].

### 2.2    Agent-Mediation in Personal Intelligence

Most authors agree that 'personal intelligence' (PI) is electronic based or technologically enhanced facility that extends human capabilities. Solachidis et al. [6] defined personal intelligence as a layer that constitutes collective intelligence and deals with "enabling users to both upload and access multimedia information submitted to the intelligent services using a range of devices". Solachidis et al. [6] looked at personal intelligence as a contribution to collective intelligence, which they believed will be extensively used in the year 2015 across the diverse user trends on technology that may change by that year. This has brought to the attention of having an intelligent 'being' (i.e. a personal intelligence agent) with the ability to 'roam around', communicate with other systems, and perform its tasks across these trends of technology and platform. Solachidis et al. [6] also mentioned about the restriction factors in personal intelligence (i.e. event, user, content capture, terminal and network), and these factors are similar and applicable to the elements in agent environment in agent technology.

Personal intelligence involves the abilities to: (a) recognise personally-relevant information from introspection and from observing oneself and others, (b) form that information into accurate models of personality, (c) guide one's choices by using personality information where relevant, and (d) systematise one's goals, plans and life stories for good outcomes [7]. In applying these abilities into agent-mediation, it is more of targeting the agent to be rational as part of being an intelligent system. In supporting this, Mayer [8] justified with his theoretical analysis that personal intelligence has the abilities to mediate individual's cumulative life decisions to help in that person's well-being.  Following this justification, this paper looks at the concept of how this 'cumulative decisions' can be animated by software agents, from a nodal approach point of view.

## 3    A Nodal Approach to Agent-Based Mediation

While we concede that there are methodologies useful in providing the guidelines for building multi-agent systems, such as MaSE [9], Prometheus [10] and Gaia [11], a meta-level approach to building agent systems requires a comprehensive analysis of humans' and agents' functions. Consequently, we propose here a novel technique for building multi-agent systems based on the concept of nodes.

We conceive a human entity, in this case a knowledge worker, working cooperatively with a software agent in a virtual workspace called a node.  A node consists of a knowledge worker, and one or more agents, also known as role agents, to perform some roles of the knowledge worker.  The knowledge worker has a set of functions, some of which could be delegated to the agents.  Our analysis of human-human interactions mentioned by Ahmed, Ahmad, and Mohd Yusoff [12] reveals that there are two types of function of a knowledge worker: common functions (e.g. open document, create/edit document, upload/download document, delegate role, request, request response); and unique functions based on the knowledge held (e.g. analyse problem, propose solutions, response-to-request).  These common and unique functions can be further distinguished into online and offline modes.  Offline modes are usually the physical tasks, e.g. attending a meeting to make a crucial decision, which the knowledge worker needs to do to complete a current online work cycle.

The knowledge worker in a node needs to perform these functions with his or her knowledge to drive some work process.  There are also mundane housekeeping functions, which the knowledge worker needs to mind and for which a software agent could be deployed.  Based on our analysis of human-agent and agent-agent interactions [12], these functions assist or supplement human functions, e.g. receive and forward documents, communicate with other agents, remind deadlines, calculate and award merit/demerit points, log actions, inform process delays, track process, and search. A node can thus be specified as follows:

- a knowledge worker, KW;
- a set of the knowledge worker's functions, F;
- a set of one or more agents, A;
- a set of agent's functions, G;
  i.e. a node can be represented as a four-tuple structure, N = <KW, F, A, G>.

With such specification, a node can be easily replicated to instantiate another virtual workspace consisting of a knowledge worker, agents, and their functions. A significant parameter which is necessary to complete information and knowledge flow between two nodes is the exchange of information and resources link between them. Such function can be tasked to a software agent and implemented by manipulating the link between the nodes. Figure 1 shows a typical node and an interaction between two nodes.

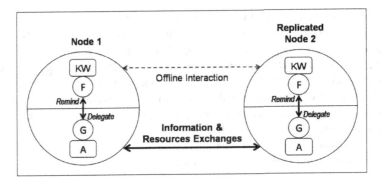

**Fig. 1.** A Node and an interaction between Nodes

In this figure, we assume that the agent itself has its own belief-desire-intention (BDI) architecture including the environment which it shares with its human counterpart. Figure 1 also displays a model in which an agent is delegated to assist the knowledge worker in performing diverse mundane tasks for all roles of the knowledge worker. While the agent performs the necessary communication, information searches, and resources exchanges, offline interactions between knowledge workers are also necessary to resolve issues of the task at hand. Belief-desire-intention (BDI) architecture is a mature and commonly adopted architecture for intelligent agents [12]. FIPA agent communication language (ACL) message use BDI to define their semantics [13], which the communication between agents is based on. The BDI semantics gives the agents the ability to know how it arranges the steps to achieve the goal [12], for example agent has the 'desire' in achieving the goal that it will never stop until it has achieved the goal.

An alternative structure shown in Figure 2 depicts two agents A-1 and A-2, in Node 1, where each performs mundane tasks for some role of the knowledge worker. In the figure, agent A-2 performs the task for one role of the knowledge worker, in this case, exchanging information and resources with Node 2. Presumably A-1 is delegated to perform other mundane tasks for other roles of the knowledge worker.

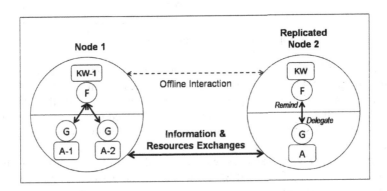

**Fig. 2.** A Node with Two Agents

We could extend this concept to an organisational setting in which an ecosystem of multiple nodes connected in a web of  human experts, non-experts, internal knowledge resources (databases, knowledge bases, and repositories) and external knowledge resources (WWW). A knowledge worker, while performing his or her job, is likely to communicate and exchange information and knowledge between these nodes. Consequently, multi-structured (structured, semi-structured, and unstructured) patterns of work processes emerge that contribute to the achievement of some objectives and the subsequent fulfillment of some goals. Such patterns of work processes represent the personal intelligence of the knowledge worker. Figure 3 shows such ecosystem involving multiple nodes.

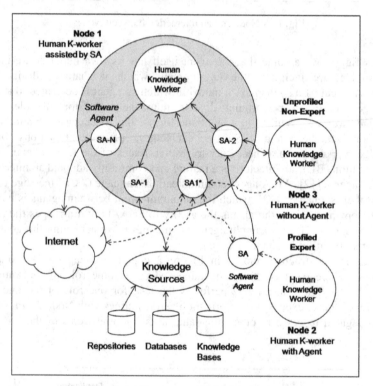

**Fig. 3.** An Ecosystem of Multiple Nodes

Figure 3 also shows how a knowledge worker connects to three nodes with the help of a software agent (SA) for each of these different nodes. As an example, SA-2 refers to 'Software Agent 2' that connects the knowledge worker to another knowledge worker who is either an unprofiled non-expert or a profiled expert, who may or may not be assisted by a software agent. Similarly, SA-1 is another software agent that mediates the searching, connecting to the knowledge sources, and retrieving the required knowledge. Figure 3 emphasises the possibilities of having different software agents (referred to SA-1, SA-2 and SA-N) to exhibit the personal intelligence of the knowledge worker. All the N agents communicate and collaborate with each other so that their human counterpart is served cooperatively to manifest an efficient mediation process. In order to achieve such efficiency, each agent should be able to match its human counterpart's profile with the relevant knowledge resources and knowledge experts.

## 4   Discussions and Conclusion

The conceptual model of agent-mediated personal knowledge management is conceived based on the people-oriented KM strategy, in which individual knowledge workers and their software agents drive the cycle of task performances while using and exchanging information and knowledge resources to complete the tasks. Knowledge workers are the central core of organisational processes, who create, use, exchange, and share knowledge with other knowledge workers to achieve some common goal. Each individual knowledge worker, while interacting with other knowledge workers, performs structured, semi-structured and unstructured tasks. The variety of ways in performing these tasks constitutes their personal approaches in managing their knowledge.

A knowledge worker and his or her agents can be abstracted as a node, which consists of the knowledge worker, his or her functions, the agents, and their functions. The agents perform their functions to assist the knowledge worker in performing his or her main functions, the completion of which supports the achievement of some organisational goal. Such symbiosis between human and agents can be construed as the intelligence in personal knowledge management. This concept can be extended to an ecosystem of multiple nodes, which interacts to manifest multi-structured processes of collective intelligence in organisational knowledge management.

The nodal approach to agent-mediated personal intelligence enables the replication of nodes where each node represents a function of an individual knowledge worker's intelligence in personal knowledge management. Such intelligence overlaps to reveal tasks for a common goal, the integration of which manifests an agent-mediated collective intelligence in organisational knowledge management.

In our future work, we will analyse the technical requirements to create a node and implement the interactions between nodes. We will also explore the possibility of modeling information and knowledge resources as nodes in which a software agent resides to mediate the update and retrieval processes.

## References

1. Bradshaw, J.M., Carpenter, R., Cranfill, R., Jeffers, R., et al.: Roles for Agent Technology in Knowledge Management: Examples from Applications in Aerospace and Medicine. AAAI Technical Report SS-97-01, pp.9-16 (1997)
2. Nwana, H.S.: Software Agents: An Overview. Knowledge Engineering Review 11(3), 205–244 (1996)
3. Jennings, N.R., Faratin, P., Lomuscio, A.R., Parsons, S., Sierra, C., Wooldridge, M.: Automated Negotiation: Prospects, Methods and Challenges. International Journal of Group Decision and Negotiation 1–30 (2000)

4. Ali, G., Shaikh, N.A., Shaikh, A.W.: A Research Survey of Software Agents and Implementation Issues in Vulnerability Assessment and Social Profiling Models. Australian Journal of Basic and Applied Sciences 4, 442–449 (2010)
5. Paprzycki, M., Abraham, A.: Agent Systems Today: Methodological Considerations. In: International Conference on Management of e-Commerce and e-Government, pp.1–7 (2003)
6. Solachidis, V., Mylonas, P., Geyer-Schulz, A., Hoser, H., Chapman, S., Ciravegna, F., et al.: Collective Intelligence Generation from User Contributed Content. In: Fink, A., et al. (eds.) Advances in Data Analysis, Data Handling and Business Intelligence, Studies in Classification, Data Analysis, and Knowledge Organization, pp. 765–774. Springer, Berlin (2010)
7. Mayer, J.D.: Personal Intelligence. Imagination, Cognition and Personality 27, 209–232 (2008)
8. Mayer, J.D.: Personal Intelligence Expressed: A Theoretical Analysis. Review of General Psychology 13, 46–58 (2009)
9. Wood, M.F., DeLoach, S.A.: An Overview of the Multiagent Systems Engineering Methodology. In: Ciancarini, P., Wooldridge, M.J. (eds.) AOSE 2000. LNCS, vol. 1957, pp. 207–221. Springer, Heidelberg (2001)
10. Padgham, L., Winikoff, M.: Prometheus: A Methodology for Developing Intelligent Agents. In: Giunchiglia, F., Odell, J.J., Weiss, G. (eds.) AOSE 2002. LNCS, vol. 2585, pp. 174–185. Springer, Heidelberg (2003)
11. Zambonelli, F., Jennings, N.R., Wooldridge, M.: Developing Multi-agent Systems: The Gaia Methodology. ACM Transactions on Software Engineering and Methodology 12(3), 317–370 (2003)
12. Ahmed, M., Ahmad, M.S., Mohd Yusoff, M.Z.: A Collaborative Framework for Human-Agent Systems. International Journal of Computer and Network Security 1(2), 69–76 (2009)
13. FIPA Communicative Act Library Specification SC00037J 2002/12/03

# Benchmarking the Performance of Support Vector Machines in Classifying Web Pages

Wein-Pei Wong, Ke-Xin Chan, and Lay-Ki Soon

Faculty of Information Technology, Multimedia University,
Cyberjaya, Selangor, Malaysia
lksoon@mmu.edu.my, {chankexin,pei_424}@hotmail.com

**Abstract.** In this paper, we benchmark the efficiency of support vector machines (SVMs), in terms of classification accuracy and the classification speed with the other two popular classification algorithms, which are decision tree and Naïve Bayes. We conduct the study on the 4-University data set, using 4-fold cross validation. The empirical results indicate that both SVMs and Naïve Bayes achieve comparative results in the average precision and recall while decision tree ID3 algorithm outperforms the rest in the average accuracy despite. Nevertheless, ID3 consumes the longest time in generating the classification model as well as classifying the web pages.

**Keywords:** web classification, support vector machines, Naïve Bayes, decision tree.

## 1 Introduction

Web classification is part of web mining and text mining where it applies data mining techniques to discover interesting and non-trivial patterns from the web. Some Web classification task are performed only on the textual content, while others may take into consideration the HTML tags, anchor texts of hyperlinks, URLs, and other metadata of Web pages, such as the page title [2, 5, 6, 8-11, 13]. Considering the emerging popularity gained by SVMs in recent years [3, 4, 11], we have implemented SVMs, decision tree and Naïve Bayes algorithms in Weka on 4-university data set[1]. The aim of this project is to compare the results produced by these three algorithms. Due to the space limitation, we only compare the results in terms of precision, accuracy and the time taken by each algorithm to perform the classification. Section 2 of this paper briefly explains the data set, the process flow of our experiment. Section 3 discusses the results and findings. Lastly, we conclude this paper in Section 4.

## 2 Data Set and Experimental Setup

The 4-university data set contains web pages collected from computer science departments of a few universities in January 1997 by the text learning group in

---

[1] http://www.cs.cmu.edu/afs/cs/project/theo-20/www/data/

D. Lukose, A.R. Ahmad, and A. Suliman (Eds.): KTW 2011, CCIS 295, pp. 375–378, 2012.
© Springer-Verlag Berlin Heidelberg 2012

Carnegie-Mellon University (CMU), under their World Wide Knowledge Base (Web->kb). There are a total of 8,282 web pages that were manually classified into seven categories.   Training and testing on pages from the same university are not recommended because each university's web pages have their own idiosyncrasies. Thus, as recommended, we have conducted the training on three of the universities plus the miscellaneous collection in order to generate the classification model, and testing on the pages from a fourth held-out university is performed to obtain the performance of the classification algorithms.  Our classification model is multi-class classifier, where one classification model is generated by each classification algorithms, for all the seven classes.   Although some works focused on only four categories, namely *Student*, *Project*, *Staff* and *Faculty*, we shall work on all the seven categories in this empirical study [10-11].

Given the labeled web pages, we have done the data pre-processing by de-tagging the HTML files and tokenizing the web content into words.   Subsequently, the processes of words selection, word-stemming and stop-word-removal have been conducted.  For word selection, we have used the pre-processing functions provided by Weka [12], where we have selected 300 most influential words for each category. Note that 300 are chosen due to the time constraint of this project, other numbers will be attempted in the future work.  After these pre-processing steps, we are left with 717 words to represent the web content.  Before sending these web pages to Weka, we have split them into the training and test data for 4-fold cross-validation.  Each word is deemed as a feature to represent the web pages, where "1" indicates the existence of the work in the web page and "0" otherwise.  The performances of these algorithms are compared in terms of precision and classification accuracy.

## 3     Results and Discussion

Figure 1 and 2 illustrate the precision and recall achieved by all three algorithms in each category.  It is notable that ID3 performs exceptionally poor in classifying the *Staff* category, achieving 0 precision and recall respectively.  The reason could be that the *Staff* category has the least number of web pages, which is only 1.65% of the data set.  Based on this observation, we may conclude that ID3 does not work well when the given dataset is relatively small, as compared to SVMs and Naïve Bayes.  In terms of precision for each category, Naïve Bayes performs consistently except in the *Student* category where SVMs gives the best precision of 0.54.  As shown in Fig. 2, ID3 achieves the best accuracy in most of the categories except *Course*, *Faculty*, *Other*, and *Project*.  We are not able to detail the comparative study of these algorithms in terms of time taken to construct classification models and perform prediction here due to the limited space.  Nevertheless, in short, ID3 takes the longest time while Naïve Bayes needs the least time due to the nature of the algorithm [1].  In average, SVMs consumes less than 50% of the duration taken by decision tree algorithm.

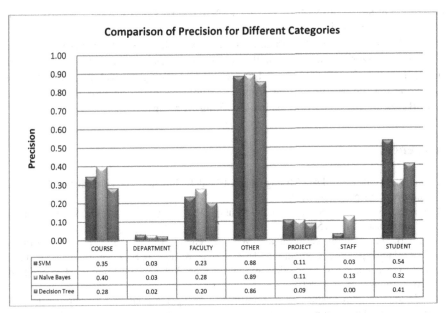

**Fig. 1.** Comparison of precision for different categories by three algorithms

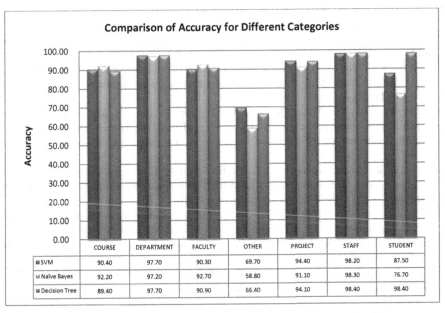

**Fig. 2.** Comparison of accuracy for different categories by three algorithms

# 4    Conclusion

Our empirical study shows that SVMs and Naïve Bayes perform relatively similar in terms of average precision when classifying the 4-university data set. On the other hand, ID3 achieves the best average accuracy. Nevertheless, our results indicate that ID3 does not work well with small data set and it takes the longest duration in constructing the classification model as well as classifying the web pages. Given the results and the observation of our empirical study, we may conclude that SVMs and Naïve Bayes are the better options in classifying web pages.

# References

1. Han, J., Kamber, M.: Data Mining Concepts and Techniques, 2nd edn. Morgan Kaufmann Publishers, Elsevier (2006)
2. Kosala, R., Blockeel, H.: Web Mining Research: A Survey. ACM SIGKDD Explorations 2(1), 1–15 (2000)
3. Liu, B.: Information Retrieval and Web Search. In: Liu, B. (ed.) Web Data Mining, Exploring Hyperlinks, Contents and Usage Data, pp. 18–236. Springer, Heidelberg (2007)
4. Liu, B., Chang, K.C.C.: Editorial: Special Issue on Web Content Mining. ACM SIGKDD Explorations 6(2), 1–4 (2004)
5. Peleato, R.A., Chappelier, J.C., Rajman, M.: Using Information Extraction to Classify Newspapers Advertisements. In: Proc of the 5th International Conference on the Statistical Analysis of Textual Data, Lausanne, Switzerland (2000)
6. Qi, X., Davison, B.: Web Page Classification: Features and Algorithms. ACM Computing Surveys 41(2), 1–31 (2009)
7. Quinlan, J.R.: C4.5: Programs for Machine Learning. Morgan Kaufmann, San Francisco (1993)
8. Shanks, V., Williams, H.E.: Fast Categorisation of Large Document Collections. In: Proc of the 8th International Symposium on String Processing and Information Retrieval, pp. 194–204 (2001)
9. Shen, D., Chen, Z., Yang, Q., Zeng, H.J., Zhang, B., Lu, Y., Ma, W.Y.: Web-page Classification through Summarization. In: Proc of the 27th Annual International ACM SIGIR Conference on Research and Development in Information Retrieval, pp. 242–246 (2004)
10. Soon, L.-K., Hwang, K.-B., Lee, S.L.: An Empirical Study on Harmonizing Classification Precision using IE Patterns. In: Proc. of the 2nd International Conference on Software Engineering and Data Mining, Chengdu, China (2010)
11. Sun, A.-X., Lim, E.-P., Ng, W.-K.: Web Classification using Support Vector Machine. In: Proc of the 4th Workshop on Web Information and Data Management, pp. 96–99 (2002)
12. Witten, I.H., Frank, E.: Data Mining: Practical Machine Learning Tools and Techniques, 2nd edn. Morgan Kaufmann, San Francisco (2005)
13. Yang, Y.: An Evaluation of Statistical Approaches to Text Categorization. Journal of Information Retrieval 1(1-2), 69–90 (1999)

# Conic Curve Fitting Using Particle Swarm Optimization: Parameter Tuning

Zainor Ridzuan Yahya, Abd Rahni Mt Piah, and Ahmad Abd Majid

School of Mathematical Sciences, Universiti Sains Malaysia,
11800 USM, Pulau Pinang, Malaysia
zainor.usm@gmail.com, {arahni,majid}@cs.usm.my

**Abstract.** We solve curve fitting problems using Particle Swarm Optimization (PSO). PSO is used to optimize control points and weights of two conic curves to a set of data points. PSO is used to find the best middle control point and weight for both conic curves to provide piecewise conics that preserve $G^1$ continuity. We present numerical result using parameter changes in PSO scheme. We obtain appropriate parameter values of PSO that provide best error and fastest time to solve curve fitting problem.

**Keywords:** Curve fitting, particle swarm optimization, parameter tuning.

## 1 Introduction

It is generally known that there is a trade-off between exact methods and soft computing (heuristic) methods. Exact methods guarantee accurate/optimal solutions but require a relatively long computational time. Soft computing methods, on the other hand, do not guarantee optimal solution but require a relatively short computational time. For most medical imaging purposes, which often require constructing empirical models which preserve the shape and match data samples or approximate data samples and unknown entity at intermediate location, short computational time is preferred. Thus, we intend to explore the ability of Particle Swarm Optimization (PSO) methods in curve fitting for this purpose.

PSO is an optimization technique proposed by Kennedy and Eberhart by means of particle swarm [9]. PSO incorporates swarming behaviours observed in flocks of birds, school of fish, swarm of bees and even social behaviour, from where the idea emerged. PSO is a population-based optimization tool, which could be implemented and applied easily to solve various function optimization problems, or problems that can be transformed to function optimization problems. As an algorithm, its fast convergence compares favourably with many global optimization algorithms like Genetic Algorithm [3], Simulated Annealing [6] and other global optimization algorithms. To apply PSO successfully, one of the key issues is finding ways to map problem solution into PSO article, which directly affects its feasibility and performance.

D. Lukose, A.R. Ahmad, and A. Suliman (Eds.): KTW 2011, CCIS 295, pp. 379–382, 2012.

Galvez et al. [4] use PSO to perform surface fitting with a Bezier surface to a given cloud of data points. Delint et al. [1] use PSO for curve fitting using NURBS. But these studies do not discuss the parameter tuning in particular even though the process is important to obtain good results. In this regard, we wish to study the effect of PSO parameter changes for curve fitting problem.

Parameter tuning plays an important role for any evolutionary computing algorithm. Parameter changes in each soft computing scheme will affect its effectiveness and computation time. For PSO, the setting of the parameter determines how the algorithm optimizes the solution. Therefore, we will determine the best setting of parameters in curve fitting problem.

## 2     PSO for Curve Fitting

Our proposed PSO algorithm starts with having an initial swarm position of swarm size = 25. PSO is used to find the best control point and weight which provide the best fitted curve to given data. For our proposed PSO scheme, we set inertia $\omega = 1.0$ and $C_1 = C_2 = 1.494$. The iterations will stop after 250 iterations. The pseudo code of our proposed technique is:

**Proposed PSO Curve Fitting Algorithm**
Get data points;
Find middle point from chord length parameterization;
Use Least Square to approximate intermediate point for the left curve;
Determine search space to find intermediate control points;
**Initialization**
       Initialize intermediate control point for each particle;
       Initialize weight (randomly chosen from [0, 1]) for each particle;
**For** i = 1 until last iteration
     **For** j = 1 until number of particle
          Update left control point of the particle;
          Calculate right control point to achieve $G^1$ continuity;
          Update left conic weight;
          Calculate right conic weight;
          Evaluate solution (newSolution);
          **If** newSolution is better than bestSolution
               bestSolution = newSolution;
               bestControlPoint = newControlPoint;
               bestWeight = newWeight;
          **end If**
          Update velocity for control point and weight;
     **end For;**
  **end For;**

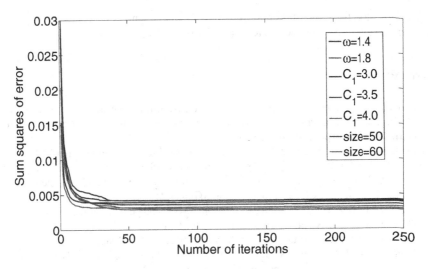

**Fig. 1.** The PSO parameters effect

## 3    Parameter Tuning

Parameter tuning plays an important role in PSO scheme. Effects of parameter changes on the PSO have been discussed by many authors [11, 10, 5, 8, 7]. In this section, we discuss the impact of changes to inertia ($\omega$), coefficient factors ($C_1$ and $C_2$) and swarm size. The inertia, $\omega$ controls the momentum of the particle [2]. From the literature, it suggest that $\omega$ is within the range [0,1]. On the other hand, $\omega$ can be greater than 1 but must be done with care. Therefore, in our research we determine the value of $\omega \in [0, 2]$. For acceleration coefficients, the usual choice for $C_1$ and $C_2$ is $C_1 = C_2 = 1.494$. However, other setting was also used in different papers and ranges between [0, 4] [2]. Therefore, we will search the coefficients factors within this range. For swarm size, researchers limit the number of particles to between 20 and 60 [2]. Therefore, we will tune the swarm size to be within this range. The result of the proposed parameter tuning is given in Fig. 1. From the figure, a combination of $\omega = 1.8$, $C_1 = C_2 = 4.0$ and swarm size = 60 gives the best error. We then run the combination of the same problem and we get time = 1.25 seconds with Sum Squares of Error (SSE) = 0.0026 which is in an acceptable region.

## 4    Conclusion

Instead of using a direct technique, we use an indirect technique to perform curve fitting. To find the best solution, we map the curve fitting problem to the best PSO scheme. Changes in the PSO parameters affect computational time and error. A combination of $\omega = 1.8$, $C_1 = C_2 = 4.0$, and swarm size = 60 gives the best error in curve fitting problem. In the future, we will try to use similar algorithm to solve surface fitting problem. Even though PSO does not give a very good solution, the error obtained is within an acceptable region set by intended user.

**Acknowledgments.** The authors would like to thank Universiti Sains Malaysia for supporting this research under its RU grant 1001/PMATHS/817049.

# References

1. Delint I.S.A, Shamsuddin, S.M., Ali, A.: Particle Swarm Optimization for NURBS Curve Fitting. In: Sixth International Conference on Computer Graphics, Imaging and Visualization, CGIV 2009, pp. 259–263 (2009)
2. Das, S., Abraham, A., Konar, A.: Particle Swarm Optimization and Differential Evolution Algorithms: Technical Analysis, Applications and Hybridization Perspectives. In: Liu, Y., Sun, A., Loh, H.T., Lu, W.F., Lim, E.-P. (eds.) Advances of Computational Intelligence in Industrial Systems. SCI, vol. 116, pp. 1–38. Springer, Heidelberg (2008)
3. Eiben, A.E., Smith, J.E.: Introduction to Evolutionary Computing. Springer (2007)
4. Gálvez, A., Cobo, A., Puig-Pey, J., Iglesias, A.: Particle Swarm Optimization for Bézier Surface Reconstruction. In: Bubak, M., van Albada, G.D., Dongarra, J., Sloot, P.M.A. (eds.) ICCS 2008, Part II. LNCS, vol. 5102, pp. 116–125. Springer, Heidelberg (2008)
5. Kennedy, J.: Small Worlds and Mega-Minds: Effects of Neighborhood Topology on Particle Swarm Performance. In: Proceedings of the 1999 Congress of Evolutionary Computation, pp. 1931–1938. IEEE Press, New York (1999)
6. Kirkpatrick, S., Gelatt, C.D., Vecchi Jr., M.P.: Optimization by Simulated Annealing. Science 220, 671–680 (1983)
7. Lvbjerg, M., Rasmussen, T.K., Krink, T.: Hybrid Particle Swarm Optimizer with Breeding and Subpopulations. In: Proceedings of the Third Genetic and Evolutionary Computation Conference (GECCO 2001), pp. 469–476 (2001)
8. Ratnaweera, A., Halgamuge, S.K., Watson, H.C.: Self-Organizing Hierarchical Particle Swarm Optimizer with Time-Varying Acceleration Coefficients. IEEE Transactions on Evolutionary Computation 8(3), 240–255 (2004)
9. Weise T.: Global Optimization Algorithms, Theory and Application (Version: June 26, 2009) (e-book), http://www.it-weise.de/projects/book.pdf (accessed January 13, 2010)
10. Shi, Y., Eberhart, R.C.: Fuzzy Adaptive Particle Swarm Optimization. In: Proceedings of the Congress on Evolutionary Computation 2001, pp. 101–106. IEEE Service Center, IEEE, Seoul, Korea (2001)
11. Shi, Y., Eberhart, R.C.: A modified particle swarm optimiser. In: IEEE International Conference on Evolutionary Computation, Anchorage, Alaska, May 4-9, pp. 69–73 (1998)

# Author Index

Abd Majid, Ahmad 379
Abdullah, Rusli 343
Aghassi, Hadi 133, 356
Ahmad, Abdul Rahim 364
Ahmad, Azhana 82, 226
Ahmad, Faisul Arif 113
Ahmad, Mohd Sharifuddin 82, 226, 272, 368
Ahmad, Rashidi 272
Ahmadi-Abkenari, Fatemeh 123
Ahmed, Moamin 82
Ainon, Raja Noor 62
Albert, Dietrich 24
Annamalai, Muthukaruppan 272
Anthony, Patricia 303, 313
Arvin, Farshad 113
Atan, Rodziah 343

Bahari, Anita 216
Behrouzifar, Mohammad 143, 184, 234
Beseiso, Majdi 364
Bong, Chin Wei 103
Boo, Vooi Keong 303, 313
Bulgiba, Awang M. 62

Caracciolo, Caterina 11
Chai, Lian En 352
Chan, Ke-Xin 375
Chan, Weng Howe 323
Chin, Yin Fai 333
Chong, Chuii Khim 352
Choon, Yee Wen 352

Danyaro, Kamaluddeen Usman 164
Dengel, Andreas 1
Deris, Safaai 352
Doraisamy, Shyamala C. 113, 262

Eftekhari Moghadam, Amir Masoud 194
Ektefa, Mohammadreza 262

Halim, Shamimi 272
Hassan, Rohayanti 333
Hockemeyer, Cord 24

Ishak, Ros'aleza Zarina 293
Ismail, Roslan 364
Ismail, Shahrinaz 368

Jaafar, Jafreezal 164
Jalilvand, Ali 348
Javani, Mehran 194
Jörke, Jan 154

Kamarulzaman, Hamzah 103
Keizer, Johannes 11
Khalid, Atifah 72
Kickmeier-Rust, Michael D. 24
Kiu, Ching-Chieh 283

Lahsasna, Adel 62
Lee, Chee Kiam 93
Liew, Mohd Shahir 164
Lim Sien Niu, Michelle 216
Liwicki, Marcus 1
Lodemann, Michael 206
Lukose, Dickson 72, 93, 216, 283, 293
Luttenberger, Norbert 154, 206

Mahmoud, Moamin A. 226
Marnau, Rita 206
Memar Kouchehbagh, Sara 262
Moghadam Charkari, Nasrollah 143, 184, 234, 348
Mohamad, Mohd Saberi 52, 174, 323, 333, 352
Mohamed Salleh, Abdul Hakim 52
Mohd Hashim, Siti Zaiton 244
Mohd Sharef, Nurfadhlina 254
Mohd Yassin, Norlidza 72
Mohd Yusof, Norfadzlia 93
Mohd Yusoff, Mohd Zaliman 82, 226
Moorthy, Kohbalan 174
Mozafari, Kourosh 184, 234
Mt Piah, Abd Rahni 379
Murad, Masrah Azrifah Azmi 343
Mustaffa, Supiah 293
Mustapha, Aida 226
Mustapha, Norwati 262

Nader Abadi, Sedigheh    356
Neubert, Joachim    37
Nussbaumer, Alexander    24

Ramli, Abdul Rahman    113
Ranaivo-Malançon, Bali    360
Rapa'ee, Suriani    72
Roghanian, Emad    356

Samsudin, Khairulmizam    113
Selamat, Ali    123, 244
Selamat, Md Hafiz    244
Selan, Nor Ezam    93, 216
Shayegh Boroujeni, Hamidreza    143,
    184, 234
Sheykhlar, Zahra    133
Soon, Lay-Ki    360, 375

Steiner, Christina M.    24
Suriani Affendey, Lilly    262

Taheri Makhsoos, Poonia    143
Talib, Amir Mohamed    343
Tan, Sieow-Yeek    283
Tochtermann, Klaus    37

Weber, Markus    1
Wong, Wein-Pei    375
Wu, Sheng-Chuan    46

Yahya, Zainor Ridzuan    379
Yong-Jin Tee, Jason    360
Yoong Lam, Hong    103

Zainuddin, Roziati    62
Zedlitz, Jesper    154
Zolfpour-Arokhlo, Mortaza    244